Applied Science

For almost two centuries, the category of 'applied science' was widely taken to be both real and important. Then, its use faded. How could an entire category of science appear and disappear? By taking a longue durée approach to British attitudes across the nineteenth and twentieth centuries, Robert Bud explores the scientific and cultural trends that led to such a dramatic rise and fall. He traces the prospects and consequences that gave the term meaning, from its origins to its heyday as an elixir to cure many of the economic, cultural, and political ills of the United Kingdom, eventually overtaken by its competitor, 'technology'. Bud examines how 'applied science' was shaped by educational and research institutions, sociotechnical imaginaries, and political ideologies and explores the extent to which non-scientific lay opinion, mediated by politicians and newspapers, could become a driver in the classification of science.

Robert Bud is Emeritus Keeper at London's Science Museum. He has led science, medicine, and curatorial research at the Museum, writing and editing books across chemical, biotechnological, and scientific instrument history.

SCIENCE IN HISTORY

Series Editors

Lissa Roberts, University of Twente
Simon J. Schaffer, University of Cambridge
James A. Secord, University of Cambridge

Science in History is a major series of ambitious books on the history of the sciences from the mid-eighteenth century through the mid-twentieth century, highlighting work that interprets the sciences from perspectives drawn from across the discipline of history. The focus on the major epoch of global economic, industrial and social transformations is intended to encourage the use of sophisticated historical models to make sense of the ways in which the sciences have developed and changed. The series encourages the exploration of a wide range of scientific traditions and the interrelations between them. It particularly welcomes work that takes seriously the material practices of the sciences and is broad in geographical scope.

A full list of titles in the series can be found at: www.cambridge.org/sciencehistory

Applied Science

*Knowledge, Modernity, and Britain's
Public Realm*

Robert Bud

Science Museum, London

CAMBRIDGE
UNIVERSITY PRESS

CAMBRIDGE
UNIVERSITY PRESS

Shaftesbury Road, Cambridge CB2 8EA, United Kingdom

One Liberty Plaza, 20th Floor, New York, NY 10006, USA

477 Williamstown Road, Port Melbourne, VIC 3207, Australia

314–321, 3rd Floor, Plot 3, Splendor Forum, Jasola District Centre, New Delhi – 110025, India

103 Penang Road, #05–06/07, Visioncrest Commercial, Singapore 238467

Cambridge University Press is part of Cambridge University Press & Assessment, a department of the University of Cambridge.

We share the University's mission to contribute to society through the pursuit of education, learning and research at the highest international levels of excellence.

www.cambridge.org
Information on this title: www.cambridge.org/9781009365239

DOI: 10.1017/9781009365260

First published 2024

A catalogue record for this publication is available from the British Library

Library of Congress Cataloging-in-Publication Data
Names: Bud, Robert, author.
Title: Applied science : knowledge, modernity and Britain's public realm / Robert Bud.
Description: Cambridge ; New York, NY : Cambridge University Press, 2024. | Series: Science in history | Includes bibliographical references.
Identifiers: LCCN 2023035631 (print) | LCCN 2023035632 (ebook) | ISBN 9781009365239 (hardback) | ISBN 9781009365253 (paperback) | ISBN 9781009365260 (epub)
Subjects: LCSH: Technology–Great Britain–History. | Science–Great Britain–History.
Classification: LCC T26.G7 B835 2024 (print) | LCC T26.G7 (ebook) | DDC 609.41–dc23/eng/20230822
LC record available at https://lccn.loc.gov/2023035631
LC ebook record available at https://lccn.loc.gov/2023035632

ISBN 978-1-009-36523-9 Hardback

This book is dedicated to the staff of the Science
Museum Library and Archives, past and present.

Contents

Figures

Acknowledgements

In the period of more than a decade this book has taken to research and write, and the even longer period of planning, I have benefitted from the support of many professionals, institutions, scholars, and friends. Research was made possible by libraries and archives. Too rarely are librarians publicly appreciated, and I should like to acknowledge a special debt to colleagues in the Science Museum Library and Archives service who helped me over more than four decades and to the numerous other archivists and librarians who went out of their way to help. In several cases, the access and use of unpublished material required the permission of descendants of the original authors or their agents. I acknowledge their speedy and enthusiastic assistance with gratitude.

The Science Museum, London, unstintingly supported research towards this project over a long period. I need to thank, in particular, Ian Blatchford, director of the Science Museum Group, and Tim Boon, head of Research and Public History. The Arts and Humanities Research Council made the project possible through two grants for fellowships, grants number AH/I027177/1 and AH/L014815/1. I am also grateful to the Science History Institute for a travel grant to study the remarkable archive in their care. I should like to acknowledge the intellectual engagement and privileges made possible by the Department of Science and Technology Studies at University College London through an attachment as 'Senior Honorary Research Fellow', and by the Department of History and Philosophy of Science at the University of Cambridge through an attachment as 'Affiliated Scholar'.

Amongst the numerous individuals who provided general advice and support, I need to acknowledge the co-founders of the international research network CASTI ('Conceptual Approaches to Science, Technology and Innovation'), Tim Flink, the late Benoît Godin, David Kaldewey, Eric Schatzberg, Désirée Schauz, and Apostolos Spanos. CASTI proved a most stimulating context for scholars with differing but interacting interests. At various stages, the manuscript and parts of it have been read, criticised, and responded to by generous colleagues.

I am indebted to them for their careful and frank criticisms and helpful suggestions. Those colleagues include Jon Agar, Dominic Berry, Philip Gummett, Roger Highfield, Daniel Kevles, John Langrish, Roy MacLeod, Peter Reed, and Audra Wolfe. Conversations with such colleagues as Bernadette Bensaude-Vincent, Peter Collins, Stephen Davies, Philip Gummett, Frank James, and Helmuth Trischler are much appreciated. I am grateful too for the chance to talk to those who have made or are making British science policy in the twentieth and twenty-first century, including Sir Geoffrey Allen, former secretary of the Science Research Council; Jonathan Pearce, associate director, Biological Medicine, Medical Research Council; the late Sir Toby Weaver, former deputy secretary of the Department of Education and Science; David, Baron Willetts, former UK Universities and Science Minister; Louise Wood, formerly deputy CEO of the National Institute of Health Research; and the participants and speakers in the meeting on the role of the Science Minister organised by the Mile End Group of Queen Mary, University of London, and the Science Museum. A wider circle of supporters made possible the writing of this book through the COVID-19 crisis. I am most grateful for the moral support provided by regular online and physical meetings with Alan Banes, Geoffrey Cantor, P. Thomas and Nan Carroll, Paul Greenhalgh, Roy MacLeod, David and Margaret Miller, Lewis Pyenson, Tony Rose, and Becky and Jeffrey Sturchio.

I have valued the comments made by physical and virtual audiences at the following presentations: the 7th Three Society Meeting, Philadelphia; Science in Public conference, London; the 24th International Congress of Science, Technology and Medicine, Manchester; the International Conference on the History of Concepts, Bielefeld; the Being Modern conference, London; the 'Re-energising Ideology Studies' conference, Nottingham; the 'Universal Histories and Universal Museums' project, Paris; the Kranzberg Lecture at the 45th ICOHTEC conference, St Étienne; also at the numerous conferences and workshops held by the British Society for the History of Science, CASTI, European Social Science History Association, the History of Science Society, the Society for the History of Alchemy and Early Chemistry, and the Society for the History of Technology; and at the departmental workshops of the Department of History and Philosophy of Science at Cambridge University; the Department of Science, Technology and Society at University College London; the Science Museum; and the Science, Technology and Society Circle, Harvard University.

I must thank publishers for their permission to reuse short sections previously published in articles and chapters. Chapters 1–3 are derived in part from an article published in *History and Technology* on 30 May 2014 © 2014 Trustees of the Science Museum, available online: www .tandfonline.com/10.1080/07341512.2014.921416, reprinted by permission of Taylor & Francis Ltd, on behalf of Trustees of the Science Museum. The journal's website is www.tandfonline.com. I am grateful to the Royal Society for permission to republish material in Chapter 1 from 'Kantianism, Cameralism and Applied Science', *Notes and Records of the Royal Society* 72, no. 3 (2018): 199–216. Berghahn Books has kindly allowed me to draw upon material I first published as 'Categorizing Science in Nineteenth and Early Twentieth-Century Britain', in David Kaldewey and Désirée Schauz (eds.), *Basic and Applied Research: The Language of Science Policy in the Twentieth Century* (New York: Berghahn Books, 2018), 35–63, particularly in Chapters 6 and 7. In addition, parts of Chapter 6 were first published in my article, 'Modernity and the Ambivalent Significance of Applied Science: Motors, Wireless, Telephones and Poison Gas', in Robert Bud, Paul Greenhalgh, Frank James, and Morag Shiach (eds.), *Being Modern: The Cultural Impact of Science in the Early Twentieth Century* (London: UCL Press, 2018), 95–129. I am grateful to the Science Museum for permission to republish this material. Quotations from Crown Copyright public sector information held at the National Archives and published © Crown Copyright material are licensed under the Open Government Licence v3.0 www.nationalarchives.gov.uk/doc/open-government-licence/version/ 3/. Quotations from Parliamentary material contain Parliamentary information licensed under the Open Parliament Licence v3.0 www .parliament.uk/site-information/copyright-parliament/open-parliament-licence/. Other permissions for individual quotations are indicated with the relevant references. I am grateful to all the many organisations that have been helpful and to Mr Richard Haldane for his generous support and permission to quote from his grand-uncle's writing.

The advice of the editors and referees of articles I have published on the topic of applied science, including Bernard Lightman, Martin Collins, Larry Holmes, and Michael Freeden, is deeply appreciated. I also owe a deep debt of gratitude to the commitment and support of Jim Secord and Simon Schaffer and the commissioning editor, Lucy Rhymer.

This book is informed by my experience and interactions over four decades as a curator at the Science Museum in London and particularly by my initial ten years of experience as curator of the Industrial Chemistry collection. During that period, I was privileged to work with Gerrylynn K. Roberts. Together, we authored *Science versus Practice* (Manchester: Manchester University Press, 1984), exploring the development of ideas about pure and applied chemistry in Victorian Britain. I should like to acknowledge her contribution to the longer-term development of my thinking. For the errors that survive, I alone am responsible. Throughout the process, Lisa Bud-Frierman and Alexander Bud have engaged with repeated enthusiasms over discoveries, small and large, maintaining interest and support throughout.

Introduction
The Biography of a Concept

With a swoop of the imagination, the 1956 promotional film 'Atomic Achievement' transports viewers from the awesome phenomenon of the sun's energy to the launch of the world's 'first' full-scale nuclear power station – and onward to the future.[1] Beginning with a dramatic reminder of the cosmic potential of 'the atom', the film zooms in on Britain's need for a new source of electricity. It shows the solution: the country's new power station and the nation's great Harwell laboratory behind the research it embodied. Repeatedly, the film returns to Harwell, for, the narrator reminds us, 'no new success on the application of theory to practice is regarded as final'. In sharp contrast to the laboratory's small pilot reactor, the viewer also sees the large, visually arresting factories that produced the uranium fuel for electricity generation, generated the electricity in the full-scale power station itself, and made the plutonium needed for atomic weapons. The sponsors intended the spectacular industrial development to seem the result of scientific research and a symbol of the nation's modernity. They were drawing upon the familiar framework of triumphant tales of 'applied science'.

By contrast, the very existence of applied science as a separate and important category has often been questioned. Criticisms have come from two directions. To some, there is only a single 'science', and any distinction between pure and applied is arbitrary and mistaken. For others, the application of science alone is utterly inadequate as a prescription for industrial success. In their eyes, different skills, expertise, and processes are critically important to innovation, and scientific research should be neither privileged nor isolated. Both arguments have been frequently and articulately voiced, and in some countries, such as France, the concept has not achieved a stable interpretation. However,

---

[1] 'Atomic Achievement' (1956). Online at https://media.nationalarchives.gov.uk/index.php/atomic-achievement/ (consulted January 2020), including a transcript of the script. See Chapter 7 of this volume for a more extended treatment of atomic research and of this film.

1

elsewhere, such as in the United Kingdom and the United States, applied science served as an important concept from the mid-nineteenth to the late twentieth century. Policymakers, politicians, journalists, and public debate deployed it within such different contexts as educational curricula, professional development, the bureaucracy of research policy, and the interpretation of modern times. Applied science was credited not only with facilitating change but also with shaping a civilisation. Indeed, national and local issues so enriched the concept that its historians need to choose their geographical focus.

In Britain, the concept has been associated with knowledge, policies, and attitudes adopted to keep the nation 'modern'.[2] This book, therefore, addresses the question: What made applied science seem such a potent economic, cultural, and political elixir in the United Kingdom for many decades, and then saw it superseded? The term has had epistemic connotations; it has been promoted, and blamed, for its science policy implications, and once weighed heavily as a cultural reality. These meanings have not been separate. The narrative shows how the category's confection has followed from interactions between organisational and professional promotion, the state's ambitions and their limitations, public debate on the prospects for British industry and civilisation, and familiar allegorical stories.

Addressing the long-term history of a concept, even in a single country, is an ambitious task. Britain has been continually susceptible to overseas ideas and models, particularly from those countries that seemed to pose commercial, military, or cultural challenges, such as France, Germany, and the United States. Though political cultures, institutional forms, traditions of knowledge, and intellectual contexts have differed, concepts have circulated internationally, countries have held problems in common, and they have shared diverse influences. Britain has been influential, and focusing on its single context should contribute to understanding the international picture. Yet this approach also suggests that to understand the range of meanings and uses of applied science, even in English-speaking countries like the United States, one will need to turn to their own local traditions and specific concerns.

---

[2] This book follows the parlance of the period under study, and historians' practice, to use the terms 'Britain' and 'British' interchangeably with 'United Kingdom', its people and adjectives. Technically, the term 'Great Britain' refers just to the union of England, Wales, and Scotland. After 1801 the term 'United Kingdom' referred to the further union of this body with the whole of Ireland, and, from December 1922, with Northern Ireland. This legal term, however, was rarely used in the vernacular during most of the period covered by this book. It should be emphasised that the scope of this volume relates to the whole United Kingdom.

The book builds on several generations of scholarship investigating British science as construed by people outside the scientific elite as well as within it. Furthermore, newly available resources make it possible, as never before, to incorporate the active contribution of laypeople. Digitised newspapers and accessible government archives allow the researcher to study the circulation between specific policy contexts and the mass media of newspapers, museums, and broadcasting. So, modern resources permit us to follow the past interactions of specialists, politicians, and public opinion in thinking about the very structure of science. There is also a contemporary resonance: I hope this volume will support today's thinking about the roles of science in our own divided societies.

Timespan and Issues

As the eighteenth gave way to the nineteenth century, suddenly, the implications of the advances in knowledge seemed enormous. On the same day, 1 January 1818, the public met two important characters for the first time. 'Frankenstein' was the inventor who drew on advanced science to produce a 'creature', in the new novel by Mary Shelley. The misguided professor and his monstrous invention would appear as repeated metaphors in warnings over the role of science.[3] The other character introduced that day was 'applied sciences'. This was the creation of Samuel Taylor Coleridge, the great poet and Shelley family friend.[4] It, too, came to serve as a companion through turbulent times, both characterised by enduring canonical stories and repeatedly modified in meaning.

Therefore, the early years of the nineteenth century demarcate one end of this book; developments at the beginning of the twenty-first century mark its end. Though there has been no history of applied science over this period, we can benefit from a rich scholarship suggesting narratives of the times, appropriate questions, and newly available sources. Accordingly, this introductory chapter explores the contemporary literature on whose shoulders the volume stands. It also highlights the issues at stake and outlines the book's structure.

---

[3] See Turney 1998; Shelley 2017.
[4] Coleridge had used the term 'applied sciences' in the prospectus for *Encyclopaedia Metropolitana*, which was published late in 1817, but it was introduced to the public in the 'Treatise on Method' used as an introduction to Part One of the *Encyclopaedia Metropolitana*, published on 1 January 1818. Mary Shelley was a contemporary of Coleridge, daughter of his friend William Godwin, and in *Frankenstein*, she quoted lines from Coleridge's 'The Rime of the Ancient Mariner'.

Many of the earlier discussions were shaped by controversy over applied science's importance. Typically, economists once accorded little attention to the scientific origins of inventions when they were accounting for industrial change.[5] Meanwhile, historians of science tended to foreground the discovery of the most general knowledge of nature. Such condescension of others has long irritated its enthusiasts. Thus, in the mid-nineteenth century, Scottish natural philosopher and inventor David Brewster objected to the absence of 'balloons, and steam-boats, and steam-guns, and gas illumination, and locomotive engines, and railways' from Whewell's multi-volume history of science.[6] A century later, the status of applied science lay at the heart of debates over the rebirth of British prosperity after World War Two. The published text of C. P. Snow's much-cited 1959 Rede lecture *The Two Cultures* uses the term eight times and three times on the last two pages. Snow insisted that the disdain too often shown even by pure scientists was unjustified, for the methods of thinking were the same across the whole scientific spectrum.[7]

For many, there was nothing to be doubted about the Rede Lecture.[8] Nor was the wide range of references, including education, culture, research and industrial benefits, and the timescale ranging between the early nineteenth century and the future unusual. Yet, three years later, the Cambridge academic F. R. Leavis published a famously biting critique, warning that Snow was not talking scientifically, or even exclusively about science, but about education, values, and international competition.[9] Ironically, Snow may have agreed about the broad implications of his approach, but, within a generation, treatment of such a network of issues through the concept of 'applied science' would seem old-fashioned. In later writings, many authors have silently translated former uses as 'technology'. Time and again, 'technology', with its somewhat different meanings, competed to describe the space between science and practice. Throughout this book, we will see a jostling between the two interpretations in policy and rhetoric.

---

[5] On the attitudes of economists, see Godin 2017, 66.
[6] David Brewster, 'Whewell's History of the Inductive Sciences', *Edinburgh Review* 66 (1837): 110–51, on 147.
[7] Snow 1993.
[8] Stefan Collini, 'Introduction', in Snow 1993, vii–lxxiii; Ortolano 2009. Both Collini and Ortolano reflect on the ways Snow expressed views of the time, and indeed pre-war years.
[9] Leavis 2013, 59. Ortolano has explored the differing views of Snow and Leavis, and of their friends towards both society and science. See Ortolano 2009, 182–93. Snow's friends in the 1950s and 1960s were proponents of 'planning', a 1930s theme taken up here in Chapter 6. Leavis and his friends were defiantly against.

Technology was not a new competitor. From the mid-nineteenth century, its attraction expressed British respect for German ways of thinking and doing. As a result, this category and its German antecedent *Technik*, evoking culture and skills beyond science, are a constant presence in this account.[10] Throughout, readers will encounter attempts to find a domestic equivalent to such models of thinking about science, industry, and society. The respect for Victoria's husband, Prince Albert, a typical enlightened German ruler; his advocacy of the 1851 Great Exhibition; and the institutions created in its wake was an early sign of an enduring aspiration.

The period covered by this book was short enough for just three lifetimes of experience and, indeed, for the ragged survival of stories handed down between generations. The brevity of these two centuries in terms of memory, despite their eventfulness, shapes this story. Particularly in the 1950s and 1960s, British historians remarked on the apparent parallels between modern campaigns for the 'proper' recognition of science and the ambitions of such nineteenth-century advocates as the chemist Lyon Playfair.[11] Activists attributed the country's recent economic difficulties to its failure to heed the warnings of the previous century. Many shared the analysis that the issues of the mid-twentieth century were substantially the same as those discussed a hundred years earlier.

Late in the 1960s, despite considerably increased investment, science had seemingly failed to spawn the wealth its promoters had promised. The economies of other countries, which were, at that time, funding science less well, particularly Japan and Germany, grew faster, and the voice of engineers at home grew louder. Consequently, in both Britain and the United States, the adequacy of the category of applied science came into question. Analysts were redefining it even at the end of the twentieth century. In 1997, the American political scientist Donald Stokes explored the relationship between basic research conducted out of curiosity and applied research, expressing what modern terminology 'ought' to highlight in his book *Pasteur's Quadrant*.[12] Though the use of the term faded early in the twenty-first century, with the extraordinary growth of the biomedical sector and, specifically, health services research, a new concept of translational research became popular in policy circles, drawing on the old template.

---

[10] However, *Technik* was theorised by philosophers to a far greater degree than 'applied science' ever was. See Schatzberg 2018, 97–117. For German historical literature, see, for instance. Dietz et al. 1996.
[11] Crowther 1965; Cardwell 1972.   [12] Stokes 1997.

Methods and Historians

Categorisation within science and technology, and differentiation between the two, proved invigorating challenges for swiftly growing disciplines associated with science and technology studies of the 1960s.[13] At that time, many sought to sweep away problematic terminology. Thus, writing early in the 1970s, Otto Mayr seemed to be a little-heard voice when he called for historians' understanding of past reflections on the relationship between science and technology.[14] Only in more recent years have the roots and evolving meanings of the classification of science attracted attention.[15]

The meaning of the term 'applied science' has often been shaped by the relationship between the lay public and campaigning professionals. When, in 1995, Ronald Kline published his now-classic paper reflecting upon leading American engineers' public speeches and general articles over the past century, he looked at their search for status to explain engineers' willingness to show subservience to pure science. Kline observed the use of applied science moving from teaching to research. He found, too, a shift away from talk of applied science as a thing, characterised by empiricism and trial and error as in Edison's laboratory. Instead, the phrase came to connote 'original research conducted in industrial labs by university-trained scientists and engineers using "scientific methods"'.[16] Since Kline based his findings on just the engineers' approaches, he suggested that historians also extend his work to study the rhetoric of both other scientific groups and non-scientists.[17] Britain is an appropriate site for taking a further step in this direction, as the historiography of the engagement of the wider British society in science is particularly rich.[18]

---

[13] For the importance of science-technology relationships to the history of technology as a discipline, see Staudenmaier 1985.

[14] Mayr 1976, 671.

[15] See, for instance, the 2005 Stevens Institute of Technology conference whose proceedings were published as McClellan 2008.

[16] Kline 1995, 209. Kline sees four different connotations: the application of pure science; the application of scientific method to the useful arts; an autonomous body of knowledge which could be taught; and practices of research, teaching and innovation. The same variety of connotations can be seen in the speeches and writings of British engineers, research leaders, and scientists.

[17] Kline has also noted continuities in rhetoric from the educational to the research eras. I should like to point also to the important work of Paul Lucier, highlighting the conflicting moral economies implied by the terms 'pure' and 'applied' science in Gilded Age America. The former represented a distance from the commercial marketplace, while the latter expressed a much closer proximity. See Lucier 2012.

[18] See, for instance, from a huge literature, Nieto-Galan 2016. For a summary of literature on science and the twentieth-century British press, see Bud 2020.

Public talk about science has also signified far more than science and even engineering narrowly defined. This book frequently refers to Eric Schatzberg's study of the term 'technology' as used in America.[19] It points to a tension between a purely instrumental role and technology's identity as a broader expression of a distinctive branch of human culture. In recent years, historians have become more interested in the diversity of contributions that have made up such discussions.[20] Scholars studying public attitudes have moved from talking about 'popularising science and technology' to 'public engagement with science and technology', and historians can still do more to follow the profound implications of this shift.[21]

In daily speech, 'applied science' has often evoked the 'sociotechnical imaginaries' that Sheila Jasanoff and Sang-Hyun Kim understand as 'collectively held, institutionally stabilized, and publicly performed visions of desirable futures, animated by shared understandings of forms of social life and social order attainable through, and supportive of, advances in science and technology'.[22] In linking collective beliefs about how society functions to 'modernity's great aspirations' and 'adventures with science and technology', Jasanoff identifies four long-standing problems.[23] She reflects on the difference in outcomes of superficially similar debates in different countries, the interpretation of past and future, the meaning of space, and the relationship between collective formations and individual identity. Imaginaries, and the problems they address, appear vividly in the discussion across the whole of our period. Repeatedly, this book finds interactions between publicly shared imaginaries and planning for the use of applied science as a tool.

We find anxiety over an imminent British 'decline', threatened, and bemoaned, since the mid-nineteenth century, permeating this book's landscape. From mid-century, and particularly from the 1870s, there was concern that manufacturing industries in other countries, and Germany and America in particular, were overtaking British enterprise in both innovation and scale. Looking back, David Edgerton has shown in multiple studies of 'declinism' that interpretations of Britain's past have tended to over-emphasise economic decline and the anxiety it caused, underestimating the importance of 'techno-nationalism' and military ambition.[24] Building on such warnings that threats of 'decline'

---

[19] Schatzberg 2018.
[20] See, for example, Secord 2004; Bensaude-Vincent 2009; Bowler 2009; Nieto-Galan 2016.
[21] Miller 2001.   [22] Jasanoff 2015, 4.   [23] Jasanoff 2015, 5.
[24] For an exploration of the impact of 'declinism', British techno-scientific rhetoric, and policies, see Edgerton 2006. For the deployment of the concept after World War Two,

should be read as a rhetorical strategy, this book is concerned with following up the consequences of debate that deployed such arguments. Its treatment of institutional development during the later twentieth century can also take advantage of archives that have only recently opened. However, rhetoric based on fear of failure provided much earlier opportunities for backers of applied science. In 1874 the Earl of Derby launched the Society for the Promotion of Scientific Industry, dedicated to averting national economic disaster through its antidote, applied science.[25] The new organisation issued a medal linking Britannia and Athena (see Figure I.1). A chemical flask, the centrifugal governor of steam-pioneer James Watt, and a hammer were set against a background of belching chimneys and a train passing over a bridge. Such allegories would long outlast this Society.

In Britain, movements within civil society were also responsible for the development of educational institutions, which again gave meaning to applied science. In 1800 there were eight long-standing universities in the country. Just two were in England and Wales (Oxford and Cambridge), five in Scotland (Glasgow, Edinburgh, St Andrews, and two in Aberdeen), and one in Ireland (Trinity College, Dublin). Like many others in Europe, Oxford and Cambridge Universities were not in good health by the late eighteenth century. They were largely devoted to preparing young men for a career in the Anglican church. Several other countries achieved academic renewal through initiatives of the state allied with demands from a qualification-hungry public service. However, amongst the British, typically, private rather than public interests and institutions drove educational initiatives.

Educational innovation in nineteenth-century Britain generally came through establishing a wide variety of privately funded colleges, private medical schools, and trade schools. Numerous industrialists and politicians expressed scepticism of the vocational benefits of academic knowledge, believing primarily in the virtues of learning in the workplace. However, scholars, men of science, and educational campaigners were greatly impressed by European institutions, first French and then German. The federal University of London was formed before the mid-century, and, in the North of England, the federal Victoria University centred in Manchester emerged during the 1880s. At the

---

see Edgerton 2018. More generally, see English and Kenny 1999; English and Kenny 2000; and English and Kenny 2001.

[25] See, for instance, 'Lord Derby on the Promotion of Scientific Industry', *The Bradford Observer*, 17 January 1874, British Newspaper Archive/British Library Board. Derby called for Britain to obtain an advantage in 'applied science'. See Farrar 1972.

Figure I.1 Scientific imaginary: medal of the Society for the Promotion of Scientific Industry, 1874. © Science Museum Group.

beginning of the twentieth century, an educational revolution occurred: in some cities, private colleges amalgamated, and elsewhere component colleges of the Victoria University became independent, to form 'civic universities'. Uniquely, the complex of government-funded colleges and museums in London's district of South Kensington, building upon the physical and economic legacy of the Great Exhibition, provided an opportunity for a home-grown version of the 'continental' model.

Across the period, institutions establishing their rules and vocabulary for professional stakeholders also became publicly recognised brands.

I shall suggest that 'applied science' served as part of the language by which enthusiasts drove institutional development.[26] Through curricula, familiar names of departments, their signature achievements, and their supporters' rhetoric, such bodies fixed the meaning of 'applied science' in the public imagination. Conversely, the concept and its associated ideologies have performed work in shaping the identities of modernising social movements.[27]

The ambitions have been high: from the exorcism of British traditions of 'rule of thumb' manufacturing in the nineteenth century to industrial rationalisation in the interwar years, to defiant modernism represented by stories of penicillin, the jet engine, television, and radar in the years after World War Two.[28] The creation and policies of institutions countering decline have frequently been responses to the work of imaginaries. The subsequent fate of the concept of applied science was part of the late twentieth-century reorganisation of the institutions that embodied it.[29]

From the mid-nineteenth century, applied science also lay at the heart of British industrial policy. Even at that time, science had a reputation as 'public knowledge'.[30] Looking at a later period, the analysts Mark Harvey and Andrew McMeekin point to the orthodoxy widely shared after World War Two. In their book *Public or Private Economies of Knowledge?* they point to a distinction between science and technology that prevailed then and indeed long before: 'science was inevitably somehow public knowledge and technological knowledge was private knowledge'.[31] Applied science sat at an unstable position between these two, as Harvey and McMeekin had suggested in an earlier article: 'The boundaries are both shifting and "fuzzy".'[32] This characterisation was important because the distinction between private and public has been critical to the expectations of intervention by successive governments. As early as 1795, the political thinker Edmund Burke famously warned that one of the hardest problems in British legislation was establishing 'What the state ought to take upon itself to direct by the public wisdom, and what it ought to leave, with as little interference as possible, to

---

[26] The philosopher John Searle coined the term 'institutional act' to describe the use of language made compelling by an institution. Searle 2010.
[27] Van Dijk 2013.
[28] For more on 'defiant modernism', see Bud 1998. See also Jones 2004.
[29] See Boden et al. 2004.
[30] See Ziman 1968; Golinski 1992. On the prevalence of secrecy for military or commercial reasons, see, for instance, Galison 2004.
[31] Harvey and McMeekin 2007, 10. See also Nelson 1989. Although nominally concerned with technology, Nelson deals too with such applied sciences as metallurgy and pathology.
[32] Harvey and McMeekin 2010, 482. See also MacLeod and Radick 2013.

individual discretion'.[33] The characterisation of science as 'public' has meant that time and again, politicians accepted the provision and diffusion of such knowledge and the scientific training of the workforce as legitimate objects of support by the British state.[34] In contrast, until the second half of the twentieth century, government participation in private business ran counter to a philosophy of non-intervention in the market.[35] Even in support of the military, government-performed research was often carefully distanced from what was considered the sphere of private corporations.

Elsewhere, most notably among German states, the situation has been different. Prussia, for instance, as part of its efforts to industrialise, promoted a clustering of theoretical and practical knowledge. In a government-led endeavour to catch up with Britain, concerns about private and public boundaries were less important than across the North Sea. Over recent years, Ursula Klein has suggested that in Prussia, one saw the aggregation and teaching of heterogenous 'useful sciences' bringing together knowledge, irrespective of ownership.[36] This pragmatic aggregation, she argues, has had long-term consequences in Germany and, more broadly, has reached into the 'technoscience' of the present day. France, too, has accorded knowledge a different structure.[37]

In recent decades, historians of Britain have paid renewed attention to the effect of being part of a global empire and Commonwealth, particularly before World War Two. A large literature now explores how visionaries dreamed of finding raw materials overseas and using them in British industry and locally – to benefit empire, metropole, individual colonies, and themselves.[38] Scholars such as Michael Worboys have shown how engagement with colonial problems profoundly affected the development within Britain of a constellation of disciplines around applied biology.[39] Sabine Clarke's study of development policy in the Caribbean after World War Two explored the distinction of 'fundamental' research appropriate to the government-funded laboratory from the role of private

---

[33] Burke 1800, 45. This was originally a 1795 memorandum to the Prime Minister. On laissez-faire and state intervention, see Mandler 2006.
[34] Even during the nineteenth-century 'Revolution in Government', the British attitude to government was distinctive. See essays in Thompson 1993. Historians of science have noted the interdependence of concepts of the distinctive role of government and concepts of the role of science. See, for instance, Olby 1991; Clarke 2018; and Agar 2019.
[35] This formulation draws heavily on Dobbin 1994, 158–212.   [36] Klein 2020, 229.
[37] On French terminology, see Bensaude-Vincent 2020; on German conceptions, see Schauz 2020.
[38] See, for instance, Drayton 2000; Hodge 2011.
[39] Worboys 1979; see also Worboys 1996.

firms.[40] Moving away from an earlier language of diffusion from the centre, the metaphor of 'networks' has proven valuable. The modern literature has examined central institutions in several countries, agriculturalists, and colonial officers around the globe and the communication between them.[41] It has shown how such imperial contexts as Ireland served as laboratories, testing institutional concepts later implemented in Great Britain.[42] The biographies of many individual lives across the empire have been traced, following young graduates of such British institutions as the Royal School of Mines seeking overseas positions and colonial scientists passing part of their careers in Britain.[43] But, repeatedly, scholars have found that the significance and political meaning of educational and research trajectories remained locally specific, and have warned against assuming the simple transposition of cultures across the world.[44]

From the end of World War Two, British conceptions of applied science were linked closely to American policymakers' interpretations of the innovation process, which came to dominate the entire 'West'. The term 'applied research', often used interchangeably with 'applied science', has often been counterposed to basic or pure research within broader strategies of science policy. Because historians have recently come to be interested in the history of such categories, they have sought out new tools for their exploration.[45] Those tools proved useful in the sophisticated study of the category of 'technology', incorporated first in the idea of 'technological change' and, after World War Two, in 'technological innovation'.[46] Building on such literature, this book explores how the language of 'applied science' came to seem outdated.

The wealth of public usage is an important thread running throughout this study. It links the raging arguments over institutional change and the commitment of vast resources to educational, military, and industrial advancement. Disputes employing 'applied science' occurred in newspapers, in Parliament, and in public speeches. At times, they were to be heard first in the discussions between intellectuals or the interplay within universities and government departments.[47] Campaigners summoned up

---

[40] Clarke 2018, see, particularly, chapter 4.
[41] On the roles of colonial officers and others as hybridisers, see Beinart et al. 2009; and Tilley 2011.
[42] Macleod 1997.   [43] See, for instance, MacLeod 2009b; Roberts and Simmons 2009.
[44] Chambers and Gillespie 2000.
[45] Schauz 2020. See also Bud 2013b; Kaldewey and Schauz 2018; Müller and Schmieder 2018.
[46] Schatzberg 2018; Godin 2019; Wisnioski 2012.
[47] For a classic study of talk by German physicists on grand public occasions, see Carson 2010.

'applied science' to encourage supporters, denounce opponents, and argue for a course of action such as a new educational policy. Even within a single language such as English, each context has its vocabulary, repertoire of allegorical stories, and appropriate podia.[48]

Such important terms have been explored under a variety of designations. 'Keywords and 'ideographs' have been proposed, for instance, but the historiography around 'concepts', formulated in Germany during the 1960s, has been the most developed.[49] Here, it is worth emphasising three points. First, concepts differ from mere words by the multiplicity of their applications and meanings. These coexist in multiple layers so that new connotations do not oust old ones but instead enrich them. Second, they have also helped define our sense of time. Third, conflict over meaning is normal, and all interesting concepts are what Gallie called 'essentially contested'.

Terms often win their connotation through stories traditionally told to embody morals and connect with tradition. Quivers of canonical stories can provide vital sustenance to social movements.[50] Even if the retelling of triumphs, such as the development of penicillin, appears tritely formulaic, accounts of the past can reveal ideology. Why ideology, and not just 'frame of thought' or some less tendentious word? I suggest that the familiar allegories were cited together to convince and press for changing priorities in vigorous political debate. Such interpretations of the past to guide future action have given significance to their key terms. Their very familiarity gives them the cohesive power to hold together social movements intended to recast the future. They lie at the heart of the language of appeals to established authority in the search for power.[51] Michael Freeden also relates their analysis closely to conceptual history, suggesting that ideologies serve as clusters of concepts.

The history of the phrase and the concept behind it are inseparable, for, as this book shows, there was no pre-existing idea expressed through a series of linguistic devices.[52] Nor, inversely, was the term used to describe completely unrelated concepts. Instead, until the late twentieth century, its use in argument served to refine the ideas it expressed. At the

---

[48] Explored in more detail in Bud 2013b; see, particularly, two classic articles: Georges 1969 and 1994.

[49] Introducing a huge literature on concepts, see for instance, Koselleck 2002. For alternatives to 'concepts', see Gallie 1955; McGee 1980; and Williams 1976.

[50] Fine 1995; and see Klandermans and Roggeband 2007. On the concept of 'techno-tale', see Sumner 2014.

[51] Freeden 1996, 22–23.

[52] The possibilities of approaching conceptual history by starting with concepts or with language are reviewed by Richter 1986. The argument here is of course related also to the 'language act' methodology of Quentin Skinner; see, for instance, Tully 1989.

same time, it is necessary to study the deployment of related terms and, above all, to look at the competing concept of 'technology'.

The instruments are now on the table: institutions, narratives, sociotechnical imaginaries, concepts, and ideologies. They come from different disciplinary backgrounds. Yet combining them offers a useful and coherent family of analytical techniques. The jargon will not often appear in the history that follows, yet its work permeates the text.

The Digitised Mass Media as Resource

Concern with the language used in public makes appropriate the use of newspapers' recent large-scale searchable digitisation.[53] Between the early nineteenth century and World War Two, newspapers printed speeches by politicians in Parliament and around the country at full length. They often also reported the proceedings of meetings held, for instance, to discuss local universities' development, voicing their own opinions and publishing readers' letters. In addition, the numerous advertisements provided a record of the vernacular. A wide distribution of papers, by region and by class orientation, is available for both the nineteenth and the twentieth centuries through the online 'Britishnewspaperarchive' corpus.[54]

With such a wealth of data, it would be tempting to seek rigorously quantitative results. Nevertheless, dependence on numbers alone would presume a false indication of an unachievable accuracy.[55] The quality of photography and the accuracy of optical character recognition are variable, and the papers provide an uneven representation of different views. One cannot rely on such bases, and needs additionally to turn to diverse forms of support.[56] Combining quantitative analysis with careful inspection of the content has served to reveal significant trends in public speech.

Journalists won an additional platform late in the 1920s when broadcasting, first radio and later television, supplemented print as opinion-former and mirror. Even in earlier periods, the printed word did not have a monopoly, for performances, celebrations, theatre, and film have each proved persuasive media. Temporary exhibitions and

---

[53] The term 'public realm' has been used interchangeably in this book with 'public sphere'. Jürgen Habermas' work on the public sphere deeply informs this work, but not his idealisation of a particular era. See Mah 2000.

[54] See www.britishnewspaperarchive.co.uk, British Newspaper Archive/British Library Board.

[55] For the use and limitations of the corpora of digitised newspapers, see Milligan 2013; Manuel 2021.

[56] Sewell 2005, 370.

permanent museums could attract hundreds of thousands or even millions of people, making their leaders into influential agents. In exploring the language through which founders argued for universities, colleges, exhibitions, and museums, this study has turned to the private archives of institutions. We shall follow the publications and other media that used the term 'applied science', the institutions and associations' inner lives giving it meaning, their buildings embodying it, and the stories interpreting it.

The Book's Structure

The narrative is structured in three 'stages': 'Origins and Pedagogy in the Nineteenth Century', 'Research in the Early Twentieth Century', and 'After World War Two'. In Stage 1, three chapters explore the term's introduction, its acquisition of a rich range of significance with several layers of meaning, and the early relationship to 'technology'. Above all, references to the proper and necessary subjects of teaching came to dominate. In Chapter 1, we see the introduction of the term in the *Encyclopaedia Metropolitana* and its diffusion and development through the initiatives of King's College London, debates over agriculture and the repeal of the Corn Laws, and the enterprise of publishers. Chapter 2 shows how civic colleges provided the context for the term's mass use and its enduring claim to the public sphere. It constituted the teachable, open knowledge that complemented private proprietorial expertise learned on the job. Concluding this section, Chapter 3 deals with the complementary emergence of 'technology' – a word linked more closely to practice and to German and American usage, and a concept perennially competing with applied science.

In Stage 2, three chapters pursue the term across the early twentieth century. The dominant reference changed from pedagogy to a style of research applying 'pure science'. Beginning the exploration, Chapter 4 takes the narrative through the crucial years of the national efficiency movement leading up to World War One. It puts the talk then in the context of the emphasis on research across the early twentieth century. New institutions such as the National Physical Laboratory and a new political landscape shifted the emphasis from the previous focus on pedagogy, yet there was continuity in the use of the term and in the underlying ideological constraints. Chapter 5 takes the narrative into the interwar years during which the concept's meaning of applied pure research was firmly established, in close association with new institutions, such as large laboratories and the Department of Scientific and Industrial Research founded during World War One. Here the book describes in more detail the ambitions of key research programmes. They were important both to scientists and to the lay community, whose attitudes are the focus of Chapter 6. This addresses the broader cultural standing of applied science. In the aftermath of World War One, optimism was challenged by people worried about the human ability to cope with the changes brought about by science to modern society. The enthusiasms of proponents and the anxieties of sceptics clashed over modernity and rationalisation, animating the discourse of applied science.

In Stage 3, two chapters deal with the post–World War Two years. Again, governments and industry adopted scientific research to solve enduring social and economic problems. Increasingly, however, there was concern about the adequacy of applied science as the basis for industrial policy, and the concept competed, ultimately unsuccessfully, with 'technology'. Chapter 7 shows how the two coexisted in the 1950s in educational and research policy. It explores the experience of the UK Atomic Energy Authority, analysing tensions between the science-oriented research centre at Harwell and the design team in the North of England. Chapter 8 explores the shift towards technology in government policy during the 1960s. Building on community and political interest in an impending 'industrial revolution', an interventionist government introduced the 'Ministry of Technology'. This chapter analyses the tensions within the public service over applied science and technology at a time when more interpenetration between state and private interests was becoming the norm. As new governments moved responsibility for applied science to users and the private sector, the concept's significance faded. Yet the chapter shows how in the case of biosciences supporting health care, the expectations of applied science still inform the new category of 'translational research'.

Stage 1

Origins and Pedagogy in the Nineteenth Century

1 Applied Science Conceived
The Early Nineteenth Century

Introduction

Even in the twentieth century, users of the phrase 'applied science' drew on connotations developed through the turbulence of the industrial revolution and its aftermath. Then, numerous individuals and schools sought to interpret and change the relations between science, prosperity, and civilisation.[1] They addressed such problems as the connections between industrial success, the practical arts, and the sciences. How was the vastly increased scale of knowledge to be organised conceptually and institutionally? And how should society educate the young in an age in which past and future were in danger of being torn apart?[2] Talk of applied science expressed both the hopes for and anxieties about the new civilisation and contested solutions to the problems it was creating. This chapter addresses the question: How did the term become familiar in society? It explores, in detail, the earliest uses because even these demonstrate a close integration of knowledge classification and determined engagement with large audiences. Their legacy lived on into the twentieth century.

Whereas the industrial revolution propelled British manufacturers to global dominance, the country had paid a high price, as many contemporaries complained. Rapid social and cultural change allowed unlettered manufacturers to become wealthy without amassing or even valuing traditional cultural capital. In the 1830s, the philosopher John Stuart Mill reflected upon the polarisation of culture. He counterposed the enthusiasts for 'our enlightened age', who could point out such qualities as its advancement of knowledge and growing physical comforts, to those fixated on the cultural cost, the loss of individual energy, and submission to artificial wants. Mill sensed the growth of what he diagnosed as a reaction against eighteenth-century patterns. Those had been experimental, innovative, infidel, abstract, and matter of fact,

---

[1] Klancher 2013. On the arts and technology, see Schatzberg 2018. [2] Wellmon 2015.

whereas the response, what he called the 'Germano-Coleridgean' doctrine, was 'ontological', 'conservative', 'religious', 'concrete and historical', and 'poetical'.[3] The commitment of Samuel Taylor Coleridge to prioritising Christian faith, enduring truths, and the poetic, feeling mind would continue to be set against visions of progressive change and celebration of the material world. On the other side, Mill identified Jeremy Bentham, the philosopher and founder of utilitarianism, as a recent standard-bearer for the celebration of progress.

In the heat of such dissent, the term 'applied science' would gain its early meanings. Strikingly, the term's first use was intended to correct excessive 'enthusiasm' for material improvement. It also became a symbol of the knowledge needed to sustain that enthusiasm. This chapter will explore, therefore, the dialectically opposed origins of our modern term. While many layers of meaning would subsequently be superimposed, the early nineteenth century's ambivalent experience would have an enduring legacy. It underlay the aspiration for a common route to technical knowledge and liberal education.

Structuring the Arts and Sciences

'Applied science' began its English life as a quintessential 'Germano-Coleridgean doctrine', a translation from German and a contribution to the understanding and promotion of 'Civilisation'. Having been created by an intellectual, it was disseminated through the power of the book market and advertising. The early nineteenth century was a time of neologisms, and Coleridge himself coined many, including 'psychoanalytical' and 'subconsciousness'.[4] Through words, he was seeking to interpret, and to refine, a new world order. During the second decade of the nineteenth century, he endeavoured to integrate managed change in knowledge with a vision of the economy and society. Three key influences shaped this social programme: the chaos of post-revolutionary France, the economically harsh aftermath of the French wars at home, and Coleridge's German encounter through study in Göttingen.[5] The turmoil experienced in France after the Revolution, in social order and

---

[3] The essay on Coleridge was published as 'Art. 1', *London and Westminster Review* 33 (March 1840): 257–302. For the reference to 'our enlightened age', see 261, and for comparison of the 'Germano-Coleridgean doctrine' with eighteenth-century philosophy, see 263–64. The essay on Bentham was first published as 'Art XI the Works of Jeremy Bentham', *London and Westminster Review* 29 (August 1838): 467–506.

[4] See Shapiro 1985. On Coleridge's use of language as a political act, see Smith 1986, 202–51.

[5] Edwards 2012.

religious belief, would be a decisive influence on his thought – ultimately as a threat against which to react in horror. Coleridge linked it to the economic upheaval Britain was experiencing in a period of post-Napoleonic Wars distress and industrial revolution. Thus, in his 1817 *A Lay Sermon*, he warned that a reborn Plato would be seen as less worthy in modern London than an innovative 'handicraftsman from a laboratory'.[6] Coleridge feared '*the overbalance of the commercial spirit in consequence of the absence or weakness of the counter-weights*' (emphasis in the original).[7] Those counter-weights would have to include a dutiful moral leadership holding correct beliefs and capable of rightful thinking, what he called a 'clerisy' – a 'permanent, nationalized, learned order', and an appropriately educated and cared-for working class.[8]

During the period Coleridge was wrestling with social concerns, the publisher Rest & Fenner invited him to edit a new encyclopaedia. Responding positively, Coleridge felt that the education promised to the attentive reader by the publication, and the method it would promote, could contribute to the broader goal of a stable civilisation. It was in proposing an overall plan for his model of knowledge that he introduced the term 'applied sciences'. Therefore, he developed the concept's philosophical meaning together with thinking about its utility to the publishing project.

The term's use in the proposal and then the introduction to the *Encyclopaedia* also had a more international context.[9] German literature, in particular, directly influenced the writing of Coleridge.[10] Consequently, to interpret the structure of the *Encyclopaedia Metropolitana*, it is helpful to bear in mind four related factors: the construction of useful knowledge, particularly in Germany; the genre of the encyclopaedia; the role of Kantian categories within it; and the specific utility and trajectory of the term 'applied sciences'.[11] These came together in cameralism, the

---

[6] Coleridge 1852a, 192.   [7] Coleridge 1852a, 188–89. Emphasis in original.
[8] The definition here of 'clerisy' is from Coleridge's 1830 *On the Constitution of the Church and State According to the Idea of Each*. See Coleridge 1852b, 78. On Coleridge's thinking about clerisy, see Prickett 1979. Coleridge's thinking within romantic reflection on political economy, and the wide dissemination of organic metaphors, has been widely discussed. Gallagher 2009, 13.
[9] This prospectus was reproduced by Snyder 1940.
[10] See, for instance, Class 2012 and Keanie 2012.
[11] There is a long tradition of pointing to Kantianism as marking a new era in encyclopaedia structure. It can be found as early as Meyer's *Conversations-Lexikon*'s first edition, *Das grosse Conversations-Lexicon für die gebildeten Stände* (1846), vol. 8, 581. For a recent reflection on the role of Kantian methodology, see Wellmon 2015. See also Yeo 2001 and Klancher 2013.

eighteenth-century German study of governance.[12] At its heart lay the concept of *Wissenschaft*, a category generally translated as 'science', in flux and formation at the time.[13] It included a broad range of state-deployed bureaucratic expertise ranging from chemistry to agriculture and even the study of fortifications. Denise Phillips has suggested that the term denoted 'not just a body of knowledge; it was a form of communicative practice'.[14]

Throughout the eighteenth century, publishers sought to help curious Europeans wishing to access the increasing knowledge of the age. In France, between 1751 and 1772, the *Encyclopédie* was published in twenty-eight volumes by enlightenment scholars intent on providing a compendium of the world's secular knowledge. This achievement was widely revered in its time and is familiar today. However, during the early nineteenth century, Coleridge leant much more towards German culture and was reacting against the French *philosophes*. Germans, too, were endeavouring to encapsulate the huge and growing tree of knowledge in major educational enterprises.[15] For instance, the *Oekonomische Encyklopädie* of Johann Georg Krünitz grew to 242 volumes. At this scale, it seemed that the encyclopaedic assembly of all knowledge had reached its limit. Increasingly, analytical techniques were needed, and more structural maps of what was known would be required.

The restructuring of knowledge developed by Immanuel Kant at the end of the eighteenth century rescued the encyclopaedists. A Kantian analysis offered a widely accepted philosophical structure for an emerging genre of guides and treatments. These shorter, albeit complex, route maps to knowledge, structured around a pure and applied dichotomy, managed to embody the understanding of the time and its ambitions.[16] Moreover, because they covered natural philosophy, morals, rights and responsibilities, history, and technology, they wove into intimate unity understanding, facts, and their social and administrative implications.

Kant's 1786 *Metaphysical Foundations of Natural Science* expressed concern with truth value and the link with the Almighty. It also distinguished between a pure science (*reine Wissenschaft*), based on first principles, and applied rational cognition (*angewandte Vernunfterkenntnis*),

---

[12] On Coleridge and cameralism, see Bud 2018b. On cameralism, see Lindenfeld 1997; Meyer and Popplow 2004; Wakefield 2009; and Seppel and Tribe 2017.
[13] Phillips 2012.   [14] Phillips 2012, 220.
[15] On the encyclopaedia as a primary form of education, see Wellmon 2015. See also Garfield 2022.
[16] The first two commonly cited Kantian encyclopaediae are Eschenburg 1792 and Schmid 1810. See Wellmon 2015, 91–95, and Bud 2018b. Also see Van Miert 2017 for a more comprehensive look at German classification systems.

dependent upon empirical knowledge. Quickly, admirers substituted the word *Wissenschaft* for *Vernunfterkenntnis*. Though there had been the occasional isolated earlier use in German, by the late 1790s in Germany, the Kantians had made *angewandte Wissenschaft* (applied science) their own.[17]

Coleridge laid out his personal vision in his essay 'Treatise on Method', published on that fateful first day of 1818. Here he explained his commitment to understanding things in terms of their relations with others, rather than in isolation. Within this overall schema, the usage of 'mixed and applied sciences' served the same role for Coleridge as the German original was doing for other Kantian encyclopaedists at the time.[18] It demarcated those areas of knowledge that depended largely on empirical evidence, contrasting with the synthetic a priori of the pure sciences.

For Coleridge, 'Method' sustained social and intellectual structure. 'From the cottager's hearth or the workshop of the artisan, to the Palace or the Arsenal, the first merit, that which admits neither substitute nor equivalent, is, that *every thing is in its place.*'[19] This resulted from the integrated vision of industry, the military, science, society, and knowledge structured by a Kantian method. Certainty and timelessness lay at the heart of the structure Coleridge had been developing for two decades. Its consequence was a partitioning of knowledge:

There are, as we have before noticed, two sorts of relation, on the due observation of which all Method depends. The first is that which the Ideas or Laws of the Mind bear to each other; the second, that which they bear to the external world; on the former are built the Pure Sciences; on the latter those which we call Mixed and Applied.[20]

Coleridge distinguished between mixed sciences (mechanics, hydrostatics, pneumatics, optics, and astronomy) and the more particular 'applied sciences', referring to the 'application' of ideas to the study of properties. He further explained that in speaking 'of certain changes in those properties, or of properties existing in bodies partially, then we popularly call the Studies relative to such matters by the name of Applied Sciences; such are Magnetism, Electricity, Galvanism, Chemistry, the Laws of Light and Heat, &c.'[21] These categories were dependent on changeable theory and hypothesis. In turn, this reliance meant a lower status than

---

[17] I have explored the use by German Kantians of 'angewandte Wissenschaft' at greater length in Bud 2018b.
[18] For the *Encyclopaedia Metropolitana* as a Kantian project, see also Alice D. Snyder, 'Introduction', in Snyder 1934, vii–xxvii at xxii.
[19] Snyder 1934, 12, emphasis in original.   [20] Snyder 1934, 55.   [21] Snyder 1934, 60.

those pure sciences possessed of 'all the purity and all the certainty which belong to that which is positive and absolute' and headed by theology, justified by Divine Revelation.[22] Quite separately from his consideration of the 'application' of ideas to properties, Coleridge also reflected that 'pursuits of utility' were increasingly being dealt with 'scientifically, or, as the more prevalent expression is philosophically'.[23] Some of the most poetic lines in the essay emphatically argued for the inclusion of such useful knowledge as that promulgated by his old friend, the chemist Humphry Davy: 'where Davy has delivered Lectures on Agriculture, it would be folly to say that the most Philosophic views of Chemistry were not conducive to the making our valleys laugh with corn'.[24] Natural history and medicine, with such subsidiaries as pharmacy, also fitted into this category. The links between science and medicine were palpable, particularly to the frequently unwell poet.[25] Revolutionary improvements might be beneficial, but they needed to be rooted in the existing culture. Coleridge intended to anchor new knowledge in old certainties and find a way to bond 'trade' with 'literature', and 'commerce' with 'science'. Without such bonds there would be no civilisation. 'Woe to that revolution', he warned, 'which is not guided by the historic sense.'[26]

Coleridge intended his project to be structured intellectually, unlike earlier alphabetically ordered compendia, such as the *Encyclopedia Britannica*. This vision of the structure of knowledge, and of the *Encyclopaedia*, found a place for fine art as a separate category.[27] The publisher, however, overruled this extra complexity. Coleridge himself left the project because of this disagreement, together with the commercial decision to publish sections of different parts simultaneously, rather than in strict logical order.[28] Nonetheless, the thirty impressive volumes of the *Encyclopaedia* amounted to more than a publication; they were a cultural symbol and became the subject of much visible advertising. These qualities underpinned the wide dissemination of Coleridge's new term.

Encyclopaedia Metropolitana

Coleridge's influence would continue to be decisive, despite his early departure from the editor's seat, and an edited form of the *Treatise on*

---

[22] Snyder 1934, 57–58.   [23] Snyder 1934, 63.   [24] Snyder 1934, 63.
[25] Roe 2002; Vickers 2004.   [26] Snyder 1934, 68.
[27] On the meaning of the fine arts in Coleridge's vision for the *Metropolitana*, see Degrois 1991.
[28] On the complex full range of disagreements, see Snyder 1940.

Method served as the general introduction. Even when the publishers declared bankruptcy in 1821, the project and the vision survived. A consortium of publishers and booksellers rescued the *Encyclopaedia*, bringing in a new succession of editors faithful to Coleridge and his ambition. Through the Bishop of London, William Howley, the consortium was in touch with a close-knit band of high-Anglican scholars, generally referred to as 'The Hackney Phalanx'.[29] Conservative theology, Tory politics, family ties to industry and commerce, personal patronage characteristic of the time, and loyalty to Coleridge characterised these men. Emphasising the holiness of life and good works, nourished by the Sacrament, they also celebrated science as a branch of learning and insight into God's creation. From this network, in line with the promotion of their philosophy and community, Bishop Howley and his protégé, Charles James Blomfield, recommended a series of editors.[30]

The first editor of the *Encyclopaedia Metropolitana* was William Rowe Lyall. He would serve as chaplain to both Howley and Blomfield and editor of the group's magazine, *The British Critic*. Then, after just a year, the Reverend Edward Smedley, a frequent contributor to *The British Critic*, took over.[31] Smedley made the encyclopaedia project his own in an all-absorbing endeavour taking him to his death in 1836 and – one cannot but suspect – foreshortening his lifespan.[32] His successor was Hugh Rose, Hackney Phalanx leader, and another (former) chaplain to Howley and then to Blomfield, who also took on the leadership of the Anglican King's College London.[33] After just two years, Rose resigned out of exhaustion, and he, too, passed away shortly after. So, it was left to his brother and fellow cleric, Henry, to bring the project to a successful conclusion in 1845.

Despite this long sequence of leaders, the project consistently followed its early vision. It dealt with 'pure sciences', including theology and mathematics, in the first two volumes (each in folio format and about 700 pages). Six subsequent tomes covered the 'mixed and applied sciences'. The *Encyclopaedia*'s advertising achieved a high profile in the press and, becoming a powerful promoter of its key concepts, contained the earliest uses of

---

[29] See Mark Smith, 'Hackney Phalanx (act. 1800–1830)', in *Oxford Dictionary of National Biography*. On the general theology of this group, see Varley 2002.
[30] The roles of Howley and Blomfield in recommending Lyall were described by Lyall's nephew George Pearson in the introduction he wrote to a new edition of Lyall's *Propaedia Prophetica*. See Pearson 1885. I am grateful to Clive Dewey for this reference. For Lyall, see Dewey 2007.
[31] For Smedley's contributions to *The British Critic*, see Houghton 1979.
[32] For the year 1833/34 we have impatient notes from Edward Smedley to Charles Babbage, who was taking the lead on science in the *Encyclopaedia Metropolitana*, familiar to editors today. See BL Add. MS 37188, ff. 68, 69b, 148, 171.
[33] For Hugh Rose, see Burgon 1891, 62–152; and Valone 2001.

'applied sciences' in British newspapers and periodicals. Occasionally, the notices corrupted his original term (in the plural) to the singular and thus introduced, perhaps by accident, the phrase 'applied science'.[34]

Coleridge's introductory 'Treatise on Method' identified the potential audiences or, one should say, pupils. Coleridge himself had worried about pandering to 'the Reading Public', comparing the attitudes exhibited in modern prefaces to the gross flattery of wealthy sponsors of the past. He warned that watered-down general education would be disastrous, as he wrote, 'You begin, therefore, with the attempt to popularize science: but you will only effect its plebification.'[35] Coleridge anticipated the *Encyclopaedia* would appeal instead to the 'numerous and respectable class of readers' seeking 'daily reference' on points relevant to their 'desires or business', as well as those with 'the leisure and inclination to study Science in its comprehensiveness and unity'.[36] The cost of buying the encyclopaedia is revealing of its potential audience.[37] Advertisements in *The Times* show that the middle-class businessman seeking to acquire, or to flaunt, improvement could have obtained six volumes for little more than the cost of a piano. Although the publishers would not follow all of Coleridge's wishes, the great work did share the knowledge of the 'clerisy' with readers who would find cultural improvement and daily reference.

Initially, association with Coleridge himself was an essential part of the branding.[38] The promotion of a philosophy, the publication, and the intellectual concepts that structured the *Encyclopaedia* gained meaning from each other. Nonetheless, across a quarter of a century of publishing history, publishers would come to lay less importance on its origins in the sage's virtual argument with supporters of the French Revolution. Indeed, the ironic outcome would be a hybridisation of Coleridge's carefully thought-through categories with the classification of just those revolutionaries.

Most frequently, authors would interchange 'applied science' with the older English expression 'practical science'.[39] Since the seventeenth century, this term had been used occasionally to describe the intersection of

[34] See, for instance, the advertisement in *The New Times*, 31 January 1829, British Newspaper Archive/British Library Board.

[35] Coleridge 1852a, 79.   [36] Snyder 1934, 66.

[37] Collison 1966. The publishers offered the first volume at £2.2s. In the same issue of *The Times* (3 February 1829), readers would find a pianoforte advertised at £9.10s.

[38] Collison 1966, 756–58.

[39] In the Britishnewspaperarchive corpus (consulted December 2019) the term 'practical science' occurs 338 times for 1849, whereas 'applied science' was used only 90 times. By 1869 the numbers were roughly equal ('practical science' 185 times; 'applied science' 213 times), and by 1879 the balance had shifted decisively: 163 'practical science'; 313 'applied science'.

artisanal concerns with scientific achievement. It might have expressed the enduring Baconian tradition in Britain, but traditionally it did not connote the application of theory. Instead, practical science often represented a useful counterpart to 'abstract science', based on observation rather than abstractions.[40] When Francis Bacon, in the seventeenth century, had expressed an aspiration, a desideratum, of using science as the root of invention in Salomon's House, that had been but a dream.[41] The qualification still held a century later when the editor of a new edition of Bacon's works argued that 'all Arts, Inventions, and Works, should flow from natural Philosopers [sic]'.[42] However, during the later eighteenth century, campaigners conceived new strategies bringing science to bear on practice. Makers and doers were successfully drawing on multiplying experimental methods and scientific knowledge, as well as on craft skills and expertise, to reform agriculture, industry, and war. Brewers used the readings of thermometers and saccharometers to monitor processes with scarce, and variable, attention to any theoretical significance.[43] Only now are the multiple dimensions of the so-called industrial enlightenment experienced across Europe, but most intensely in Britain, becoming familiar.[44]

During the 1830s, newspaper use of 'practical science' increased substantially as the term acquired institutional status.[45] For instance, there were numerous articles about two new popular exhibition halls showing a variety of machines illustrating scientific principles and displaying curious scientific phenomena. These establishments also tended to mark a shift in the term's meaning.[46] The Adelaide Gallery was a traditional hall of remarkable machines, but the slightly later Polytechnic was more ambitious. Dedicated to improving and entertaining its audience, it

---

[40] A similar point about an earlier time is made by Westman 2011. I am grateful to Robert Westman for pointing out the parallel.

[41] See Keller 2015. On Bacon as visionary, see also Schauz 2020, 65–88.

[42] Peter Shaw 1733, vol. 2, 379, footnote 'a'. On the use of this annotation, see Golinski 1983. On the changing views of Bacon, see Yeo 1985.

[43] Sumner 2015.

[44] See Jones 2008. See also Mokyr 2004 and the critiques of the claims of this book in *History of Science* 45 (2007): 123–221. For a sceptical reflection on the significance of science to the big picture of industrialisation, see Ashworth 2017. The argument here does not need Mokyr's claim that the industrial enlightenment was the actual cause of the industrial revolution, just that belief in the linkage between knowledge and prosperity was one of the phenomena of the time. See also Klein 2016.

[45] In the 'Britishnewspaperarchive' corpus, three uses of the term 'practical science' were recorded for the first decade of the nineteenth century, sixty-two in the period 1810–1819, and the number increased from about twenty per year to several hundred in every year between 1833 and 1836, after the Adelaide Gallery had opened, and almost 500 in 1838 and 1839 once the Polytechnic Institution was open.

[46] On the two institutions, see Altick 1978; Weedon 2008. See also Berkowitz and Lightman 2017.

could boast radical gentleman-inventor George Cayley among its promoters. He had visited revolutionary France, and now attempted to shape the Polytechnic to show Londoners the interaction between abstract science and industrial arts.[47] Increasingly, the newer term 'applied science' would represent such ambitions.

France

One might assume that Britain and France's military and cultural conflicts around 1800 extended to their ambitions for science. Though the wars stretched over two decades, such an assumption would be simplistic. French institutions and accounts of French concepts would be influential in England both during and after the conflicts. The English recognised early nineteenth-century France as the centre of abstract science.[48] France's achievements during the Revolutionary and Napoleonic Wars in bringing science to industry's aid were also legendary. Heroic stories of the contemporary development of domestic potassium nitrate manufacture to replace now-unavailable imports and, through the production of gunpowder, saving the Revolution were both immediate and long-told.[49] Key accounts told of the triumph of the marriage of French science and the practical arts. They were reflected upon and encouraged by the chemical manufacturer and Napoleonic minister Antoine Chaptal. In his seminal *Chemistry Applied to the Arts* (*Chimie appliquée aux arts*), the metaphor that Chaptal used for chemistry was that of a lighthouse for the manufacturer. This guide illuminated the options, but only the practitioner could decide which to choose. He combined the French state's heritage of centralising the use of knowledge, going back a century to Colbert's government, with the contemporary cameralist approaches developed in the small German states.[50]

---

[47] Certainly, the promoters of the Institution had a variety of motives. Conflicts over the role of the Institution between Cayley and his stakeholders are manifest in a plea for support from Cayley to Babbage in 1839. See Cayley to Babbage, 2 December 1839, BL ADD 37191 f. 271, British Library. A preliminary printed prospectus emphasised the importance of scientific principles; see *Prospectus of the Royal Gallery of Arts and Sciences*, Box C, Cayley papers, National Aerospace Library/Royal Aeronautical Society, Farnborough.

[48] A review of the *Encyclopaedia* reflected that over the previous half century huge strides had been made in Britain in practical mechanics, though little in theory, while in France the reverse was true: 'Art IX: *Encyclopaedia Metropolitana*', *Monthly Review* ser. 2, 89 (1819): 187–99, on 192.

[49] See Multhauf 1971; Bret 2002, 241–79.

[50] Thus, the Prussian agricultural reformer Albert Thaer wrote an admiring review of Chaptal's 1808 *Chemistry Applied to the Arts*. [Albrecht Daniel Thaer], Review of *Die Chimie in ihrer Anwendung auf Kunste und Handwerke* dargestellt von J. A. Chaptal (Intelligenzblatt, 3), *Annalen des Ackerbaues* 8 (1808): 567.

Chaptal promoted an appreciation of industry that went beyond the application of scientific principles. As he sought to improve the apprenticeship system towards the end of Napoleon's reign, he spied a possible means in the Paris Conservatoire des Arts et Métiers. Established during the Revolution's early years, the Conservatoire had become an educational centre for workers seeking to learn such trades as weaving. Chaptal envisaged three departments, dealing, respectively, with products made of wood, of metal, and precision instruments.[51] Together with the second director, Gerard-Joseph Christian, he envisioned the Conservatoire as the centre of a reformed apprenticeship system. Its disciplinary orientation would be towards 'technonomie', a French interpretation of technology, at whose heart lay the tacit skill of the artisan.[52]

A younger generation, with different intentions, was looking greedily at the same institution. The firebrand Charles Dupin (born 1784), radicalised during the early Revolutionary years as one of the first cohort of engineering students in Paris's fabled École Polytechnique, had adopted its optimistic rhetoric. He was awed by the potential of 'science appliquée aux arts' – science applied to the arts (by which he meant the practical arts). In the immediate aftermath of the Napoleonic War, he visited Great Britain. Travelling through the country, he noted its remarkable development while it had been closed to the French. On his return to Paris, under the restored monarchy, the star of Chaptal, a former Minister of Napoleon, had waned. Now, Dupin drew upon an alliance with the influential physicist Arago, and these two transformed the Conservatoire. They persuaded the King to give the institution a new mandate to promote the application of science in industry.[53] Dupin developed his model of the relationship between the expert in science and the aspirational artisan. The reconstructed body, which would come to be known by its acronym CNAM, offered workers evening lectures from its three distinguished professors and preached the benefits of science applied to the arts.[54] For the language, the founders were indebted to the work of Chaptal. However, Dupin and his colleagues, the professors of industrial economics and of industrial chemistry, were much more aggressively confident than their forbears about science's ability to change practice directly.

---

[51] Chaptal 1893, 90–92.   [52] Sebestik 1986. On Christian, see Sebestik 1984.
[53] 'Ordonnance' of 25 November 1819, article première, *Recueil des Lois, Décrets, Ordonnances* 1889, 50. For CNAM, see, from a large literature, Fontanon et al. 1994; Mercier 2018.
[54] On Dupin, see Fox 1974; Christen-Lécuyer and Vatin 2009. After the restoration of the monarchy the institution was temporarily renamed Conservatoire Royale des Arts et Métiers.

Dupin's career took him from youthful radicalism to pillar of the establishment. In the aftermath of the 1815 Restoration, he had been brave to promote the achievements of the revolutionary years, but living until the 1870s, Dupin became a baron and a national icon.[55] With his rose-tinted accounts of Britain, political roles in France, active networking, and prolific writing, he would be popular and influential across the Channel. In the 'Britishnewspaperarchive' corpus, his name appears in more than a thousand articles of the 1830s and 1840s.[56] In 1839, Dupin received the ultimate accolade: satirical condemnation in one of just five *Times* pen portraits of French leaders: 'The current of his wandering eloquence is too strong for him; he cannot restrain it, but must talk, talk, talk. He has an itch for quarto pages, and must print, print, print.'[57] Despite the scorn of the British 'thunderer' for the French scribbler, by the late 1830s, the relations between science and practice had become a matter of public discussion even in England.

As Dupin came to be known across the Channel, the term 'applied science' entered the anglophone public sphere. A new link emerged through importations from, and hybridisations with, Dupin's French phrase 'science appliquée aux arts'. Moreover, the concept came to serve as an essential part of the response to calls for academic engineer training in Britain.

Negotiating 'Applied Science' at the BAAS and King's College London

Into the early 1830s, the ageing Coleridge continued to exert a personal authority. He made his influence felt in the British Association for the Advancement of Science (BAAS). Its founders in 1831 represented his views in the sequence of sections: 'A' was mathematics and physical science, 'B' was chemistry, and engineering low down at 'G'. The historians of its early years have reflected on this hierarchy, noting that 'applied science and the mechanical arts hardly flourished'.[58] Nonetheless, the new Association proved a considerable success, with many scholars, practitioners, and enthusiasts attending the meetings. Its historians have described the leadership, counting several of Coleridge's friends and

---

[55] See Christen-Lecuyer and Vatin 2009; Bradley 2012.
[56] See www.britishnewspaperarchive.co.uk (consulted 10 December 2019). British Newspaper Archive/British Library Board.
[57] 'Characters of M. Laffitte and M. Charles Dupin', *The Times*, 12 October 1839. The same article was republished a few days later in *The Morning Advertiser*, 16 October 1839. British Newspaper Archive/British Library Board.
[58] Morrell and Thackray 1981, 258.

leading contributors to the *Encyclopaedia*, as a Coleridgean 'clerisy'.[59] The Cambridge Master of Trinity, William Whewell, president in 1841, had been one of the *Encyclopaedia*'s first authors, writing the article on Archimedes.[60] Another of the founders was Charles Babbage, a scientific consultant to the *Encyclopaedia*'s editor, Edward Smedley. He also contributed several major essays, including a substantial entry of eighty-four pages on the principles of manufactures, which would constitute the first part of his 1832 volume *On the Economy of Machinery and Manufactures*. Here, Babbage offered a definition derived from Coleridge and the *Encyclopaedia*. 'The applied sciences derive their facts from experiment; but the reasonings, on which their chief utility depends, are the province of what is called abstract science.'[61] Like Coleridge before him, Babbage saw, and indeed practised, a distinct division of labour between collaborators who were 'most skilled in the theory' and those 'with practice in the art' – and he both expressed and showed high respect for the latter.[62]

A second institution also promoted the poet's classification developed for the *Encyclopaedia*. The staunchly Anglican King's College London (KCL), founded in 1829, led the introduction of 'applied sciences' to teaching. Even for this religiously conservative establishment, survival and influence in a new world required creative interventions. During a short tenure as College Principal, the *Encyclopaedia Metropolitana* editor Hugh Rose made an innovation of long-term consequence when he added a Department of Civil Engineering to the College in 1838. Since its foundation, KCL had boasted a Medical Faculty, through which it sought to ensure doctors' moral and theological preparation.[63] The objective of the younger department would be clear. Youths entering the booming railway construction industry needed an education that would civilise and provide the understanding of theory useful to a professional career. As the prospectus explained, 'A Course of instruction in the principles of those sciences which admit of an application to the uses of life, forms the basis of the entire system of this Department.'[64] The professor of natural philosophy and astronomy, Canon Henry Moseley,

---

[59] This is a substantial theme of Morrell and Thackray 1981.
[60] The word 'scientist' was famously coined by Whewell after Coleridge had rejected the characterisation of attendees as philosophers. See Ross 1962.
[61] Babbage 1832, 307.
[62] Babbage's description of these two categories follows immediately on from his reflection on applied and abstract sciences. On the practical design and building of his engine, see Swade 2000; Schaffer 2019.
[63] See Hearnshaw 1929, 139–40. KCL was established to create an Anglican alternative to the strictly non-sectarian University College London founded in 1826.
[64] King's College London, *Department of Civil Engineering and Architecture and of Science Applied to the Arts* (London: Clay, 1840), 1.

had a French education and acknowledged his debt to Dupin in print.[65] A cleric who would be the chaplain to Queen Victoria, Moseley argued that since the Almighty made humanity, humanity's works, such as machines, possessed the spark of the divine.[66] Also passionately interested in education, Moseley can be attributed with the decision, a year after KCL's introduction of engineering, to emphasise its teaching's scientific basis. The College launched a course on 'Science Applied to the Arts', adapting the French term promoted by Dupin.[67] Correspondingly, the prospectus promised the potential student would experience 'actual application to practical purposes'.[68]

This move had been some years in the making. In 1836 the two pre-existing institutions, University College London (UCL) and KCL, were incorporated into a new federal University of London, founded to offer academic degrees. Three years later, the University's Faculty of Arts began a lengthy consultation on certificates of proficiency in three new fields: civil engineering, navigation, and hydrography.[69] While proponents gave no particular name to this class of subjects, it was treated as a specific kind of knowledge. Using the *Encyclopaedia Metropolitana*'s schema, it would have been called 'applied sciences'. Though, ultimately, the Senate did not implement the special certificates, it did accept these subjects as part of London University's scope.

The influence of the French model of knowledge was evident in a front-page news article entitled 'Education-Science' published in numerous provincial newspapers. The piece, dealing with the country's novel engineering courses, was seemingly a preparation for launching London's proposed certificates and cited the potential civil engineering qualification.[70] Furthermore, the article urged emulation of Paris's new 'École Centrale des Arts et Manufactures'. This college was intended to create a cadre of outstanding civil engineers in France and featured fiercely utilitarian and materialist rhetoric.[71] As an institution, the

---

[65] Moseley's 1843 textbook based on his College lecture course, cited Dupin on its second page; see Henry Moseley 1843, vi. On Moseley, see Layton 1974, 75–94.
[66] Moseley 1857, 20.
[67] In the 'Britishnewspaperarchive' the first use is to be found in an 1825 report of lectures given by Dupin. See 'Education for Mechanics in France', *The Morning Chronicle*, 19 November 1825, British Newspaper Archive/British Library Board.
[68] For the course, see King's College London, *Prospectus*, 1845 (n.p.).
[69] 'Committee Appointed for the Purpose of Considering the Subject of Certificates of Proficiency in Civil Engineering, and Navigation and Hydrography', Committee of the Faculty of Arts, 16 February 1839 and 12 February 1840, 'Minutes of the Committees of the University of London', Ms 3/3, Special Collections, University of London Library.
[70] 'Education–Science', in, for instance, *The Northern Whig*, 17 November 1838, British Newspaper Archive/British Library Board.
[71] Olivier 1850 and Olivier 1851, iii–xxi. See also Weiss 1982; Grelon 2000.

École Centrale was a frequently cited model. Still, as personalities, its founders were unknown in England, where the face of French scientific education was Charles Dupin, founder of CNAM.

At KCL, language changed when the ambition to teach practice failed to convince British employers. Two years after the course on civil engineering launched, it proved necessary to qualify its promotion in the College prospectus with the note that the course was seen not as a substitute for professional training but 'as an introduction to it'. In short succession, the College opened an exhibition of the magnificent, if now out-of-date, Royal collection of instruments illustrating the scientific and engineering principles. It replaced Moseley, who had left to become one of the first inspectors of schools, and welcomed its next principal, who had spent the previous thirteen years in Germany as tutor to the future King of Hannover. These developments laid the foundation for a new department committed to teaching the profession's intellectual underpinning rather than claiming to teach vocational skills themselves.

Accordingly, in 1844, the College retitled civil engineering as 'applied sciences'.[72] Descriptions of the courses offered emphasised that they would be both liberal and useful. Both the future engineer and surveyor who would go on to specialised training and 'the merchant and the man of business' needed the same basic preparation. That would be 'in those branches of knowledge which are understood to form the groundwork of a liberal education on the continent'.[73] On the other hand, the newspaper advertisement for its classes emphasised practicality. Without promising a professional education, they would offer 'a thoroughly practical education for those who are afterwards to be engaged in the business pursuits of active life ... This Department provides also (in addition to the general course) a complete system of Elementary Instruction in Engineering and Architecture.'[74] Drawing directly on contemporary German practice, Coleridge's educational principles, and the opportunities for maximising student flow, this was the first course in Britain to use Coleridge's term. KCL's search for students through frequent advertisements in the press, first appearing in August 1844, would itself lend

---

[72] See https://kingscollections.org/catalogues/kclca/collection/e/10ki4665/ (accessed March 2023).
[73] King's College London, *Department of General Instruction in the Applied Sciences 1845*, 3. Bound in King's College Calendar 1844–1845. For Coleridge's ideas on education, see Coleridge 1852b, 95–97. William Whewell at Cambridge, but closely associated with KCL, would publish his 'On Liberal Education', drawing on ideas of Coleridge and the importance he laid on truth that could be relied on to be permanent, thus reconciling Coleridgean priorities with contemporary pedagogy. See Buchanan 2013, 36.
[74] See, for instance, the front-page advertisement, 'King's College, London', *The Morning Herald*, 12 August 1844, British Newspaper Archive/British Library Board.

visibility and authority to its concept. For the College, the term 'applied sciences' served to denominate those aspects of professional education that underlay, but did not replicate, the knowledge gained in the workplace. At the same time, they would give an appropriate preparation to a well-educated subject of the crown.

High Farming and Applied Science

KCL had intended its engineering course to meet the high demand for professional expertise during the railroad building boom. However, the turning point for public thinking about science was through its role in farming. This industry's baffling problems, outstanding opportunities, and national importance gave it the highest public profile of the 1840s. From the late 1830s, the price of wheat in British cities became a major political issue.[75] A complex system of tariffs – the so-called Corn Laws – protected growers and landowners, to the cost of the impoverished, the urban poor, and the middle classes. The debate over the Corn Laws would not only be the defining event for British politics in the mid-nineteenth century but also frame the development of rhetoric about applied science.[76] Once again, press interest and advertising would spread new usage.

Since the post-Napoleonic slump, 'high farming' informed by science had been favoured as a solution for the woes of British agriculture bedevilled by low prices.[77] A variety of new fertilisers hit the market. In 1849 a model 'high farmer' with 320 acres and produce worth £3,289 spent over £500 on fertilisers.[78] Agriculturalists had to choose carefully. As *The Polytechnic Review* warned, they negotiated paths between 'blind credulity' and 'obstinate mistrust'.[79] Low prices and the threat of imports imposed both economic pressures on farmers and a new culture.

Political tensions between urban consumers and rural producers intensified, and the crisis deepened. In 1838 the Anti–Corn Law League was founded in the new metropolis of Manchester. The following year, landowners who disagreed about the corn laws but agreed that science could, in principle, contribute to the well-being of the agricultural sector founded the Royal Agricultural Society. Its motto, 'Science with Practice,' expressed sensitivity to the demands of practice, the seduction of science, and the border between the two. The caution with which the Society addressed this boundary was apparent in its scepticism

---

[75] Gambles 1998.   [76] See Jones 2016.   [77] Moore 1965.
[78] 'British Agriculture and Foreign Competition', *Blackwood's Edinburgh Magazine* 67 (1850): 94–136, 222–48, see 118.
[79] 'Agricultural Chemistry', *Polytechnic Review* 1 (1844): 1–18, 91–102, at 101.

towards the agricultural chemistry proposed by Professor Justus Liebig of Hesse's Giessen University. The first editor of its *Journal*, the practical, if advanced, gentleman farmer Philip Pusey, never accepted a single contribution by Liebig.[80]

The novelty of Liebig's theories and aggressive disdain for traditional know-how made enemies of farmers in both Germany and Britain.[81] He warned his English translator, 'You know I did not wish to write an 'Agricultural Chemistry but a "Chemistry of Agriculture". I must avoid anything bearing on practical agriculture.'[82] He was concerned to ensure the reading of his work as a radical reinterpretation of plant and animal physiology in terms of organic chemistry and not, primarily, farming. Accordingly, he had been acclaimed by many chemists, particularly in Britain, as an international hero. Nonetheless, while Liebig was a scientific celebrity, he was not universally beloved.

This professional and intellectual terrain was negotiated with agility and success by the Scottish chemist James Finlay Weir (J. F. W.) Johnston. Though today scarcely remembered, in the 1840s, he was among the most widely known of all scientists. Johnston was an accomplished self-promoter and the author of a long sequence of internationally successful books on chemistry.[83] A cosmopolitan man, he had spent time in Sweden learning chemistry from Berzelius, was fluent in German, and admired Kant's work.[84] At first, learning of Liebig's move to agricultural chemistry, he was an enthusiastic follower. Well-connected in Britain, too, Johnston was deeply involved with the British Association for the Advancement of Science's infancy alongside such men as Whewell and Babbage. Moreover, he was the rare holder of a British academic post in chemistry at the University in Durham, near the Scottish border, founded in 1832. Nevertheless, Johnston felt isolated in his small institution that had failed to win over local industrialists, and was disappointed by a paltry salary.[85] So he took an additional

---

[80] See Brock 2002, 172.   [81] Finlay 1991; Brock 2002, 170.

[82] See Liebig to Lyon Playfair, 14 August 1841, Papers of Lyon Playfair, Imperial College London, College Archives and Corporate Records Unit; translated in Reid 1899, 47. On Liebig's ambitions, see Werner and Holmes 2002. Also see Brock 2002: 145–82.

[83] *The Catechism of Agricultural Chemistry and Geology*, published in 1844, passed through thirty editions in Johnston's own lifetime. For its influence, see Snelders 1981.

[84] David Knight, 'Johnston, James Finlay Weir (1796–1855), Chemist', in *Oxford Dictionary of National Biography*. For his admiration of Kant, see J. F W. Johnston, 'Meeting of the Cultivators of Natural Science and Medicine at Hamburg [*sic*], in September 1830', *Edinburgh Journal of Science* n.s. 4 (1832): 189–244, see 206.

[85] On the failure of Durham's early engineering course on which Johnston taught, see Preece 1982. Johnston's salary as Reader of Chemistry at Durham was only £50 a year (David Knight, personal communication). I am grateful to the late Professor David Knight for advice based on long study of the Durham University archives. For

position, developing the phrase 'applied science' as a trademark. Using it, he could endear himself to customers in the agricultural sector while maintaining his chemical credentials. The term, relaunched by him, entered the press and acquired a new meaning through his indefatigable industry.

Today Durham is associated with its superb Norman cathedral and a great university. Then the county was becoming, as *The Times* described it, 'very little more than one huge colliery', with the grime, noise, disasters, and industrial spirit that entailed.[86] Like KCL, the small and struggling university, with sixty students and a dozen academic staff, was a new High-Anglican foundation with ambitious spiritual and practical goals.[87] Before advancing to the leadership of KCL, Hugh Rose had been Durham's founding professor of divinity and was, therefore, among Johnston's few academic colleagues during the 1830s. Inspired by its southern counterpart's introduction of a course in civil engineering, Durham applied to the Prime Minister for support for agricultural education in 1841. Johnston himself aspired to a chair in 'scientific agriculture' in an expanded university. Depending on an offer of money from the government, Durham's scheme was always unlikely to be implemented and ultimately failed.[88] Moreover, as a Free-Church Presbyterian, Johnston was quite unlike the ideal of KCL teachers with their mission of creating Anglican gentlemen.[89] Nonetheless, he was mindful of his stakeholders. As a chemist, he was more respectful of agriculturalists than Liebig, treading carefully between the extremes of an argument over fertilisers.

We can watch the progression of Johnston's campaign, beginning with an initial public appearance at the lavish, widely reported dinner of the Yorkshire Agricultural Society in August 1841.[90] He began his address by reminding the audience how important their art of agriculture was and that the Royal Agricultural Society's motto was 'Science with Practice'.

complaints about salary, see Johnston to Robert Blackwood, October 1842, ff. 209–212, Ms 4061, National Library of Scotland.

[86] Editorial, *The Times*, 5 October 1850.   [87] Jones 1996, 50, 68–69.

[88] See Roberts 1973, 76–77; Pusey to Sir Robert Peel, 16 January 1842, ADD MS 40500, f. 166, British Library. This was also discussed by Johnston in his letter to Robert Blackwood, October 1842. On Johnston's personal engagement in the negotiations, see Johnston to Blackwood, 'Saturday morning', [1843], ff. 156–57, Ms 4066, National Library of Scotland.

[89] Free-Church Presbyterians left the Church of Scotland in 1843 to form the evangelical 'Free Church'. Its members resisted the authority of local gentry and rich land-owners. Johnston's friend David Brewster was also an early member.

[90] 'Dinner of the Committee', *The Hull Packet*, 6 August 1841, British Newspaper Archive/British Library Board.

His motto, he told the gathering, was 'Science applied to the Arts'. This expression served to emphasise the links with KCL, which Durham, and Johnston, in particular, were hoping at that moment to emulate with the support of the Royal Agricultural Society.

Despite his public allegiance to 'science applied to the arts', Johnston changed his motto within eighteen months.[91] As a venue to make the switch, he chose *Blackwood's Edinburgh Magazine*, run by his friend Robert Blackwood. The method was a carefully worded review of the practical *The Book of the Farm* by esteemed agricultural writer Henry Stephens and published by Blackwood in April 1843.[92] Johnston's essay began with the sentence: 'Skilful Practice is Applied Science.' 'The genuinely scientific man', Johnston explained in his review:

does not despise the *practice* of any art, in which he sees the principles he investigates embodied and made useful in promoting the welfare of his fellow-men. He does not even undervalue it – he rather upholds and magnifies its importance, as the agent or means by which his greatest and best discoveries can be made to subserve their greatest and most beneficent end.[93]

To explain his meaning, Johnston compared the growing dependence of agriculture on science with the history of navigation. As sailors became more confident about travelling out of the sight of land, the more dependent they grew on the tools provided by the astronomer and hydrographer. The telling of such stories was critical to Johnston's model. It offered a grand narrative whose latest episode lay in the activities of the modern farmer and, of course, his advisor. As part of his campaign, Johnston sought to persuade Blackwood to launch a new journal of agriculture.[94]

Finding a way out of his poorly remunerated Durham post, three months after his review, Johnston took the position of chemist to the newly formed 'Agricultural Chemistry Association'. Tenant farmers created this in the hope that it could mitigate the effect of low cereal

---

[91] Johnston seems to have experimented with the new term as early as July 1841, hiding behind the anonymity of a newspaper article. In July 1841 an article, probably by Johnston himself, reported that he had been appointed an honorary fellow of the Royal Agricultural Society. Together with a verbatim transcript of the president's speech on the occasion, an editorial note said it showed 'how Mr Johnston's endeavours to illustrate this most important branch of applied science are estimated in the highest quarters'. *The Durham Advertiser*, 16 July 1841, British Newspaper Archive/British Library Board.

[92] J. F. W. Johnston, 'The Practice of Agriculture', *Blackwood's Edinburgh Magazine* 330 (April 1843): 415–32.

[93] Johnston, 'The Practice of Agriculture', p. 415. Emphasis in original.

[94] Johnston to Robert Blackwood, 'Saturday Evening', [1843], ff. 152–55, Ms 4066, National Library of Scotland.

prices.[95] Their association would provide chemical advice to members needing to know more about the fertiliser needs of their soils and the qualities of commercial products. Over the next four and a half years, Johnston and his five full-time assistants would conduct almost 2,000 analyses, send over 3,500 explanatory letters, and deliver 100 lectures and addresses.[96] The team dealt, therefore, with practical issues, showing the rational basis for agricultural decisions.[97]

Johnston's personal use of his term was amplified by his publisher, Blackwood, which used sophisticated cross-advertising to promote its various products. So, Johnston's epigram 'Applied science is Skilful Practice' featured prominently in the numerous press advertisements for Stephens' *Book of the Farm* in whose review it had appeared.[98] In addition, in 1847, a new edition of Johnston's *Lectures on Agricultural Chemistry* offered an opportunity to promote his trademark term as a characteristic of a course in chemistry. Again, he emphasised that it was useful to both the practice of agriculture and its communication.

The objections which have been urged against the introduction of general chemistry and geology do not bear upon this branch of *applied* science, – because it has a direct special relation to the art by which the proprietors of the soil are maintained, and which is daily practised by those persons with whom the clergy in the rural districts have the most frequent intercourse.[99]

Blackwood then widely advertised the new edition of the *Lectures* in local papers, reusing Johnston's terminology. The book exhibited 'a full view of the actual state of our knowledge upon this important branch of applied science'.[100]

Johnston exploited his institutional base to make his term the trademark of a social movement. In the north of Ireland, a group of 'Improvers' copied the private organisation model, founding the Chemico-Agricultural Society of Ulster in 1845. Farmers in British Caribbean colonies having to restructure after abolishing slavery formed

---

[95] For a detailed account of the foundation of the Agricultural Chemistry Association, see Bud 1980, 192–94.
[96] 'Agricultural Chemistry Association', *The Scotsman*, 13 January 1849, British Newspaper Archive/British Library Board. For Johnston's assistants, see 'Report of the Committee of Management of the Agricultural Chemistry Association of Scotland', *Journal of Agriculture* 1 (1845–47): 248–50.
[97] See, for instance, 'Edinbro' Agricultural Chemistry Association', *The Cork Examiner*, 1 March 1844, British Newspaper Archive/British Library Board.
[98] See, for example, the advertisement published in *The Morning Post*, 20 April 1843, p. 2, www.Britishnewspaperarchive.co.uk/The British Library Board, and elsewhere.
[99] Johnston 1847, 10. Emphasis in original.
[100] See, for instance, an advertisement on the front page of *The Dublin Evening Post*, 25 March 1847, British Newspaper Archive/British Library Board.

the Royal Agricultural Society of Jamaica.[101] In England, the Liebig student Lyon Playfair offered analyses on behalf of the Royal Agricultural Society. A private agricultural college (now the Royal Agricultural University) was successfully established at Cirencester to offer chemical analyses and train scientific farmers. The professorship of chemistry went to a former Johnston assistant. Enthusiastic admirers linked the use of his term and the Agricultural Chemistry Association and, in turn, publicised his cause. At the time of Corn Law repeal, supportive articles deploying Johnston's usage appeared in Scotland's *Caledonian Mercury* and Ireland's *Belfast Newsletter*. The voice of free trade, *The Economist*, published several articles vigorously promoting Johnston and his linguistic innovation.[102]

Ireland experienced the challenges to agriculture most cruelly. The infamous potato blight bringing misery and starvation to millions (1846–49) also brought to a head the long-term problems of developing the Irish economy. In the 1840s, multiple interlinked crises were demanding national attention. We can appreciate the tension between practical farming expertise and claims for authority based on knowledge of chemistry through the work of Julianna Adelman. She points to the contrast between the intensely practical Flax Improvement Society in Belfast and the Chemico-Agricultural Society established in 1845 in the same city. This Society's chemist and journal editor, John Frederick Hodges, was a former Liebig pupil who had never worked on a farm. He recruited Johnston as an early speaker to the Society – a private organisation described as an 'offspring' of the Scotch Agricultural Chemistry Association.[103]

The social, educational, religious, and nutritional issues underpinning constitutional legitimacy and sectarian conflict were already too urgent for the government to ignore. One solution, urged the popular volume *The Industrial Resources of Ireland* (1844), was science as the source of

[101] On the overseas influence of the Scottish society, see 'Report of the Committee of Management of the Agricultural Chemistry Association of Scotland', *Journal of Agriculture* 1 (1845–47): 260–61. On the Ulster society, see Linde Lunney, 'Hodges, John Frederick', in *Dictionary of Irish Biography* (online edition: Cambridge University Press and Royal Irish Academy, 2021); 'Carrickfergus Agricultural Society', *Pharmaceutical Times* 2 (1847): 52–53. On Jamaica, Bud and Roberts 1984, 58.

[102] 'Scientific Agriculture', *The Economist*, 26 August 1848, 966–68, on 967; 'Experimental Agriculture', *The Economist*, 10 November 1849, 1246–47, on 1246. [Untitled leader article], *Caledonian Mercury*, 19 October 1846, British Newspaper Archive/British Library Board, published just a few months after the Parliamentary votes to repeal the Corn Laws in June 1846; 'Royal Irish Agricultural Association', *The Belfast Newsletter*, 18 August 1846, British Newspaper Archive/British Library Board.

[103] 'Chemico-Agricultural Society of Ulster', *The Belfast News-Letter*, 21 August 1846, British Newspaper Archive/British Library Board.

'improvement'. This route to resolving such intractable political problems appealed greatly to Prime Minister Robert Peel.[104] When the British government founded new secular colleges in Belfast, Cork, and Galway in 1845, each boasted a department of agriculture, and despite his lack of agricultural experience, Hodges took the professorial post in Belfast.[105] Men with more practical farming expertise headed other departments in Cork and Galway, but all failed to attract students. Within scarcely more than a decade, the three Queen's colleges closed their departments of agriculture. They could not compete as centres for training, with model farms providing cheaper, more practical instruction.[106] Adelman argues that though their student numbers had been low, that was an insufficient explanation for their selection for closure. She suggests that, additionally, agriculture was disparaged as an inappropriately practical topic for a university. Yet, at the same time, the courses were criticised for purveying knowledge rather than farming expertise. This tension in Irish agriculture was but one early example of a perennial problem in the role of applied science.

Nonetheless, the term became popular, and it may seem that Johnston had cynically appropriated Coleridge's carefully crafted coinage to an entirely different agenda. Indeed, his acceptance of 'skilful practice' as a component of knowledge associated with science's epistemological certainty would have been unacceptable to his predecessor. Yet, like Coleridge, Johnston saw his concept in terms of a philosophically unified coherent body of knowledge. In a *Blackwood's* article of 1849, he declared the importance of ideas, and of mind: 'But it is after these first ruder though more imposing conquests over nature have been made, that the demand for mind, for applied science, becomes more frequent, and the results of its application less perceptible.'[107] Public opinion was moving with him. The radical *Daily News* newspaper, founded by Charles Dickens, suggested in 1857 that a new industrial mentality symbolised by applied science was taking over the countryside because of the end of protection.[108] The article's casual use of the term indicated the penetration of the usage that Johnston had instigated.

---

[104] See Adelman 2015.
[105] In addition to inviting Johnston to speak at the meeting of the Chemico-Agricultural Society, Hodges also cited him frequently in his volume *First Steps to Practical Chemistry*.
[106] Miller 2014, especially chapter 2, 41–62.
[107] [J. F. W Johnston], 'Scientific and Practical Agriculture', *Blackwood's Edinburgh Magazine* 65, no. 401 (March 1849): 255–74, on 260, attributed to Johnston by the *Wellesley Index*.
[108] Editorial, *The Daily News*, 3 April 1857, British Newspaper Archive/British Library Board.

Virtually every use of 'applied science' in the British press between 1817 and 1838 referred to the *Encyclopaedia Metropolitana*.[109] But after that, the connotation changed as Johnston effected a radical revolution. During the years 1843–48, references to his work dominated the public use of the term. The number of advertisements for the KCL course was significant, but much more widespread was Johnston's usage promoted both by press articles it inspired and by Blackwood's advertising.

Beyond Johnston

By the late 1840s, the term appeared increasingly in the daily press and the public sphere. Excitement over the telegraph and chemistry showed its use in discussions of a broadening range of industries. Rather than denoting merely individual enthusiasms, it had become a symbol of increasingly popular educational and industrial agendas. For example, an article referring to an exhibition of manufactures suggested 'applied science' would have a popularity with the public, which went beyond that of mere 'abstract science'.[110] Here, the term had slipped further to describe machines, rather than any form of knowledge or understanding. Publishers, advertisers, journalists, and the authors of letters to the editor drew upon it with increasing frequency.

Whether one favoured the philosophy and classification of Coleridge, Dupin, or Johnston, the academically trained chemist's expertise was the epitome of applied science. By the 1840s, businesses across such industries as mining, brewing, glassmaking, gunpowder making, and dyeing were employing chemists.[111] In the Manchester area, four calico-printing houses had used at least one of these new professionals. Lyon Playfair worked at the second largest of these, Thomson's of Clitheroe, which employed 1,400 people.[112] Though the company afforded him a luxurious laboratory, library facilities, and a high salary, Playfair complained to a friend of his concern lest his interest in 'pure science' be discovered.[113]

---

[109] Of the thirty uses to the end of the 1820s in the 'Britishnewspaperarchive' corpus, all but one occur in advertisements for the *Encyclopaedia Metropolitana* (consulted 10 December 2019). At the end of the 1830s a very few non-*Encyclopaedia* references to the category are to be found. Several are reprints of each other.

[110] 'The Birmingham Exhibition of Manufactures and Art', *The Daily News*, 29 September 1849, British Newspaper Archive/British Library Board.

[111] Bud and Roberts 1984.

[112] See 'A Day at a Lancashire Print Works', *Penny Magazine* 12 (1843): 289–96; also Bud 1980, 198–206.

[113] See Playfair to Andrew Ramsay, 26 March 1842, and Playfair to Ramsay, 21 May 1842, NLW MS 11574D, National Library of Wales. Playfair to Liebig, 29 June 1841, Liebigiana 58, no. 3, Papers of Justus von Liebig, Bayerische Staatsbibliothek, Munich.

Chemistry was a discipline long associated with industrial advances, and its progeny, the study of electricity, offered radically new opportunities and demands. Thus, the Electric Telegraph Company, incorporated in June 1846, launched the public development of a new mode of communication. In an 1849 article on the telegraph, a friend and contemporary of Playfair, the Scottish chemist George Wilson, took his subject more broadly. He was concerned that the public was a 'little crazed' at that moment 'on the subject of applied science', with the inevitable consequence of disappointment and reaction.[114]

In many contexts, the new term's use had replaced the earlier 'practical science' and implied there was now an avenue to improvement employing theoretically informed 'abstract science'. This programmatic connotation attracted sympathetic users. For example, from 1845, the newly founded, non-conformist *British Quarterly* used the phrase to classify news, probably on account of the enthusiasm of George Wilson, an active contributor to the journal. Similarly, book publishers and booksellers, such as the well-known firm of Samuel Highley, also with a Scottish association, classified books as 'applied science'.[115] Typically, these dealt with engineering, often encompassing those works previously seen as 'practical science'. Such commercial interpreters of the public sphere were promoting the use of the term as a natural category.

Certainly, such individuals as Coleridge, Dupin, and Johnston were important pioneers. However, this chapter has suggested the importance too of the *Encyclopaedia Metropolitana* and institutions such as KCL, CNAM, the Agricultural Chemistry Association, and Blackwood. These acted as powerful amplifiers and transformed subjective opinion into apparent factual reality. It is striking that though they were the agents of quite different philosophies and conjured up dissimilar sociotechnical imaginaries, they identified generally similar activities within their interpretations. Though inspired by various models, uses of 'applied science' merged within the general hubbub, and former distinctions soon faded.

Conclusion

The concept of applied science, as it emerged by the late 1840s, had several roots. The *Encyclopaedia Metropolitana*, inspired by Coleridge;

---

[114] [George Wilson], 'The Electric Telegraph', *Edinburgh Review* 90 (1849): 434–72, see 442; authorship ascribed by the *Wellesley Index*. Of course, in telegraphy, innovation could come often from the machine shop, as Israel 1992 has shown. With rhetorical verve, Wilson argued here that, while valuable, scientific theory was itself an inadequate guide to practical possibilities.

[115] See, for instance, Samuel Highley advertisements, *Athenaeum*, no. 1351 (17 September 1853): 1112; *Literary Gazette*, no. 1912 (10 September 1853): 896.

practical science; the 'science appliquée aux arts' of Dupin; and Johnston's ambitious campaign would each have a role. The influence of their shared legacy would last for more than a century. The term had come to embody the belief in science's power to inform or even transform industrial practice and the culture of the people deploying it. Coleridge, the editors of the *Encyclopaedia Metropolitana*, and KCL believed that the promoters of new techniques 'needed' the broadening influence of science. It would be both civilising and practical. Even Coleridge would emphasise that chemistry could make valleys 'laugh with corn', as proof of the value of linking practice to science. KCL was careful to emphasise that its education would imbue students with the appropriate culture and not claim to be a complete engineering training. To Johnston, the need was more practical than cultural. He was careful to respect farmers' expertise as he emphasised the relevance to them of applied science. Still, even he suggested it would provide a useful cultural underpinning for the pastor in frequent communion with his agricultural parishioners.[116] It was striking that both strains would survive in the connotations of the concept.

These uses became familiar across society through the press in articles and advertisements promoting the *Encyclopaedia Metropolitana*, major institutions such as KCL, and the analytical laboratories addressing the complexities of fertilisers and the challenge of Corn Law reform. So, by the late 1840s, secondary transmission of the term coming to be cited in newspapers and used in publishers' catalogues was well established. Although the historian might decode the intentions behind innovators' usages, such distinctions would not have been evident to the contemporary reader. Adept users were successfully addressing public ambivalence over the practical ability of science to make a difference. Their classification of science, educational ambitions, and visions of society were intimately interconnected. In this early period, users emphasised the term's social functions. The meanings of 'applied science' obtained their significance from the social goals they could forward, and their subsequent development would depend upon the creation of new institutions.

---

[116] Johnston 1847, 10.

2 Applied Science Institutionalised
The Liberal Science College

Introduction

Change in the talk about science was confirmed by the Great Exhibition of 1851 mounted within the dramatic Crystal Palace in London's Hyde Park. The occasion marked the peak of Britain's domination of the mass production of goods by machine. Not just a spectacular event, the Exhibition also inaugurated a period of rapid organisational change. Setting British achievement within an international perspective, unlike previous shows, this presented products from around the globe. While celebrating British industrial pre-eminence, the exhibition also launched a discussion about the risks of its loss. From that time on, campaigners for a variety of causes, particularly science, would point to decline, to its roots, to the means of mitigation through their favourite measures, and to the importance of new bodies – such as those promoting knowledge of applied science.

The Great Exhibition's success and rich diversity of industrial achievements stimulated debate about the factors that had made these qualities so abundant in the modern age. It also informed worrying arguments about British ways and their relationship to foreign competition. As a means of defence that was both powerful and culturally acceptable, 'applied science' became a taken-for-granted reality and national need. Newspaper usage of the term, which had jumped in 1849, stayed at the new higher level through the 1850s, and a network of new colleges gave it further significance. This chapter will explore the implications of this concatenation of public knowledge and authority.

The Great Exhibition attracted visits and support from across the country. The city of Liverpool provided vigorous help thanks to the vision of customs officer and keen botanist Thomas Archer. Subsequently, having coordinated a presentation of useful imported plants at the Crystal Palace, he opened a 'Museum of Applied Science' in the city's Royal Institution. The term entered regional newspapers through frequent reports of this locally high-profile establishment. When the Museum opened in 1857, it was devoted to showing the industrially valued products of plants, but this would be too limited an interpretation of its meaning.[1] He had looked forward to leading an institution offering a wider opportunity to elucidate applied science.[2] His personal dream was realised in 1860, as he ascended to the directorship of Edinburgh's Industrial Museum of Scotland.

Elsewhere, the Great Exhibition also prompted institutional change. The religious and the scholarly celebrated the new levels of industrial achievement made possible, at least in part, by science.[3] Across the country, including Manchester, Bury, Newcastle, and Glasgow, lectures were advertised, and organisational experiments attempted in the name of applied science.[4] Announcements of classes deploying the phrase, either in the collective sense or in the plural, made it more familiar. So did advertisements by schools offering preparation to boys preparing for engineering and agriculture, the army, and the East India Company.[5] Even Cambridge professor William Whewell was convinced that something had changed in the effect of science upon industry.[6] Before the Great Exhibition, he had seen the arrow of influence as going from machines to science. He now proclaimed the significance of the other direction, from science to invention. After Whewell's death, the University of Cambridge brought in a new course for its ordinary degree. In May 1865, teaching began under the rubric of 'mechanism and applied science', superseding the earlier draft title 'mechanism and practical science'.[7] Meanwhile, professors associated with the new colleges had developed a rhetoric of applied science that sustained their institutions and their own positions in the broader community and within

---

[1] In 1847 Kew opened a Museum of Vegetable Products, renamed the Museum of Economic Botany in 1852. See Cornish 2013. On the imperial context, see also Drayton 2000.

[2] Thomas Archer to President of the Liverpool Royal Institution, 24 July 1860, LRI 1/3/4.1, 109, LRI papers Special Collections and Archives, University of Liverpool Library. On the Liverpool Museum, see Ormerod 1953, 56. On Archer, see A. Galletly, 'Obituary Notice of the Late T. C. Archer, F.R.S.E., Director of the Museum of Science and Art, Edinburgh', *Transactions of the Botanical Society of Edinburgh* 16 (1886): 272–76.

[3] The religious response to the exhibition is discussed by Cantor 2011.

[4] One can see benefits from Playfair's energetic endeavours in the press. See, for instance, a report of Dr Wylde's lectures, *The Glasgow Herald*, 12 August 1857, British Newspaper Archive/British Library Board.

[5] See, for example, an advertisement for Hoddesdon Scientific School, *The Times*, 6 January 1851.

[6] The scholarly response of Whewell, who concluded that science did now play a role in invention, is reviewed by Mertens 2000.

[7] Report revised by the Syndicate, 24 May 1865, Guard Book, CUL UA CUR 28.5.1, University Library Cambridge. Quoted by kind permission of the Syndics of Cambridge University Library. See Marsden 2004.

their profession.[8] In the aftermath of 1851, therefore, one finds increasing use of the expression to describe the age.

Machines on display in the Crystal Palace had crossed continents, and science was international. Yet the exact role and negotiation of the place of scientific theory and practice in the training of engineers and industrial development remained specific to individual countries. In Britain, employers and professional engineering institutions insisted that science had to be a subordinate part of the next generation's training, however valuable it was as background. Only by serving an apprenticeship to a master engineer could the young acquire important trade skills and secrets.[9] Elsewhere, governments supported academic training at various levels for managers and engineers. In Britain, where an overreaching state was deeply suspect and at a time when the government's role in maintaining the health of industry was being beaten back, the balance between public and proprietary expertise was deeply problematic. The value of science to the future engineer had to be secondary, but its encouragement was an acceptable role for the government in support of industry.

A new kind of institution, the great civic liberal science colleges, would provide a home and physical embodiment of the concept, its ambitions, and constraints. By offering courses in 'applied science', by directly promoting the category, and by serving as local foci for talk about it, these colleges became the principal vehicles through which the term was naturalised in the public sphere. They taught science with the twin promise that it would both civilise the student and underlie success in the outside world. Thus, at Owens College in Manchester, founded in 1851 (ancestor of the University of Manchester), Edward Frankland, professor of chemistry, promoted the understanding of scientific chemistry. At the same time, his examinations addressed students' likely interests: 'What is the composition of bleaching powder? Explain in chemical language the processes of bleaching and of calico printing in the discharge style.'[10] Combining an emphasis on the moral virtues of chemical reasoning and the promise of utility, Frankland's successor Henry Roscoe was particularly successful. In 1864/65, his chemical

---

[8] This chapter develops arguments for the importance of a new liberal science college model of education put forth by Bud and Roberts 1984, and in Bud 2014a, 30–36. Donnelly 1986 argues the significance of the professional ambition of chemists for the development of the applied science rhetoric. For all the differences in emphasis, the fundamental arguments of Bud and Roberts and of Donnelly are complementary.

[9] The term 'tacit knowledge' and exploration of its importance in science was developed by the chemist Michael Polanyi. See Polanyi 1958.

[10] Owens College, 'Examination Papers', in *Calendar for the Session 1851–52* (Manchester: T. Sowler, 1851): 26. For a more extensive treatment of Owens College, see Bud and Roberts 1984, 81–86.

lectures enrolled a hundred students, at a time when the College's total was 128. Many were not fully matriculated but had registered especially for Roscoe's class, which promised to fit the attendee 'for applying the science to the higher branches of Art, Manufactures and Agriculture'.[11] That promise was to be fulfilled by a course dominated by basic organic and inorganic chemistry, including one lecture per week on the 'chemical principles involved in the most important chemical manufactures'.

Applied science also served as a powerful ideological bond for a diverse scientific community. Many practical men, chemical manufacturers, and consulting chemists were members of the Chemical Society of London, founded in 1841, though the leadership was academic. Even if, day to day, members' concerns were enormously varied, the community shared an assumption of their science's industrial utility.[12] The term 'applied science' described a space between the poles of private expertise and public knowledge.

The newspaper corpus provides a rich insight into public discourse. Newspapers printed the term in such diverse contexts as advertisements, news reports, editorials, and the verbatim text of speeches. In recent years, historians have taken an interest in the resonance between politicians' speeches and the press's needs and opportunities. News was required to fill the columns, but the selection was often deeply political, aimed at institutional change, whether national or municipal. Accounts of meetings close to the editor's interest served both political platform and the paper, and mid-century politicians enjoyed useful relations with individual newspapers.[13] The interaction between the book trade and periodicals has also attracted attention.[14] Such publishers as Blackwood exploited the potential of cross-media ownership by, for instance, vigorously promoting the press's books in their magazines.[15] Thus, in the case of applied science, published comments could bestow trade advantages in one context and, elsewhere, support organisational goals. The press was, of course, not the only agency of public culture. Speeches inside and outside Parliament, exhibitions, and ritual occasions provided the opportunity to promote ideas as truths, anecdotes as representative, and concepts as realities, allowing these identifications to be widely shared.[16]

---

[11] Owens College, *Calendar for the Session 1862–63* (Manchester: T. Sowler and Sons, 1862), 35.
[12] See Donnelly 1987, 51, for a discussion of the ideology this betokened.
[13] Brown 2010.   [14] Brake 1997 and Dickens 2011.
[15] On Blackwood's, see Finkelstein 2006.
[16] The evocative term 'representative anecdote' was coined and explored by the literary theorist Kenneth Burke; see Burke 1960.

Lyon Playfair

Change was announced in London. Even before the Great Exhibition, the government was convinced that its existing Museum of Practical Geology should offer a broader service to the state. As part of plans to upgrade it to be the 'Government School of Mines', the government hired the ambitious young Scottish chemist Lyon Playfair.[17] Obtaining his position in 1843, Playfair reported his carefully negotiated cultural place to a close friend, emphasising his commitment to bringing science to bear on practical issues.[18] Indeed, he became the government's consulting chemist on problems ranging from selecting the navy's coal to Poor Law establishments. Playfair's political acuity was rewarded. Prince Albert recruited him as an organiser of the Great Exhibition.

Late in 1851, Playfair achieved another coup. The transformation of the geological collection into a national educational institution may have been shaped partly by the personal presence of Charles Dupin in London as the head of the French Commissioners to the Great Exhibition. *The Times*' announcements of events, and their dutiful listing of participants, show he met Playfair, now an advisor to Albert and confidante of the Prime Minister, at least twelve times during the summer and early autumn of 1851. When, in November 1851, the new college opened, it was with a title evoking French terminology: 'Government School of Mines and of Science Applied to the Arts'.[19] This official title did not, however, stick. In its two-page coverage of the inaugural lectures, the *Athenaeum* referred to the new institution as the 'School of Mines and Applied Science'.[20] If only by accident, the press was continuing a cycle of mutual 'infection' between discussion of cultural change and the interpretation of an institutional initiative. This hybridisation also occurred in Dublin, where the contemporary 'Government School of Science Applied to Mining and the Arts' was referred to as the 'School of Applied Science', even in official documents.[21]

In early 1852, after an indecisive election, the Tories formed a weak minority government. That November's Queen's Speech from the Throne, anticipating the government's plans, included an announcement of a comprehensive scheme for the 'advancement of Fine Arts

[17] On Playfair's ambition, see Blatchford 2021 and more generally Reid 1899.
[18] Playfair to Ramsay, 25 February 1843; NLWMS 11574D MSRAMSAY, National Library of Wales.
[19] This is dealt with in more detail in Bud and Roberts 1984, 88–89.
[20] 'Industrial Education', *Athenaeum*, no. 1255 (15 November 1851): 1209.
[21] See Department of Science and Art 1857, xviii. On the school, see Cullen 2009.

and Practical Science'.[22] A few days beforehand, Playfair capitalised on press interest to give a widely reported inaugural lecture to the Government School of Mines and of Science Applied to the Arts. Here he established one part of a rhetoric, further developed over the next two decades. Addressing technical education, Playfair reflected on a study tour of the continent. He threatened his audience with the 'superiority of the intellectual element of labour abroad' that he put down to their colleges. Playfair praised especially Paris's Conservatoire des Arts et Métiers (CNAM) and the newer École Centrale.[23]

Following the Queen's Speech announcement, despite a change in administration, the agitation of Playfair and the machinations of the design advocate Henry Cole, the new government established the Department of Science and Art within the Board of Trade. With buildings amid the western suburbs of London, it occupied land immediately to the south of the former site of the Great Exhibition, dubbed 'South Kensington'. The new department would sponsor both day and evening classes in art and science, and the vision for its area was to follow up with government-funded schools and museums. Even today, the Victoria and Albert Museum incorporates buildings that are surviving relics of the earliest plans.[24]

The department took over the recently formed Government School of Mines, which was retitled 'The Metropolitan School of Science Applied to Mining and the Arts'. And it almost came to be referred to officially as the 'School of Applied Science'. However, a clerk added 'Metropolitan' and deleted 'Applied' before a draft announcement was signed off.[25] If Playfair's most grandiose plans were then frustrated, it was a setback but not the end of his campaign. The prospectus of this school declared, 'The chief object and distinctive character of this Institution (to which everything else is made subsidiary) is to give a practical direction to the course of scientific study.'[26] Whereas such education would not in itself be a qualification to direct practical operations, it would render a graduate 'in the highest degree competent' to promote future development. Playfair took charge of science across the new department. Nonetheless, several professors at the restyled Metropolitan School of Science Applied

---

[22] For the text of the Queen's Speech, see *Hansard* HL Deb 11 November 1852.
[23] Playfair 1852, 6.
[24] The convoluted politics of the building of the Cole Building, which housed the Normal School of Science, are described in Sheppard 1975, 233–47. See also Bud and Roberts 1984.
[25] Board of Trade, 'Department of Science and Art', 5 July 1853, Ed 28/1, f. 203, the National Archives (hereafter TNA). For the history of the department, see Butterworth 1968.
[26] Department of Science and Art 1857a, 2.

to Mining and the Arts fought a long battle to prevent a move to the emerging science complex at South Kensington. Resisting assimilation within 'science', they considered their biology professor Thomas Huxley a traitor for finally leading a movement to the new district.[27]

Playfair emphasised that universities should not teach practice itself. He also believed in the crucial role of a taught knowledge intermediary between the 'abstract sciences', which should be the province of the old universities, and instruction in the workshop.[28] At first, he sought to convince his audiences of the need to create schools analogous to those on the Continent by issuing threats of national decline in the face of foreign competition. Then, during the 1850s, Playfair's argument evolved, as in his 1857 valedictory speech as Secretary of Science of the Department of Science and Art. At the time, he was a member of a network of young London-based Scots. Though sharing many of J. F. W. Johnston's assumptions and interest in science's practical benefits, these men were somewhat younger and often referred derisively to the older man.[29] Coming together in a small dining club, 'the Universal Brotherhood of Friends of Truth', this network provided the opportunity for ideas to circulate.[30] Thus, Playfair drew upon a rhetorical approach previously used by the Brotherhood's leader, the geologist Edward Forbes.[31]

In his 1857 address, Playfair no longer spoke in terms of the details of foreign practice but, instead, reflected on the unnecessary hurdles put in the way of British heroes of the industrial revolution. Here, Playfair answered what had come to be the generic challenge to scientific instruction; James Watt, George Stephenson, and Richard Arkwright had been outstandingly successful without it. The experience of Watt, the father of the modern steam engine, was a reliable standby. He had been an

---

[27] Bud 2013c, 23–24.

[28] 'Museum of Practical Geology: Government School of Mines and of Science Applied to the Arts', *The Morning Chronicle*, 4 November 1852, British Newspaper Archive/British Library Board.

[29] Playfair, for instance, wrote to his friend Andrew Ramsay, celebrating any setbacks of Johnston. See Playfair to Ramsay, January 1842, and Playfair to Ramsay, 23 July 1842, NLW MS 11574D, National Library of Wales.

[30] P. J. Hartog and R. G. W. Anderson. 'Wilson, George (1818–1859), Chemist and Museum Director', in *Oxford Dictionary of National Biography*. On the brotherhood, see Wilson 1860. See also Reid 1899, 39. So close was this 'brotherhood' that they even had a badge of red silk ribbon embroidered with the letters MEO, which were said to represent the initials of the Greek words signifying Learning, Love, and Wine. See Reeks 1920, 30.

[31] Forbes' talk was his inaugural lecture as professor of geology to the Metropolitan School of Science Applied to Mining and the Arts. See 'Museum of Practical Geology', *The Daily News*, 3 October 1853. Playfair used almost the same examples as Forbes, except he cited Arkwright rather than Forbes' example of Hugh Miller.

assistant to the eminent Edinburgh professor of chemistry Joseph Black but had not benefitted from formal education.[32] Accredited with the industry-changing Rocket locomotive, together with his son Robert, George Stephenson was another well-known autodidact. Arkwright, a barber by training, was considered the pioneer of spinning by powered machinery. The stories of such great, self-helping British inventors had become part of the canon of Victorian culture and were celebrated by Samuel Smiles.[33] To such models, Playfair countered, 'Call such men as witnesses; hear the struggles of their life to overcome the deficiency of early education; their toilsome ascent in steps cut out, one by one, in the mountain of knowledge.'[34] In other words, they had used science, but for lack of proper training, they had been forced to teach themselves. Such narrative support, provided by heroic British allegories, would be evoked for years to come.

Building the Institutional Framework: The Irish Model and the Samuelson Committee

During the 1860s and 1870s, applied science became central to the mounting debate about threats to Britain's industrial position and aspirations for technical education. Uses of the expression in the press mounted rapidly. The 1855 abolition of stamp duty on newspapers encouraged a host of aspiring literary entrepreneurs to create new serials offering an understanding of radical industrial change. As Ruth Barton has pointed out, the ambitious War Office clerk and amateur astronomer Norman Lockyer, founder of *Nature* in 1869, was the most successful of a clutch.[35] Other editors, such as Peter Lund Simmonds, editor of *The Technologist* magazine, entitled after 1870 *The Journal of Applied Science*, and William Crookes and his business partner James Samuelson also made their living by reporting on the boundaries of science and practice. Even the old-established *Mechanics' Magazine*, taken over by a new publisher, advertised itself from October 1861 as 'devoted to the applied sciences, the mechanical arts, inventions, engineering, agricultural implements, scientific discoveries and societies &c'.[36] The term was now part of the language.

---

[32] The most recent treatment of James Watt's reputation is Miller 2019.
[33] Samuel Smiles' biography of George Stephenson had been published just a few months before Playfair gave his speech. On Smiles and invention, see MacLeod 2007.
[34] Playfair 1857, 22.   [35] Barton 1998. See also MacLeod 1969.
[36] The *Mechanics' Magazine* was relaunched in 1859 as the *Mechanics' Magazine and Journal of Engineering, Agricultural Machinery, Manufactures, and Shipbuilding*. This title was maintained even while the advertisements described it in terms of applied sciences. See the classified advertisement, *Dundee Courier and Daily Argus*, 31 October 1861, British Newspaper Archive/British Library, and subsequently.

The 1840s organisation of Irish science education had established a Museum of Irish Industry with some teaching responsibilities. Two decades later, byzantine negotiations led to the proposal to further enrich Dublin's institutions by converting the Museum into a College of Science.[37] Rather than imparting practical knowledge, a committee chaired by the astronomer Lord Rosse, and including many of the London science grandees (including Lyon Playfair, Thomas Huxley, and John Tyndall) suggested that it focus upon sciences,

which are more immediately connected with and applied to all descriptions of industry including Agriculture, Mining and Manufactures and that it should in this way supplement the' elementary and scientific instruction already provided for by the Science Schools of the Department [of Science and Art] towards the training; of teachers for which schools it may also give considerable assistance.[38]

This College's curriculum would have a common two-year core followed by a specialised 'applied science' third year dealing with agriculture, mining, or engineering. Notice that this model was intended to be suitable too for teachers. Devised for Dublin in 1866, this would become the template for scientific education in the whole of Britain.

Playfair facilitated the translation from Dublin to the rest of the country. As his role in the Irish Commission showed, the 1858 move to Edinburgh as professor of chemistry did not mean total retirement from the politics of science. In his new position, Playfair raised funds for agriculture and engineering chairs, which, in turn, challenged existing organisational patterns in the University. By February 1868, he was a member of a small committee that proposed a faculty of 'applied science'. In its submission to the University Court, the Senate urged haste in approving the scheme while this was a topic of national interest.[39] Visibility in the public sphere became a means to pressure the University.

The immediate context for the urgency in Edinburgh was a renewed campaign associated with narratives about education and the

---

[37] On the negotiations behind the proposals for the transformation of the Museum of Irish Industry, see Bud and Roberts 1984, 126. On the Royal College of Science in Dublin, see Kelham 1967.

[38] Miscellanies, 13, f. 179, Papers of Sir Henry Cole, National Art Library, Victoria and Albert Museum. Quoted courtesy of the Victoria and Albert Museum. Similar, though not identical, wording is found in Commission on the College of Science for Ireland 1867, 778.

[39] For the Senate Committee's report and injunction to the Court to hasten, see Secretary of the Senate to the Court, 24 April 1868, Minutes of the Court 1 (1859–70): 359–60 and the draft report on pp. 361–65, Centre for Research Collections, University of Edinburgh. In 1869, Edinburgh's *University Calendar* advertised a division of study entitled 'applied science', incorporating agriculture, engineering, and veterinary surgery. See *The Edinburgh University Calendar 1869–70*, 140.

comparative performance of British companies and their overseas competitors. The 1860s saw the development of concerns about elementary, secondary, and technical teaching in Britain. As in Ireland, 'applied science' would benefit from the myriad of proposed institutional solutions. Typically, accounts of the campaign have started with the March 1867 polemical letter from Playfair, a juror at the Paris 'Exposition universelle d'art et l'industrie', in a letter to Lord Taunton then leading an educational enquiry, decrying the inferiority of British exhibits.[40] This letter quickly won support from a few others who had attended. Another juror, former schools inspector and distinguished churchman John Norris, wrote to Lord Taunton strongly agreeing with Playfair. Subsequently, in his 1869 book *Education of the People*, Norris would fully flesh out his appeal. He suggested that the category of 'applied science' bridged the interval between theory and manual skill. He claimed that the physical sciences taught in universities were 'too abstract, their professors too fastidious for our purpose'.[41] Norris linked his recommendation to glowing accounts of development overseas, which contrasted with British education's well-known limitations. A lengthy *Manchester Guardian* editorial cited this argument, and so did *The Liverpool Mercury* a few weeks later.[42] The sixty reports of teachers awarded subsidised visits to the Paris exhibition, generally published in local newspapers, reinforced the message. One expressed his conclusion as 'we want more "applied sciences"'.[43] Another report from a 'specialist' calling himself 'a knight of the quill' described such international events as 'really, at bottom, great competitions among the nations in art, and skill, and applied science'.[44] Further support came from a few local chambers of commerce, which disseminated their message widely through the press in February and March 1868.[45] At national levels,

---

[40] Playfair's letter and the letters of support were published in Schools Inquiry Commission 1867. On the significance of Playfair's letter, see Gooday 2000.
[41] Norris 1869, 106.
[42] [Editorial], *The Manchester Guardian*, 13 August 1867; 'Compulsory Education', *The Liverpool Mercury*, 2 September 1867, British Newspaper Archive/British Library Board.
[43] T. Brown, 'The Paris Universal Exhibition of 1867: Suggestions on Scientific Instruction', *Chorley Standard*, 9 November 1867, in 'Volume of Reports by British Science Teachers on the 1867 Paris Exhibition', MS/0210, Dana Research Centre and Library, the Science Museum.
[44] Special Correspondent, 'The Great Exhibition', *Glasgow Morning Journal*, 14 August 1867, in 'Volume of Reports by British Science Teachers on the 1867 Paris Exhibition'.
[45] 'Chambers of Commerce on Technical Education', *The Manchester Guardian*, 12 February 1868. See also 'Associated Chambers of Commerce', *The Birmingham Daily Post*, 5 March 1868, British Newspaper Archive/British Library Board; *The Leeds Mercury*, 5 March 1868, British Newspaper Archive/British Library Board.

England's Society of Arts and the Royal Scottish Society of Arts also held widely reported meetings promoting science education.[46] This build-up provided an opportunity for the German-born iron master, engineer, and Member of Parliament Bernhard Samuelson.[47] He campaigned for a parliamentary investigation of 'Provisions for giving instruction in theoretical and applied science to the industrial classes'. In moving the formation of an investigating committee, Samuelson began his speech to the House of Commons explaining that the term 'technical education' was ambiguous.[48] Confusingly, it could mean either education in the factory itself, which Samuelson thought was in any case well done in Britain, or by the school. To avoid this ambiguity, he preferred the term 'applied science'. Building on half a century of use, Samuelson emphasised a very particular connotation. In this context, it entailed practical but 'not workshop' education. This approach to academic teaching would resolve the tension between foreign success stories and the celebration of the great British tradition. Samuelson was successful, at least in obtaining a 'select committee' of Members of Parliament to look at technical education. The so-called Samuelson Committee proved to be the breakthrough event in the public history of 'applied science' in Britain. Over just the four days following Samuelson's speech, the sample of newspapers in the 'Britishnationalarchive' published over 200 articles using the term.

The category of 'applied science' ran through testimony to the committee. Lyon Playfair proposed that it be the speciality of the new colleges under discussion around Britain, leaving Oxford and Cambridge to specialise in pure science.[49] A different tack was taken by the former KCL professor Henry Moseley, responsible for establishing the Bristol Trade School for skilled workers' sons and daughters.[50] He felt applied science was too removed from practice for his clients. It often had no direct industrial application and dealt with unrealistically idealised situations such as friction-free motion. Instead, the pupils with whom he was concerned were interested in immediate personal utility and needed to learn what he called 'industrial science'. In the next chapter, we will explore the implications of such arguments.

---

[46] For more on the complex interplay of reports, see Bud and Roberts 1984, 127–30.
[47] In contrast to the born-insider Lyon Playfair, whose grandfather had been principal of the University of St Andrews, Bernhard Samuelson, born Jewish in Hamburg, made a fortune in manufacturing industry. There is no extended published biography.
[48] *Hansard* HC Deb 24 March 1868.
[49] Select Committee 1867–68, p. 59, qq. 1028–30.
[50] Select Committee 1867–68, pp. 192–99, qq. 3711–97. On Moseley's proposal, see Charles 1951.

Beyond the parliamentary committee, activists frequently associated the emulation of overseas' institutional models with appeals to end the age of the traditional 'rule of thumb'. An early example was the 1868 book of school inspector and poet Matthew Arnold, *Schools and Universities on the Continent*, based on his official investigations published the same year.[51] Within a short time, well-known engineers, such as William Fairbairn and William Rankine, were repeating Arnold's mantra.[52] The evocation of 'applied science' as the superior modern alternative to 'rule of thumb' would be an enduring rhetorical move, well into the twentieth century.

Local Initiatives

While the Samuelson Committee report urged state action, a remarkable amount of private money was spent developing a new generation of provincial colleges.[53] Local industrial needs inspired several, such as those in Birmingham and Leeds, but these also sought to display their commitment to liberal education. At formative moments, campaigners deployed stories of revered heroes to promote both their institutional objectives and the concept of applied science.

Birmingham was the heartland of the nation's metals manufacturing, of non-conformist educational enthusiasm, and of the campaign for improved education.[54] Among the educational movement's leaders was the industrialist and future politician Joseph Chamberlain. So was the local newspaper editor, *The Birmingham Daily Post*'s James Bunce, who would also serve as an executor and biographer of another enthusiast, Josiah Mason, a wealthy electroplate manufacturer and a philanthropist.[55]

---

[51] Arnold 1868b, 278. The book was adapted from a report written for the Schools Inquiry Commission; see Arnold 1868a. For Arnold's own complex relations to science, see Dudley 1942.

[52] Marsden 1992, 330; Kargon 1977, 47. Denunciations of the 'rule of thumb' were also incorporated in similarly themed speeches by politicians such as Lord Alexander Churchill, referring to the prospects for chemistry in his annual address as chairman of the Society of Arts. 'Chairman's Address', *Journal of the Society of Arts* 24, no. 1200 (1875): 4–13, see 10.

[53] In addition to the expansion of Owens College (founded 1851) and of the old universities, within just one decade, colleges were established in Newcastle in 1871, Aberystwyth in 1872, Birmingham and Bristol in 1873, Leeds in 1874, and Nottingham in 1876. In Scotland, Dundee University was founded in 1881. See Vernon 2004.

[54] See Rodrick 2004. The Birmingham Education League, of which Chamberlain was a founder in 1867, became the National Education League two years later.

[55] Leighton 2000; McCulloch 2004. On Mason, see Bunce 1970. In addition to work on electroplating, Mason led Britain's largest pen-nib manufacturer, at a time when Birmingham was the world's centre for the industry.

In 1870, Mason, already seventy-five years old, set up a science college in Birmingham, donating over £200,000 (equivalent to about $1 million at the time) to the cause. In line with local Congregationalists' principles, the college would be separate from any church, and Mason specified that theology not be taught.[56] This example highlights the tangle of alliances around the concept of applied science: a philanthropist, a local newspaper, a scientist, and other vital individuals, including a lawyer and a politician.

Preparing for his foundation, in 1867, Mason commissioned the local consulting chemist and teacher George Gore to tour the universities and laboratories of other European countries.[57] Returning from these travels to promote applied science locally, he was campaigning at the same time as the parliamentary moves by Samuelson and his supporters. Gore's overseas tour seems to have resulted in two articles published two years apart. The first, published in two parts, was entitled 'On the Relation of Science to Birmingham Manufactures' and appeared in *The Birmingham Daily Post* in January 1868. Here, Gore defined 'applied science' as 'those portions of pure science as have been adapted to practical purpose'.[58] The final quarter of the two-column polemic was devoted to the model offered by an adopted Birmingham hero. Here, Gore emphasised the debt owed by James Watt to the Scottish chemistry professors Joseph Black and John Robison. Once again, the allegory of Watt's career was used to point to the importance of applied science.[59] In the second article, Gore returned to his theme with 'Practical Scientific Education' in the *Quarterly Journal of Science*, co-edited by James Samuelson, brother to Bernhard. As a chemistry teacher at the local grammar school, Gore emphasised the need for artisans to learn not just pure science but also *'how such knowledge is applied and operates in their several occupations'*.[60]

---

[56] 'Sir Josiah Mason's Scientific College Birmingham', *The Birmingham Daily Post*, 24 February 1875.

[57] 'George Gore F.R.S.' 1909. Gore referred to this tour in his letter to Richard Norris, 1 July 1880, US41/7/22/32, University of Birmingham, Cadbury Research Library.

[58] George Gore, 'On the Relationship of Science to Birmingham Manufactures. First Article', *The Birmingham Daily Post*, 13 January 1868, British Newspaper Archive/ British Library Board; Gore claimed the paper in his list of publications, in an application for a post at the College. See Gore, 'Statement of Qualifications of Dr George Gore for the Professorship of Chemistry at Sir Josiah Mason's College Birmingham, December 1879', Cadbury Research Library. This was followed up a week later by [George Gore], 'On the Relationship of Science to Birmingham Manufactures. Second Article', *The Birmingham Daily Post*, 20 January 1868, British Newspaper Archive/British Library Board.

[59] On the use of Watt as chemist, see Miller 2004.

[60] George Gore, 'On Practical Scientific Education', *Quarterly Journal of Science* 7 (1870): 215–29, on 221 and repeated on 227. Emphasis in original.

SIR JOSIAH MASON'S SCIENCE COLLEGE, BIRMINGHAM.

Figure 2.1 Applied science embodied: Josiah Mason's Science College at the time of its opening in October 1880. *The Illustrated London News*, 9 October 1880. Courtesy Illustrated London News/Mary Evans Picture Library.

Through his use of 'applied science', Gore promoted its cultural distinctiveness. It was not just Gore to whom it was important. The concept proved to be centre-stage in October 1880, on the occasion of Mason College's opening of a magnificent new building (see Figure 2.1).[61] At this grand affair, Thomas Huxley was the invited speaker. Famously, he announced,

I often wish that this phrase, 'applied science', had never been invented. For it suggests that there is a sort of scientific knowledge of direct practical use, which can be studied apart from another sort of scientific knowledge, which is of no practical utility, and which is termed 'pure science'. But there is no more

---

[61] The vaguely thirteenth-century-styled architecture of Mason's College was described in 'Sir Josiah Mason's Scientific College', *The Birmingham Daily Post*, 13 September 1880, British Newspaper Archive/British Library Board. Birmingham University moved progressively from the centre of Birmingham to a much larger suburban site in Edgbaston, and the building of Mason College was demolished in 1964. See Gillian Thomas, 'College Is No More, but Memories Live On', *The Birmingham Post*, 24 August 1964, British Newspaper Archive/British Library Board.

complete fallacy than this. What people call applied science is nothing but the application of pure science to particular classes of problems.[62]

Huxley seems to have derived his exclamation from his close friend, the versatile and distinguished scientist John Tyndall. As early as 1873, Tyndall had translated a passage from a short French article published two years earlier by Louis Pasteur, concluding, 'There exists no category of science to which the name applied science could rightfully be given.'[63] This sentiment had subsequently become well known, not least because of its well-publicised use by Tyndall in numerous lectures, including one given twice at the Royal Institution and widely published just a year before Huxley's Birmingham address.[64] However, such arguments for the autonomy of science teaching and research from the scientists' corner were not matched by the funder's view. Mason's lawyer and old friend replied to the indictment of 'applied science' at the subsequent luncheon (in the absence of the now very aged philanthropist). 'Perhaps Mr Huxley', he said, 'would regard that phrase with more favour if he (the chairman) told him that it was probably to that phrase that they owed the existence of the Mason College.'[65]

Historians can impute differences between those whose focus was on a branch of knowledge itself and those for whom the concept constituted the application of pure science.[66] It is, nonetheless, striking that, at the time, the term 'applied science' was rarely explicitly either decoded or disputed in this way. Gore's formulation, 'such portions of pure science as have been adapted to useful purposes', carefully avoided a decision. Historian of education James Donnelly has shown the difficulty chemists found when they sought to teach courses that distinguished between scientific and technical chemistry.[67] Indeed, the flexibility of the meaning served widely as a strength.

---

[62] 'Sir Josiah Mason's Science College', *The Birmingham Daily Post*, 2 October 1880, British Newspaper Archive/British Library Board. For a reflection on Huxley's intention, see MacLeod and Radick 2013, 197.

[63] Pasteur 1871, quoted in Tyndall 1873, 221. Tyndall's own distinctive views on the primacy of pure science are discussed by Cantor 1996 and Barton 2018.

[64] Tyndall used Pasteur's quotation in his Royal Institution address of 17 January 1879, repeated on 20 January. See John Tyndall, 'The Electric Light', *Notices of the Proceedings at the Meetings of the Members of the Royal Institution of Great Britain* 9 (1879–81): 1–24, on 23. The lecture and the quotation were also published in the *Fortnightly Review* 31 (January–June 1879): 197–216. It was even quoted at length in New York, 'Tyndall on Illumination', *New York Times*, 16 February 1879.

[65] 'Sir Josiah Mason's Science College'.

[66] An attempt to distinguish these meanings is made by Donnelly 1986, 198.

[67] Donnelly 1997.

In Birmingham, the term was an essential means of describing the distinctive interface between science and practice in the setting up of Mason Science College. Through the support of the College and the publication of Gore's articles, the local newspaper had promoted the dissemination of 'applied science'. The explicit and immediate riposte to Huxley highlights the importance of the concept to the College's identity. What it meant, epistemologically, was much less apparent. When the College established its first chairs (in chemistry and physics), the teaching focused on basic principles. Nevertheless, the chemistry professor did promise to organise some factory visits to promote 'practical knowledge of some branches of Applied Science'.[68] The College offered courses in metallurgy, practical metallurgy, and civil and mechanical engineering from its second year. With the implementation of such specialisms, the rhetoric then faded.

In Leeds, at the opening of the Yorkshire College in December 1880, a few months after Birmingham's Mason College, applied science's ascendancy was acclaimed, together with the end of the age of the rule of thumb – that emblem of craft-based knowledge and production.[69] Particularly through the 1887 incorporation within a new northern federal university centred in Manchester, requirements for liberal education would balance pressures to give practical training. Such founders of the Yorkshire College as the MP and newspaper owner Edward Baines had found support for their own views across the Pennines, from a well-developed romantic narrative in which Manchester Man bettered himself morally as well as commercially through science.[70] Moreover, Manchester had already proved to be the successful home of perhaps the most significant of the provincial colleges.

During the mid-1860s, an enthusiastic campaign backed by wealthy local industrialists urged the conversion of the city's small Owens College founded in the previous decade into a great centre worthy of a modern metropolis. The six Trustees co-opted for the committee responsible for enlargement included the editor-proprietor of *The Manchester Guardian*.[71] His newspaper reported a presentation by the college principal proclaiming there could be no doubt that as the nation's premier manufacturing centre, Manchester's special strength should be

---

[68] Mason Science College, *Calendar, 1880–1881* (Birmingham, 1880), 98.
[69] 'The Yorkshire College', *The Glasgow Herald*, 4 December 1880, British Newspaper Archive/British Library Board.
[70] Joyce 1994, 163–76. Manchester scientific culture was explored by Kargon 1977, by Thackray 1974, and in the essays collected in Cardwell 1974. For the history of the Yorkshire College and Leeds University, see Gosden and Taylor 1975.
[71] Thompson 1886, 317.

'the study of experimental and applied science'. The article noted the response from the audience: 'Hear, Hear.'[72] After all, argued the principal, this was the city of Dalton and Joule. The architect of atomic theory, John Dalton, was already a municipal patron saint, a stained-glass picture of whom would adorn the new town hall. Brewer James Joule, who had identified the 'mechanical equivalent of heat', was a still-living hero. Later, statues of the two men would flank the entrance hall of the town hall.

When calls for government funding failed, local philanthropy realised municipal ambition, leading to the magnificent building still at the heart of the University of Manchester campus. Winning the active support of two of the world's most revered engineers, the industrialists Joseph Whitworth and William Fairbairn, the campaign published a fund-raising circular expressing the vision in terms of 'applied science'.[73] The concept linked the aspiring middle class's practical needs in industry with culture represented by science.

North of the Scottish border, the engineering tradition was even stronger. Fleeming Jenkin, the new professor of engineering, appointed to the University of Edinburgh in 1868, allowed that applied science would be a component of the engineer's education but only following rigorous training in the basics of science and in alliance with practical engineering.[74] As he said to the Samuelson Committee, 'I should say that technical instruction meant scientific instruction, with the addition of a small portion of applied science; science applied to the special profession which the man intended to follow.'[75] Jenkin was promoting the model hammered out by the Royal College of Science in Dublin report. At the University of Glasgow, there was the same recognition of a level in between 'pure science' and 'pure practice' as W. J. M. Rankine put it in his 1855 presidential address to the mechanical science section of the

---

[72] 'A North of England University: Proposed Extension of Owens College', *The Manchester Guardian*, 2 February 1867.

[73] Thompson 1886, 325. For the focus of Whitworth on chemistry, metallurgy, and engineering and the eschewing of 'applied science', see the *Engineering Professorship Committee*, for instance, 'Report of the Engineering Professorship on the Terms of Agreement Settled with Mr Reynolds, on the Accommodation Required for the Engineering Department and on the Scheme of Study to Be Pursued in the Engineering Classes', 13 May 1868, OCA 5/3/1, University of Manchester Library, Manchester, Special Collections Division.

[74] Fleeming Jenkin, 'Civil Engineer: A Two Years' Course', Appendix, 'Committee on Technical Education', *Journal of the Society of Arts* 16, no. 818 (1868): 634. On Fleeming Jenkin, see Hempstead and Cookson 2000 and Morrell 1973.

[75] Select Committee 1867–68, p. 138, q. 251.

British Association for the Advancement of Science.[76] Rankine and close colleagues redesignated and widely advertised the 'Breadalbane Scholarships' for 'the encouragement of pure and applied science'.[77] But, again, when it came to the content of what he taught, Rankine emphasised his vision of 'engineering science'. In analysing Rankine's work, Ben Marsden has interpreted this formulation as 'science regulated by economy'. He suggests that, for Rankine, it comprised an aggregate of physical sciences 'modified in a manner specifically designed to treat quantitatively the economic and financial constraints placed upon the practical engineer in the commercial environment'.[78] It was thus broader than merely 'applied science'.

This treatment has concentrated on the colleges established in Manchester, Birmingham, and Leeds and Scotland's old universities. There were parallel movements in a host of industrial towns, including Newcastle (1871), Aberystwyth (1872), Bristol (1873), Nottingham (1876), Dundee (1881), Liverpool (1881), and Sheffield (1884). Although several struggled initially, Liverpool, together with Leeds and Owens College Manchester, became part of the federal Victoria University. Newcastle was a college of Durham University. Again, 'applied science' resolved tensions between the vocational teaching sought by many sponsors and the liberal education demanded by the University parent. Other colleges could follow what was becoming a nationwide trend. Thus, the founders of University College Bristol, writing to their local newspaper in 1875 and 1876, cited developments in Birmingham, Leeds, and Manchester in detail to justify their approach to technical education.[79]

The foundation of these colleges was, of course, part of an international movement. Parallel processes were happening across Europe and America. In the United States, the land grant colleges date from the 1860s, and the most famous, the Massachusetts Institute of Technology, was founded in 1866. Such well-known European centres as Switzerland's centre for technical education, the Eidgenössische Technische Hochschule – known as the ETH (1854) – and Prussia's

[76] 'The British Association for the Advancement of Science', *The Morning Post*, 15 September 1855, British Newspaper Archive/British Library Board.

[77] Marsden 1992, 271.

[78] Marsden 1992, 320. I am grateful to Ben Marsden for permission to quote from his unpublished thesis. Also, Channell 1982.

[79] William Lant Carpenter, 'The School of Science & Literature or University College, Bristol', *The Western Daily Press*, 13 March 1875, British Newspaper Archive/British Library Board; J. Percival, 'University College Bristol, for Whom Is It Intended?', *The Western Daily Press*, 9 October 1876, British Newspaper Archive/British Library Board.

Technische Hochschule in Berlin (1879) come from the same era. Across the empire, too, colleges resembling their English counterparts were being established. In Sydney, Australia, the university dates from 1850. McGill University in Montreal established its 'Faculty of Applied Science' in 1878. As they grew, many of these universities would negotiate their own international standing. McGill's faculty would be outstandingly well resourced and an imperial leader in its field.[80] Circulating through the empire, New Zealander Ernest Rutherford taught physics there from 1898 to 1907. Its dean, Henry Bovey, was appointed the first head of Imperial College, London. Moreover, many graduates of such colleges as the Royal School of Mines found positions in the colonies. Nevertheless, in both the nineteenth and twentieth centuries, press conversations about applied science within Britain focused on just a few competitor countries, such as France, Germany, and the United States.

Pure and Applied Science and the Engineering Profession

However parochial the British press of the time might often seem, the historian can notice how much the relationship between 'pure' and 'applied science' was a matter of international, national, and regional discourse. Coleridge had imported both terms from the German into the early nineteenth-century British consciousness, and for him, pure science, Kant's *reine Wissenschaft*, had a particular truth value. Certainly, with their culture of *Wissenschaft*, the German universities were a constant presence for other leading scholars. These cathedrals of learning provided a vivid embodiment of the meaning and potency of 'pure' science.

Within Britain, there were also regional variations. Many scientists, particularly the Scots, found that truth and utility went hand in hand. Even the Cavendish physics laboratory at Cambridge, under its Scottish leader James Clerk Maxwell for its first five years (1874–79), was known for its contribution to establishing standards useful in telegraphy.[81] Recently, Ruth Barton has reflected on the contrasting views of the members of the London-based X Club. This included Thomas Huxley and John Tyndall, whom she found characteristically '[p]raising science

---

[80] 'Dr. Henry Taylor Bovey, F.R.S.', 1912; 'Physics and Engineering at the McGill University, Montreal', *Nature* 50, no. 1301 (1894): 558–64. The world's leading institutions are discussed in 'The Position and Work of the Central Technical College', *Nature* 55, no. 1421 (1897): 284–86.

[81] Schaffer 1992; Crosbie Smith 1998.

for its truth rather than its utility and denigrating the money-making aspects of science'.[82] Yet that did not mean they believed science to be useless. Rather, its inherent potential for utility justified an individual focus on its truth. Similarly, Graeme Gooday has reflected on the emphasis put by Huxley on the primacy of pure science.[83]

Whereas to the X Club scientists, distinctions within science were anathema, for engineers, particularly those impressed by German models, they were most attractive. Enthusiasts included such promoters of an academic preparation for a career in engineering as Owens College promoters Josiah Whitworth and Henry Fairbairn and their ally Sir William Siemens (brother of Werner). In the 1870s, Siemens would work hard, if unsuccessfully, to establish a home for all the engineering societies. The title of his grand centre would be 'House of Applied Science', and he offered to contribute £10,000 to the costs of its building.[84] His ambition complemented the oft-told narratives about such past luminaries of Britain's industrial revolution as James Watt and George and Robert Stephenson. The recounting of their experiences, the publication of meetings at which they were retold, the support of the press, and the use of 'applied science' were closely interconnected.

Pure and Applied Science on Display

The most widely experienced promotional event for science in late nineteenth-century Britain was a museum exhibit visited by almost 300,000 people during six months in the summer of 1876.[85] This helped establish the sense of applied science as a grouping, as an interpretation of technical development, and as a source of future growth. Explicitly planned to promote the cultural appreciation of science, it grew from a recommendation in the fourth report of the Royal Commission on Scientific Education that had followed on the Samuelson Committee. Norman Lockyer, Secretary to that Commission and editor of the campaigning *Nature* magazine, was the secretary and curator.[86] Among the 20,000 objects on show, Lockyer moved the great engines of Watt and the Stephensons from the nearby Patent Office Museum, where they had

---

[82] Barton 2018, 414.   [83] Gooday 2012.
[84] See, for instance, Siemens 1889. It is a nice coincidence that Siemens had begun his career working with Mason on electroplating.
[85] The most extensive modern analysis of the exhibition in English is the four part group of articles by de Clerq 2002–3. See also Cho 2001. I am grateful to Dr Cho for sharing an English translation of a draft.
[86] Bud 2014b.

embodied invention, to the exhibit where they now became exemplars of 'scientific apparatus'.

In his opening speech, Sir William Siemens, then president of the mechanical sciences section, discussed the exhibition incorporating 'The Department of Applied Science' closely linked to pure science.[87] Progress was illustrated through an overwhelming visual onslaught and by the labels, which emphasised the scientific principles underlying the machines.[88] The exhibition displayed the story of applied science as cosmopolitan and quintessentially British, both a distinctive part of European culture and useful for industrial development.

Conclusion

This chapter has shown how the concept of 'applied science' provided a language for the public debates about science in education and state policy. Above all, the result was that it came to underpin and, in turn, be institutionalised in the new colleges. Remember the immediate riposte to Huxley's criticism: 'it was probably to that phrase that they owed the existence of the Mason College'.

The term moved through several different discourses in Britain between the 1840s and the 1870s. Whether in journalism, lecturing, testifying, or exhibiting, it proved to be a means of linking individuals' claims and giving them the authority of an increasingly vibrant and real concept. The recounting and reprinting of heroic narratives of achievement cemented alliances between 'practical men' and 'men of science' by proclaiming a respectable subject of common interest. The term expressed respect for the British constitutional resolution, which placed a strict boundary between the legitimate scope of the public sphere and the privacy of the commercial enterprise.

The campaigners had developed three characteristics that would endure. First, applied science was a category incorporating, as a natural set, ingredients that might otherwise seem disparate: an approach to training, a genre of books, and the machines themselves. The most prominent usage was as a form of education necessary for future prosperity. Second, based on science, it was public rather than proprietary knowledge, though it remained hard to agree upon the detailed content of this education. Finally, 'applied science' starred in widely shared historical allegories. These represented the relationship between

---

[87] Sir C. William Siemens. 'Opening Address: Section – Mechanics', *Nature* 14, no. 342 (1876): 56–59.
[88] 'The Queen at South Kensington', *The Observer*, 14 May 1876.

science, praxis, and international leadership in Britain's triumphant past and served as models for contemporary practice assuring civilisation and prosperity. They were encoded in the new, privately endowed colleges educating the next generation and reassuring current elites.

From the mid-nineteenth century, 'applied science' linked a selective rendering of the nation's past to fears about a future perceived as full of risk. Newspapers living on the combination of familiar history and unknown future found this account particularly attractive. The 'romances' of Watt and Stephenson acquired their force from their moral significance and compelling narrative. The new colleges capitalised on this hope and its compatibility with industrialists' expectations that training in manufacturing skills would still be safe in the privacy of the factory.

3 Competing Concepts of Applied Science and Technology

Introduction

In the era of the colleges' foundation around 1880, 'applied science' dominated the space between science and practice in British education. Politicians' speeches, newspaper editorials, and novels invoked its power. Over succeeding decades, however, other interpretations increasingly contested this space. The terms 'industrial science' and 'applied research' would find niches, but the dominant competitor was 'technology'. Different kinds of organisation were attached to these concepts, and the choice of language was a socially significant act. Between the 1870s and World War Two, the usage of the two was complementary with a carefully negotiated boundary that revealed the pressures on terminology. So, this chapter asks how 'technology' was used in this period, and how it related to 'applied science'.

At the heart of the account lies the manoeuvring to develop a conception of engineering training outside both the industrial and the university sectors to meet the nation's needs for skills and correspond with instruction offered in Germany and the United States. This process underpinned both opportunities and uncertainties for 'technology', as an anecdote from 1920 made clear. In the wake of the centenary commemoration of James Watt's death, a small band of enthusiasts sought to create a society concerned with the history of engineering more broadly. They delegated the sensitive question of name to two of their most respected members. One, Henry Winram Dickinson, a Manchester graduate who had gone on to an apprenticeship, rarely used the term 'technology'; but the other, L. St L. Pendred, more professionally trained, with Belgian and French engineering experience, would speak on 'technological history' to the society.[1] The domain these two men eventually identified for the new organisation was a compromise: its

---

[1] Titley 1920 and Titley 1941. Schatzberg has pointed to American tensions between history of engineering and history of technology. See Schatzberg 2018, 207.

subtitle gave its subject as 'engineering and industrial technology'. As the main title, the two men picked 'The Newcomen Society' to steer away from the more controversial figure of James Watt, although it had been his anniversary that had propelled the foundation.[2] At about the same time, A. P. M. (hereafter Arthur) Fleming, director of a leading industrial research laboratory, at the electrical engineering firm Metropolitan Vickers (also known as Metrovick), co-wrote a book-length history of 'engineering', only once mentioning 'applied science'.[3] These examples illustrate how mobile and negotiable were uses of such terms in Britain.

In the United States, Eric Schatzberg sees the term 'technology' as having three lives: the first, during the nineteenth century, came out of the category of industrial arts.[4] A second came at the end of the century when Thorstein Veblen used the same word to represent the engineers' claim for their own culture, adopting as his meaning the translation of the German word *Technik*.[5] A third generation was associated with material progress. It was only in the 1930s, Schatzberg has suggested, that the concept came into vogue and, by the end of the decade, had become a standard way of referring to the 'material means of production, transportation, and communication' and as a means of explaining the onset of mass unemployment.[6] It acquired the twin connotations of practical means to an end and the application of science. Previously, he suggests, it had been little known in the United States. Britain did not experience such earlier silence. From the 1870s, its campaigners promoted the term to accommodate a class-based educational hierarchy.

Technology in Britain

The term 'technology' had entered British English as a direct translation from the German *Technologie*. This was an established term within cameralism, and its travel across the North Sea was also facilitated by the needs of an encyclopaedic work. The publication of Crabb's *Universal Technological Dictionary* in 1822 had an impact through its press coverage quite comparable to the publication of the *Encyclopaedia Metropolitana*. A few years later, the American Jacob Bigelow published

---

[2] I am grateful to David Philip Miller for pointing out to me the enduring controversies over Watt's standing within the engineering community. See Miller 2019, 307.
[3] Fleming and Brockelhurst 1925.
[4] Schatzberg 2018 also refers to the Society of Arts innovation, though not to its subsequent development.
[5] Schatzberg 2018, 104. [6] Schatzberg 2018, 152

his lectures in the *Elements of Technology*.[7] Lyon Playfair's friend, George Wilson, also proved to be influential. Though an early adopter of 'applied science', in 1855, he dismissed that term in favour of 'technology'. Taking his usage from the German writer on chemical technology, F. L. Knapp, Wilson described it as the 'science of utilitarian art'.[8]

Wilson was always concerned with the non-scientific factors that underpinned industrial success. As early as his 1848 article on telegraphy, he emphasised the importance of such non-scientific resources as capital and 'bold, sagacious, and often reckless empiricism' that were needed to make men the 'conquerors of physical nature'.[9] Ambitious himself, Wilson became the first director of the new industrial museum in Edinburgh in 1855, and later that year, the city's university appointed him to a specially created chair in 'technology'.[10] Intellectual, religious, and politically astute, Wilson developed a model of technologies that built on the German cameralist tradition. A modern biographer has emphasised its complexity, extending beyond ideas to incorporate their relationship to people and things.[11] However, the category posed a challenge to other professors' intellectual and economic space, and on his early death, the University discontinued his chair.[12]

Thus, it was through another route, also inspired by the different German term, *Technik*, that 'technology' entered English usage more permanently. In the German states, the period after the failure of the 1848 revolutions saw the ambitions of youth move from effecting immediate political change to dreams of engineering.[13] Across the region, states established technical schools, drawing on the French example. Preceding large-scale industrial employment, ambition to develop trumped the concerns about safeguarding proprietary knowledge, so important in Britain. Technical education came, therefore, to encompass a distinctive blend of theoretical and practical knowledge.[14] It was associated too with a new category of organisation. Whereas science was

---

[7] Before the appearance of Crabb's *Dictionary* one finds very few uses of the word 'technology'; the 'Britishnewspaperarchive' corpus (consulted May 2019) shows only five uses in the previous decade, and by contrast 142 uses in the year 1823 immediately after publication. On the significance of Bigelow's lectures, see Schatzberg 2018.

[8] Wilson 1855, 4. Wilson ascribed his use of the term 'technology' to the 1848 translation of Knapp's *Chemical Technology or Chemistry Applied to the Arts and Manufactures*. See Wilson 1855, 3–4. Knapp, in turn, had attributed his use of the term to the cameralist Johann Beckmann. See Sebestik 1983.

[9] [George Wilson], 'The Electric Telegraph', *Edinburgh Review* 90 (1849): 434–72, see 442; authorship ascribed by the *Wellesley Index*.

[10] Anderson 1992.   [11] Swinney 2016, 180.

[12] Grant 1884, 354–61; see also Donnelly 1989.   [13] Gispen 1989, 47.

[14] Klein 2020.

taught and researched in great German universities, these generally excluded engineering, which became the domain of the separately organised *Technische Hochschulen*. A lower level of trade schools, *Fachschulen*, offered training to artisans.[15]

Playfair's successor at the Department of Science and Art, John Donnelly, was impressed by German institutions, but he lacked the authority to copy them directly. Instead, he redeployed George Wilson's term 'technology' with a new meaning and purpose. Looking outside the academic sector, which was institutionalising 'applied science', in 1871, Donnelly proposed that a parallel movement adopt the alternative category to describe the training of artisans.[16] His own government department had recently squeezed out the privately funded Society of Arts from its role as the provider of science examinations.[17] Now, speaking as a citizen rather than a civil servant, he urged the Society's council to establish a system of 'examinations in the science and technology of the various arts and manufactures of this country'.[18] Accordingly, the following July, the Society hosted a conference on 'technological education'. This meeting approved the proposal for qualifications based on three levels of competence: workman, foreman, and manager, and three discrete components. These were, first, knowledge of science, examined by the Department of Science and Art; the second in 'technology' defined as 'the special application to each manufacture of the various branches of science that have to do with it'; and, third, practical knowledge based on an employer's certification.[19] Playfair, a friend of the late George Wilson and Donnelly's former colleague, seconded the motion. He reflected on the apparent threat of technology, which reached as close to practice as was acceptable: 'the word Technology sometimes frightened persons who did not quite understand what it meant. It was not at all intended by the Society to teach the practice of the arts; that was not the meaning of technology,

---

[15] For a good English-language review of German technical education, see Inkster 1991, 101–3.
[16] On Donnelly, see Armytage 1950.   [17] Wood 1913, 428.
[18] Society of Arts, 'Minutes of Council', vol. 17 (June 1870–July 1872), 11 December 1871, RSA/AD/MA/100/12/02/18. By kind permission of the RSA London.
[19] 'Conference on Technological Education', *Journal of the Society of Arts* 20, no. 1027 (1872): 725–52. The linkage between the Science and Art Department and the vocational examinations managed by the Society of Arts/City and Guilds, intended to be tight, was in fact much weaker than envisaged. In the early 1880s, out of nearly 2,000 candidates, only one in five had taken examinations in the requisite science subjects through the Science and Art Department. See question of Philip Magnus, Royal Commission on Technical Instruction 1884, vol. 3, p. 339, q. 3068. See Foden 1961.

which rather meant the science which underlie [*sic*] the art, and on which it rested.'[20]

The Society of Arts intended its examinations to build demand for an education equivalent to that of German trade schools. In 1873 the Society launched a qualification system beginning with 'cotton', 'paper', 'steel', 'silk', and 'carriages'. The politics of technical education ensured that, in public declarations, technology was defined entirely in terms of the application of science. The details of the examinations revealed, however, a more subtle meaning. For instance, Donnelly's technology proposal recommended a curriculum on paper manufacturing. This included such topics as 'The characteristic properties of the fibres of hemp, flax or linen, cotton, esparto, straw, wood, &c., when made into paper' and 'The general arrangements for water supply throughout the works, and the conditions to be attended to as respects its quality'.[21] In his 1895 presidential address to the Society, Donnelly distinguished between 'more or less pure sciences' such as mathematics, physics, and mineralogy; the 'applied sciences' such as mechanical drawing, applied mechanics, and metallurgy; and the 'technological' sciences such as machine construction, principles of mining, and hygiene.[22] Later in his lecture, he instanced plumbing, electrical and mechanical engineering, and textile fabrics as technological subjects encompassed within the examination system.[23] Technology was still neither intended by teachers nor expected by professionals to equate with engineering. At most, it was that subsidiary non-tacit component that could be examined.

Therefore, technology entered the vocabulary of British bureaucracy driven by the need for skilled workers, respect for the knowledge obtained in the machine shop, and jealousy of German educational achievements. At first, there was little take-up under the sponsorship of the Society of Arts and the gaze of suspicious employers. After a few

---

[20] 'Conference on Technological Education', 732. The observation that British bureaucrats were 'frightened' by the use of the very word 'technology' was repeated twenty years later by Playfair's friend Donnelly, testifying to a government hearing about the 'Technical Instruction Act' of 1889. Donnelly, by then Secretary of the Science and Art Department, said that 'to be correct, the word used ought to have been "Technology" and not "Technical Instruction", only I suppose people would have been frightened at the word "Technology". "Technical Instruction" has, in some places, been really carried into technical instruction, that is to say into the art itself, which is forbidden by the Act.' Royal Commission on Secondary Education 1895, Testimony of Sir John Donnelly, p. 137, q. 1235.

[21] 'Outline Scheme of the Technological Examinations Proposed by Major Donnelly, R.E.', *Journal of the Society of Arts* 20, no. 1027 (1872): 726–31, on 730.

[22] Major-General Sir John Donnelly, 'Address', *Journal of the Society of Arts* 44, no. 2244 (1895): 7–19, on 10.

[23] Donnelly, 'Address', 13.

years, though, the extremely wealthy City and Guilds of London, needing to prove their charitable status, took over running the examinations. The number of examinees multiplied ten-fold in the four years 1879 to 1883 alone, and the attraction kept growing. By 1905, almost 12,000 students passed the technological examinations, with 20,000 papers in sixty different subjects sat by 55,000 students.[24] Generally, these clients studied in the evening while employed in the daytime. The many new institutions sustained by this system also took a few day pupils and erected impressive buildings. As was typical, the School of Science on Merseyside became the Liverpool School of Science and College of Technology in 1888 and acquired a large and elegant new home in 1901.[25]

The City and Guilds also established its own teaching centres, including a college at Finsbury in North London, which offered day classes. The principal, Sylvanus Thompson, emphasised that all his lecturers were 'men who learned their work from the industrial side ... they are trained workers, endeavouring to teach the principles of their craft, not schoolmasters trying to teach trades'.[26] To produce teachers for the system, City and Guilds also opened the impressive Central Institution in South Kensington in 1884. Although the investment was considerably less than for the much-envied Technische Hochschule in Berlin ($2.5 million), it was substantial: the new body's building and equipment cost about £100,000 ($0.5 million).[27]

At the helm of the City and Guilds was a missionary clergyman, but not an Anglican vicar. The German-trained Philip Magnus was an early Reform Rabbi who, in his religious role, had also balanced aspirations for change with respect for established ways.[28] Magnus would be a doughty promoter of 'technology' for fifty years, navigating between British tradition and the German model with determination and political skill.

Magnus was quite sure it was on account of the government's overly scrupulous adherence to laissez-faire that his private organisation had to promote 'courses of instruction directly applicable to the different trades in which artisans were engaged'.[29] In an entry on 'technical education'

---

[24] Horn and Horn 1984. See also Gay 2000. For the number of subjects, see Smalley 1909, 220. For the number of students in 1883 and 1905, see Foden 1961, appendix E, xxii.
[25] 'Liverpool School of Science', *Liverpool Mercury*, 28 June 1888. For technical education in Liverpool, see Roderick and Stephens 1972.
[26] 'Professor Sylvanus P. Thompson on Technical Education', *The Electrician* 20 (1887): 19–20, on 19.
[27] A cost of £101,000 pounds for building and equipment of the Central Institution is given in 'The Position and Work of the Central Technical College', *Nature* 55, no. 1421 (1897): 284–86, on 285.
[28] Foden 1970. For the context, see Sharot 1979. Also see Sebag-Montefiore 1994–96.
[29] Magnus 1910, 107.

for the *Encyclopedia Britannica*, he explained, 'The terms "technical" and "technological" (Gr. τέχνη [techne], art or craft) as applied to education, arose from the necessity of finding words to indicate the specialised training which was needed in consequence of the altered conditions of production during the 19th century.'[30] He pointed out that the education was one which the state was too timid, tied to strict ideology preventing state intervention in the market, or, in his words, too 'pedantic' to offer.[31]

The Royal Commission on Technical Education of the early 1880s identified the practices of international competitors, the opportunities and challenges of artisan education, and the meaning of 'technology' within it. Bernhard Samuelson, who had launched the parliamentary investigations at the end of the 1860s, was the chair, and with him were other enthusiasts, including Philip Magnus and the chemist Henry Roscoe. Their report drew upon a vast mass of detailed literature, including prospectuses and syllabi and descriptions of in-depth on-site interviews of both companies and academics on the continent.[32] Among those interviewed was John Donnelly, now a lieutenant-colonel and still leading the science division of the Department of Science and Art. His testimony provides a window into the distinctions he drew between the branches of science and technology.

Donnelly explained that the mood had changed from twenty years previously when an attempt to teach agriculture would have been 'one stone more to throw at the Department'.[33] Now, the department found no problem aiding the teaching of 'applied science'. It was the analogue of 'design' on the arts side, and he gave the examples of 'metallurgy, applied mechanics, steam, nautical astronomy and agriculture'.[34] He had, nonetheless, to avoid encouraging the teaching of the art itself. So Donnelly talked about 'technology', whose education he had promoted through the private system, first supported by the Society of Arts and then the City and Guilds. He described how, lecturing on its importance in a town in the north of England, he was received silently during the meeting itself, and afterwards accosted by the angry local manufacturers:

---

[30] Sir Philip Magnus, 'Technical Education', in *Encyclopedia Britannica* (Cambridge: Cambridge University Press, 1911), vol. 26, 487–98, on 487.

[31] Magnus 1910, 107.

[32] Stored offsite, many of these are now an accessible archive within the Library and Research Centre of the Science Museum, London (Ref RCTI). For the Commission, see Argles 1959.

[33] Testimony of John Donnelly, Royal Commission on Technical Instruction 1884, vol. 3, p. 287, q. 2868.

[34] Testimony of John Donnelly, Royal Commission on Technical Instruction 1884, vol. 3, p. 283, q. 2845; see also p. 285, qq. 2856 and 2857.

we were intervening in the details of manufacture, and they said they were not going to encourage something which would bring all the workmen from the different works together to discuss matters in which trade secrets were involved, and they would have nothing to say to it, and so far from assisting the scheme they said that they would do everything they could to stop it.[35]

A few minutes later, Donnelly explained to the Samuelson Commissioners that he blamed the employers rather than artisans for the weakness of technical education in Britain. Equipped with a dry wit, he even criticized philanthropists who supported local initiatives. He compared them to irreligious people in the Middle Ages who, nearing death, 'took a slice out of the inheritance of their children to endow something religious'.[36]

Other witnesses testified to the impact of overseas initiatives and concepts on the City and Guilds team. One suggested that members had the 'Technological Institute at Massachusetts' (MIT) model before them in planning the 'Central Institution'.[37] Another explained that its role would be analogous to the polytechnic schools on the continent, and students would learn the science behind brewing, dyeing, or calico printing rather than the general principles of science.[38] Interestingly, Thomas Huxley, famous for his opposition to the very concept of applied science, explained his attitude in terms of organisations. Samuelson asked him, 'Would you, or would you not, like to see a more distinct line drawn between pure science on the one hand and its applications on the other, than is drawn at present by the Department of Science and Art, if it could be shown that applied science might be provided for in some other efficient way?' Huxley replied, 'If it could be shown that the branch of applied science could be provided for equally well in other ways, I quite think that it would be desirable to draw such a line.'[39] Hence rather than dismissing the category entirely, Huxley was happy with it, just so long as it was not confused with his own interests. The term 'science and technology' came to be used at this time to talk about a broad range of examinations and teaching, quite different from the

[35] Testimony of John Donnelly, Royal Commission on Technical Instruction 1884, vol. 3, p. 286, q. 2862.

[36] Testimony of John Donnelly, Royal Commission on Technical Instruction 1884, vol. 3, p. 287, q. 2871.

[37] Comment by Mr Roberts, Clerk of the Clothworkers Company and an Honorary Secretary of the City and Guilds Institute, Royal Commission on Technical Instruction 1884, vol. 3, p. 526, q. 4464.

[38] Testimony of Sydney Waterlow, Royal Commission on Technical Instruction 1884, vol. 3, p. 527, q. 4467.

[39] Testimony of Thomas Huxley, Royal Commission on Technical Instruction 1884, vol. 3, p. 324, q. 3006.

classical languages dominant elsewhere in education, but also internally heterogeneous.[40] Technology had to be closer to the lived experience of the artisan than its competitor. However, even this category had to be considered in terms of public knowledge, from which confidential expertise was kept separate. The industrialist Sydney Waterlow, who had been the mover for the examinations within the City and Guilds of London, warned the Samuelson Commission that manufacturers would not support any loss of secrecy.[41] Indeed in 1896, a furious row broke out over a question in the chemistry examinations that a manufacturer felt infringed his proprietary knowledge.[42]

Magnus was ambitious; he was also lucky. In 1889 the government gave local authorities the right to fund technical instruction in their districts with an extra tax, again excluding the teaching of trades themselves.[43] The following year, the government provided a historic windfall. It directed revenue from a new tax on alcohol, intended initially to reimburse publicans, instead to local needs, including technical education. The rules were brought in suddenly with little prior discussion; they applied differently to England and Wales, Scotland, and Ireland. Furthermore, districts were permitted to devote all their proceeds to reducing local taxes rather than expenditure. Nonetheless, many localities, particularly London, very quickly directed a large part of the new largesse to technical training for adults, mainly on a part-time basis.[44] As a result, approaching a million pounds a year was typically available from the 'Whiskey Money' during the final years of the next decade, several times more than past Department of Science and Art expenditure on science schools.[45]

---

[40] The first use of the term 'science and technology' to express a category in the 'Britishnewspaperarchive' of largely provincial newspapers was 1872 relating to examinations for artisans. 'Notes of the Day', *The Globe*, 18 July 1872, British Newspaper Archive/British Library Board.

[41] Testimony of Sydney Waterlow, Royal Commission on Technical Instruction 1884, vol. 3, pp. 515–17, q. 4410.

[42] A furious letter from the leader of the Manchester Technical School was included in the Minutes of the Board of Examiners, 3 June 1896. Quoted by Foden 1961, 509–10.

[43] Technical Instruction Act 1889, para. 8, p. 387. The act permitted the teaching of principles and applications of science and art but excluded 'the practice of any trade or industry'.

[44] Sharp 1971.

[45] Department of Science and Art expenditure in 1898 on science schools in England, Wales, and Ireland was £169,000 (Department of Science and Art 1899, iv). Local authority expenditure on technical education was £843,000 for the year 1897–98, lxviii. It is worth pointing out that the government's fiscal year runs 6 April to 5 April, thus spanning two calendar years.

Municipally run, whiskey-money funded, and City and Guilds–examined institutions, designating their teaching as 'technology', spread around the country.[46] Across London by 1903, evening courses were available at a ring of twelve new 'polytechnics' and three branches.[47] Unlike the École Polytechnique of Paris, these were not for the few. Instead, they predominantly offered evening classes to working people who would take the technology examinations of the City and Guilds. These bodies were funded by the London Technical Education Board, for many years headed by the redoubtable Sidney Webb.

New Roles

Innovative industries and professional opportunities associated with the 'application of science' created new challenges for old categories. Whereas apprentice training and practical experience might be sufficient for competence in some fields, by the 1880s, there were novel issues in which common sense was inadequate. Thus, in the design, manufacture, and operation of electrical equipment, such as the telegraph and the telephone, the exploitation of strange effects drew on the esoteric physics of James Clerk Maxwell. As a result, the 'physicist-engineer' with mysterious and powerful expertise became commercially important and of a socially high status.[48]

The role evolved from the 1880s.[49] In 1883 the American physicist Henry Rowland denounced American obsession with applied science to the American Association for the Advancement of Science.[50] The argument has been cogently made that Rowland's speech, 'A Plea for Pure Science', was a criticism of the Thomas Edison cult. Hugely popular and a successful innovator, Edison was also known for his disregard of high theory. The reaction of physicists such as Henry Rowland was to emphasise a division between 'pure' and 'applied science'. This distinction represented an emphasis on research and business as different social roles.[51] Famous as Rowland's intervention has been, it was also in a particular context. He developed his concerns differently the next year at

---

[46] In 1910 Magnus, looking back on the growth of the City and Guilds, said that 'it was definitely restricted to the encouragement of the teaching of technology, a term now very generally understood and which was less familiar twenty years ago'. Magnus 1910, 103.
[47] See Webb 1904; Floud and Glynn 2000.   [48] Arapostathis and Gooday 2013.
[49] Jordan 1985.
[50] H. A. Rowland, 'A Plea for Pure Science', *Science* 2, no. 29 (1883): 242–50. On this famous speech, see Hounshell 1980. The British chemist A. W. Williamson had given a lecture under the same title more than a decade earlier. See Williamson 1870.
[51] See Lucier 2012.

Philadelphia's electrical conference, held in association with a path-breaking international electrical exhibition.[52] Here Rowland was joined by Britain's William Thomson, Lord Kelvin. Now he gave a more subtle interpretation. 'By the labors of the immortal Faraday, electro-magnetic induction was discovered and the modern dynamo-electric machine became a certainty.'[53] Rather than denouncing applied science as he had done previously, he urged Americans to reconceptualise it. The new role of the physicist-engineer was coming into view: 'It is not telegraph operators but electrical engineers that the future demands,' he concluded.[54]

Four years later, the British experienced their own public debate over the skills and approach needed for the new electrical industries. The meeting of the British Association for the Advancement of Science witnessed a discussion, which at the time seemed to be a bitter conflict. The protagonist for practice was the former telegraph operator and chief electrician at the British Post Office, William Preece. A first-generation telegraph man, Preece had had to be largely self-educated. For him, the simple analogy of the flow of current in a wire to water passing through a pipe generally sufficed. Against Preece were ranged the so-called Maxwellians.[55] This group of followers of the Cambridge mathematical physicist Clerk Maxwell has been identified as a community of interest by Bruce Hunt. They included the brilliant Oliver Heaviside, Sylvanus Thompson (principal of Finsbury College), his close friend Oliver Lodge, and Henry Rowland, an honorary American member. In hindsight, it has been argued that Heaviside, Preece's principal opponent, was dedicated to an attempt to link the body of Maxwellian theory to established engineering knowledge and not to either purely theory or practice. His biographer has suggested that he was engaged in what would be called basic industrial research.[56]

Lodge, too, would contribute to both the practice of radio wave detection and the theoretical interpretation of lightning protection. As the first vice-chancellor of the University of Birmingham, his adminis-trative career and work in physics would be intimately intertwined with finding an appropriate space for applied science.[57] Some would revere him, and he was confident in his lightning rod research. However, Preece was responsible for half a million lightning rods and dismissed Lodge as

---

[52] On the significance of this occasion, see McMahon 1976, 1384.
[53] Rowland 1886, 18.   [54] Rowland 1886, 28.
[55] See Hunt 2005. Also see Buchwald and Hong 2003.
[56] Hunt 1983; Yavetz 1993. On Preece, see Arapostathis 2012. For the representation of this debate at the time, see 'Practice versus Theory', *The Electrician* 21 (1888): 730–31.
[57] See Mussell and Gooday 2020.

inexperienced in practice. Not inclined to take lectures from a physicist, Preece was dismissive of the 'innumerable students' from technical institutes who 'thrust upon the electrical world conditions and conclusions arrived at by their mathematics with a coolness and an effrontery that were simply appalling'.[58] His respect instead was for electrical engineers. Chemists, too, were at the forefront of the new world. In the production of dyestuffs and the latest pharmaceuticals, German industry, with its close links to academic chemistry and new research laboratories of its own, took the lead. Scientific skills alone, however, were not sufficient to design and run a factory. Many newly introduced processes, such as using superheated steam and distilling coal tar, required engineering expertise. The question, therefore, arose: How should engineering and chemical training be combined? The former British government alkali inspector and environmental regulator, George Davis, addressed the challenge. In the late 1880s, he gave lectures at the Manchester Technical School, wrote articles, and published a book criticising the limitations of such terms as 'chemical technology' or even the more general 'applied chemistry'. At length, he discussed the tension between theory and practice and the breadth of operations to which he was alluding. His promotion of the idea of what would be called unit operations required a new concept: 'chemical engineering'.[59] Now a higher level of technological expertise was in demand. Specialised branches of engineering and the use of more sophisticated physics and chemistry in industry sustained a duality of 'higher technological education' and 'applied science' that would be important in the twentieth century.

New Institutions

British anxiety about Germany's *Technische Hochschulen* and America's MIT was intense among experts, but was not restricted to specialists.[60] When the Kaiser opened the Danzig Technische Hochschule in 1904, the very next day, his speech about the new foundation was published, in part, by the British press.[61] At the end of the twentieth century, E. P. Hennock admiringly explored the political skill with which British

---

[58] Preece et al., 'Discussion on Lightning Conductors', *The Electrician*, 21 (1888): 644–48 and 673–80, on 679. See Yavetz 1994.

[59] Davis 1904, 2–4. See Divall and Johnston 2000.

[60] American technical education aroused much interest. In 1903, there was a privately funded investigation of American technical education, the so-called Mosely Commission. See Lupton 1964; König 1996.

[61] See, for instance, 'The German Emperor on Education', *The Scotsman*, 7 October 1904, British Newspaper Archive/British Library Board.

educational entrepreneurs of the years around 1900 had used the challenge of what he called the 'THs', particularly the high-profile Berlin *Technische Hochschule* in the suburb of Charlottenburg.[62] Citing that challenge, Liberal politician and philosopher R. B. Haldane had proposed an organisation incorporating all the existing South Kensington science-teaching institutions. The amalgamation required great diplomatic skills. Haldane memorably told the University of London, itself anxious about the innovation, that his strategy was to follow the British Constitution, 'which is to change the substance, but always leaving the form intact'.[63] Initially, he described the new body as a 'College of Applied Science'.[64] He had, however, to manage the proprietorial anxieties of the City institutions for the City and Guilds 'Central Technical College' (previously known as the Central Institution) they had founded and funded. Ultimately, as 'The Imperial College of Science and Technology', the new organisation's title reflected its ancestors' diverse values. Moreover, until the twenty-first century, the City and Guilds College endured as a visible and integral constituent of Imperial College.[65]

Manchester also aspired to a 'Charlottenburg'. In the face of international competition, the city was fighting to renew its status as a global leader in manufacturing. At the end of the nineteenth century, this inland metropolis funded the Mersey and Irwell river system's conversion into a waterway bringing ocean-going merchant ships into the heart of its warehouse district. Greatly enhanced technical education was a complementary ambition. The city boasted a strong German community with its club, the Schiller Anstalt, and the Municipal Technical School enriched by several German professors. Reconstruction gave the old institution new standing. Erected on the land previously occupied by the engineering works of Joseph Whitworth, the magnificent building was inspired by the *Technische Hochschulen* to whose scale it was comparable, but the

---

[62] Hennock 1990.
[63] Departmental Committee of the Board of Education on the Royal College of Science, *Minutes of Proceedings at Conference with Representatives of 1. The Senate of the University of London, 2. The Council of University College 3. The Council of King's College, 7 July 1905*, 166, ED 24/255, TNA. Quoted by kind permission of Richard Haldane.
[64] Board of Education 1905, 6. The formation of Imperial College and its Manchester counterpart were associated with a new turn in the interpretation of 'technology'. The new term 'technological education' was defined as 'the branch or branches of Technical Education which have particular reference to the application of science to industrial processes as distinct from those branches which are predominantly manual, commercial, artistic, medical or legal, or otherwise, in this narrow sense, professional'. Board of Education 1906, 1. See Musgrave 1964. This spawned 'higher technological education', for which see Chapter 7.
[65] On the history of Imperial College, see Gay 2007.

building was in French Renaissance style.[66] Now part of the University of Manchester, its appearance is still impressive. The title was also changed, from 'Municipal Technical School' to the 'Manchester Municipal School of Technology' in 1900. The director, J. H. Reynolds, initially proposed 'Manchester Municipal Institute of Technology' (MMIT).[67] An elaborate paper had weighed the various options for a name, referring to both the German *Technische Hochschulen* and MIT: 'technology' reflected a more scholarly approach than 'technical'. The 'School' was also soon bureaucratically upgraded to the University's 'Faculty of Technology'. The city's needs, the institution's potential, the *Technische Hochschulen*, and MIT all justified the move (see Figure 3.1).

This upgrading of the School of Technology was, however, threatening to the existing university departments. Thus, regarding degree titles, the University wanted a clear distinction between the science degrees its long-standing departments would present, and 'technology' degrees, offered by the new faculty. Meanwhile, the School of Technology wanted equality. The result was a compromise, with the School preparing students for a new category of degrees in 'technical science' – evoking the qualification in *Technische Wissenschaften* chosen by the *Technische Hochschulen* in Germany.[68] In 1915, the university's professor of history reflected ironically that the relationships between the university's faculties of science and technology, their jostling for position, and concern with each other's frontiers reminded him of 'the attempted definitions by mediaeval publicists of the spheres of church and state'.[69]

Meanwhile, change was happening at a government level. From the 1890s, the government slowly stripped responsibilities from its Department of Science and Art.[70] Bureaucratic requirements came to reinforce categories. Thus, in 1902, 'T-branch' of the newly constituted

---

[66] See the description in 'The New Municipal Technical School' 1900, 293. A cost of almost £300,000 is specified in 'The Manchester Municipal School of Technology', *The Manchester Guardian*, 16 October 1902. This compares with £450,000 for the Berlin Technische Hochschule. See 'The Position and Work of the Central Technical College'. See also Short 1974.

[67] [J. H. Reynolds], 'Considerations as to the Expediency of Changing the Name of the Municipal Technical School, Subcommittee to the Sub-Committee of the Municipal Technical School by the Board of Studies', Technical Instruction, 112–14, Board of Studies TGB 1 /2/1, Special Collections Division, University of Manchester Library. Quoted courtesy of the University of Manchester Library.

[68] Manegold 1970; Short 1974.

[69] Tout 1915, 44. On the relationship between the University and its faculty of technology, see Guagnini 2010.

[70] For descriptions of the detailed bureaucratic dance, see Balfour 1903, 170; K. O. Roberts 1969.

Figure 3.1 Technology embodied: Manchester Municipal School of Technology at its opening, engraving by Hedley Fitton. Report in *The Manchester Guardian*, 16 October 1902. Copyright Guardian News & Media Ltd 2022.

Board of Education took responsibility for 'technology', now defined by a committee of the Board as 'the application of different and specialised kinds of knowledge to practical occupations'.[71] It inspected schools and allocated miscellaneous grants that the Department of Science and Art had accumulated over the years. Eight years later, it separated the special provision for higher education bodies from that it made for part-time education. The Board published an annual 'Statement of Grants Available from the Board of Education in Aid of Technological and Professional Work in Universities in England and Wales', commonly called 'The Statement'. Certainly, the reclassification of this support to universities and selected colleges gave wide recognition to the national

---

[71] Board of Education, 'Report of the Committee on the Co-ordination of Technological Education', 1901, ED 24/36, TNA. See p. 2. The committee of six was chaired by the senior civil servant Sir William Abney and included Philip Magnus.

value of the technological and professional education they were giving.[72] Nonetheless, this complex institutional origin of 'technology' through the support given to technical education had been convoluted, and would be enduringly confusing.

Universities, the University Grants Committee, and the Classification of Knowledge

Despite the new-found respectability of technology, within those new academic centres with ambitions to specifically university status, the educational offer was generally promoted as 'applied science'. Early in the twentieth century, leading liberal science colleges across the country were redesignated universities. Birmingham was the first, and half a dozen others followed shortly. Individual civic universities in Liverpool, Manchester, Leeds, and Sheffield replaced the former federal Victoria University of Manchester. In addition, the University of Bristol was incorporated; a new university of Wales founded in 1895 included colleges in Swansea and Aberystwyth; and establishments in Reading, Southampton, Hull, and Nottingham were upgraded to University colleges. Within a few years, a thorough American survey of British higher education of 1917 found that each one of the universities and university colleges in England and Scotland was teaching applied science.[73]

Each application for upgrade to university status was carefully reviewed by a Committee of the Privy Council, comprised of current and former government Ministers and similar worthies. The Committee would interrogate the proponents, who had to demonstrate that their plans met the standards of liberal education and distinguished university from technical education, warning against the tendency 'to degrade university teaching to technical teaching', in the words of the banker Lord Balfour of Burleigh, a Committee member.[74] Thus, the 1904 charter of the University of Manchester specified that the reorganised institution was entitled to offer degrees, 'provided that degrees representing proficiency in technical subjects shall not be conferred without proper security for testing the scientific or general knowledge underlying

---

[72] The ancestry of this is shown in Morant 1910 and Morant 1911. See also Eaglesham 1963.

[73] Maclean 1917, 205. See also Drummond 2020.

[74] Question from Lord Balfour to R. B. Haldane, 'Transcript of Shorthand Notes of Proceedings before the Committee of Council, on 18 Dec. 1902', ff. 89731 p. 30, PC 8/605 pt. 2, TNA. Lord Balfour of Burleigh is not to be confused with fellow Conservative politician Arthur Balfour (also later a lord).

technical attainments'.[75] The nineteenth-century colleges had seen a distinctive break from the old universities' traditions in their adoption of 'applied science'. Now the same category distinguished their successors from the technical colleges characterised by technology. The culture, which linked 'applied' to 'pure' science and the core of western civilisation, proved critical to institutional identity.

Roots in the great history of culture proved a key distinction between the two categories. Occasionally in the nineteenth century, books on the history of science had been written. Many more were published early in the twentieth century, and the trend continued into the 1920s.[76] In 1923, the University of London established a 'Board of Studies in the Principles, History and Method of Science', convened by the distinguished philosopher Alfred North Whitehead.[77] The syllabus of the course it created listed a special 'Physical and Mathematical Section'. This specified a list of eighteen topics from the history of physical and chemical discovery, followed by a final two subjects, 'The Steam Engine and Other Mechanical Inventions' and 'The Industrial Revolution: Science and the Social System'. Typically, the course interpreted applied science as the outcome of an unfolding process that had begun with the 'Ionian philosophers' and Plato and Aristotle. By contrast, technology in Britain had neither such prestigious roots in high culture nor did its historiography have an academic base.[78]

In many cases, former independent colleges became departments of, or prepared for degrees in, 'applied science'. In Glasgow, the newly regrouped 'Royal Technical College', entitled in 1912, boasted an early twentieth-century home even grander than Manchester's School of Technology. But even this prestigious centre of technical education that was descended from the early nineteenth-century Andersonian Institution offered the degrees in 'applied science' of Glasgow

---

[75] 'The Victoria University. Manchester Charter', *The Manchester Guardian*, 29 May 1903. The same article showed that an identical condition applied to Liverpool. Again, a similar clause was to be found in the contemporary Charter of Leeds University. See www.leeds.ac.uk/secretariat/charter.html (accessed February 2023).

[76] Mayer 1997.

[77] Board of Studies in the Principles, History and Method of Science, 'Proposed Syllabus for B.Sc. Course. Draft to be submitted to the Board of Studies at Their Meeting at the Institute of Historical Research, Malet Street, W.C.1. on Thursday February 14th at 6.15 p.m.' Quoted with the kind permission of the University of London. For the origins of the course, see Smeaton 1997.

[78] The emphasis on pure and applied science rather than technology in 1920s historical writing is discussed by Scheinfeldt 2003, 152–57. There is here a strong contrast with the German tradition of *Technikgeschichte*. See Dietz et al. 1996.

University.[79] In 1905, the new University of Sheffield, struggling to prove its academic status as a full and integrated scholarly centre, incorporated the former municipal technical school as the 'Department of Applied Science'. A visiting American delegation reported that the dean 'approved of the title "Applied Science" as contradistinguished from technology, which deals with the application of science to trades'.[80]

World War One would be a moment of transformation in funding, scale, and authority. In wartime, the universities and technical colleges proved themselves at the heart of national life and formative of individual lives. Some, such as the Department of Applied Science at Sheffield, became virtual munitions factories. Others, such as the Loughborough Technical Institute in a small industrial town of the East Midlands, trained munitions workers and became wealthy as a reward.[81] Elsewhere colleges and university departments had rapidly developed technical competence for which hitherto Britain had depended on Germany. Before the war, London's Northampton Polytechnic (to become City, University of London) had developed courses and expertise in optics, including training for the new job of the cinema projectionist.[82] Now it transferred its skills to the need for gunsights and gunsighting. Imperial College, too, developed an optics specialism and was the base for a new Admiralty Engineering Laboratory. Everywhere, universities and colleges could boast of their critical role in providing scarce skills. When in 1915 their leaders wrote to the government seeking financial support, they trumpeted the sector's service to the nation, training sorely needed scientists, medical professionals, engineers, and other experts.[83]

The end of the war brought new demands, opportunities, and government funds into universities and colleges but also the need for systematisation and standard criteria for allocating resources. Government funding of a new fellowship, the 'Robert Blair Fellowship in Applied Science and Technology', widely advertised and much esteemed, rewarded London's colleges for their wartime efforts and promoted the categories of achievement it celebrated. A significant shift in funding came from the increasing contribution of the public purse and

[79] The Glasgow and West of Scotland Technical College, 'Relations with the University of Glasgow, November 1909', in 'Special Committee on Relations with the University', Special Collections, University Library, University of Strathclyde, Glasgow.

[80] MacLean 1917, 126.   [81] Cantor and Matthews 1977.

[82] See 'Report of the Principal of the Northampton Polytechnic Institute from 1 July 1907 to 30 June 1908', Library, City, University of London, p. 2; Teague 1980, 52–56.

[83] 'Grants to Universities: Financial Position during Wartime', *The Scotsman*, 7 September 1915, British Newspaper Archive/British Library Board.

bureaucratic formalisation. Even before the war, the Treasury had supported universities. Now, it had to respond to popular demand for more funding, particularly for teaching science and technology.[84] The government contributed a million pounds as an annual sum, with an extra half a million to aid reconstruction. The grant was not only much more significant than pre-war subventions, but it also consolidated a variety of earlier government routes to funding. It incorporated the pre-war Treasury subsidy for universities, grants formerly awarded by the Board for agriculture and fisheries, support for teacher training, and the subvention for 'technological and professional work' known as 'The Statement'.[85] Management of this government endowment would be the responsibility of a new intermediary body intended to protect institutional freedom and ensure accountability, the University Grants Committee (UGC).

Providing insulation between supposedly independent organisations and the government, the UGC became a sponsor of terminology, an acceptable means of nationalising individual universities, and the creator of a national system. Through this committee, the government came suddenly to contribute a third of the funding of independent bodies and to determine the bureaucratic classification of sciences, albeit embodying legacy subdivisions. The UGC reported its grants to Parliament under the heads of arts, pure science, technology (as a legacy of 'The Statement'), medicine, and agriculture. As an odd consequence, 'technology' had become an enduring subcategory of university funding while remaining philosophically challenging to universities' identity.[86]

Enthusiasm for applied science training immediately after the war benefited the civic universities.[87] The year 1918 brought an enlarged new cohort of graduates of the secondary education system launched at the beginning of the century. Additionally, the government gave grants to about 26,000 demobilised officers to become post-war students.[88] Overwhelmingly, they chose subjects from the sciences and engineering side of the curriculum.[89] In October 1919, the *Yorkshire Post* reported the unprecedented demand for places to study applied science at the once

---

[84] Kidd to McKormick, 'University Grants', 6, Ed 24/1964, TNA. On this letter, see Shinn 1980, 3.
[85] See also Hutchinson 1975.
[86] For a further discussion of this ambivalence, see Shinn 1986, 198–205. Even a generation later the botanist and university leader Eric Ashby wrote a book entitled *Technology and the Academics* (Ashby 1958) highlighting the consequent tension within universities.
[87] Armytage 1953. See also Barnes 1996.   [88] Vernon 2004, 200.
[89] Sanderson 1972, 237.

more retitled 'Manchester Municipal College of Technology'.[90] The University of Sheffield, having enrolled 115 full-time students at its formation in 1905, was teaching more than a thousand in 1919.[91] The universities of Leeds and Birmingham also doubled their enrolments. Nationally, the number of students (including those at the older universities) increased by more than half from their immediate pre-war enrolment – from 23,000 students to 37,000.[92] In retrospect, the new institutions have been criticised as excessively influenced by Oxford and Cambridge's antique values, nor did they develop as quickly as their early leaders had hoped. Still, in the early post-war period, the changes seemed revolutionary.

In particular, the civic universities emphasised applied science in the training of engineers who were, as a profession, rapidly moving from their old apprentice base. Several scholars, such as Colin Divall, have followed the rise of academic training. This was brought about by new forms of preparation for engineering careers, the involvement of private and public bodies, and new forms of accreditation.[93] During the 1920s, the Institution of Mechanical Engineers and the government's Board of Education introduced the qualifications known as 'national certificates' to replace the Science and Art Department examinations. The innovation proved an enduring success.[94] In addition to the twenty or so university courses, about a hundred technical college courses offered training recognised by the new national certificates.[95] These tended to emphasise the more practical 'technology', though where the line between the offerings of types of organisation should be drawn remained controversial. In 1921 the UGC reflected on applied science taught in the university: 'the difference between a liberal and vocational training lies not so much in the subject studied as in the spirit in which it is pursued. What is essential is to maintain that breadth and proportion of view which are implied by a University standpoint.'[96]

In 1920 Dugald Clerk, who had led the Admiralty Engineering Laboratory during the recent war, wrote a paper on training for his profession and the distinction between university and technical education.[97] Rejecting his principal recommendation, a civil servant commented: 'We see little prospect of an systematic organisation of a

[90] 'Manchester College of Technology', *The Yorkshire Post*, 21 October 1919, British Newspaper Archive/British Library Board.
[91] Herklots 1928, 26.   [92] University Grants Committee 1921, 4.
[93] Divall 1990, 67.   [94] Foden 1951, 45.   [95] Divall 1990; Zeitlin 2008.
[96] University Grants Committee 1921, 10–11.
[97] Dugald Clerk, 'Note on University as Distinguished from Technical Education', 1920, ED 24/1997, TNA.

five-year course of study covering 3 years in a university and 2 years of Technology in an associated technical college.'[98] Subsequently, as a founder member of the UGC, and the standard-bearer there for engineering, Clerk led a follow-up enquiry – mandated to distinguish between the provision of universities and technical colleges and to clarify their relationship.[99] Although eminent, its committee members could agree on little; a civil servant described the document they submitted as 'colourless', and it was never published.[100] It went very little further than the UGC's published report of the previous year, which had also been positive about applied science in the university, and anxious about technology.[101]

The needs for more of a new kind of specialist were constantly repeated. For example, in his earlier memorandum, Dugald Clark had emphasised the importance of academic leadership in universities from engineers with expertise in physics and chemistry, though not necessarily a rich commercial background.[102] Similarly, in an article for the *Journal of Careers* in 1930, Arthur Fleming, of Metropolitan Vickers, wrote about the opportunities for engineers with physics training. To encourage advisors of potential young recruits, he highlighted his profession's prospects, emphasising the link between engineering's progress and applied science.[103] Subsequently, he led a 1937 campaign to promote the academic basis of the engineering profession.[104]

Institutions hotly debated what balance between general theory and practical application was appropriate to this interpretation of applied science. In 1933, the University of London refused to validate a syllabus on textile science that Northern colleges (not themselves entitled to offer degrees) would have taught, describing it as 'a heterogeneous set of

---

[98] V. R. Davies to A. Abbott, n.d. (July 1928), ED 24/1997.
[99] University Grants Committee, 'Minutes of Meeting', 3 February 1921, UGC 1/1, TNA.
[100] The committee included included Clerk himself; J. H. Petavel, then director of the National Physical Laboratory (in succession to Glazebrook); Cecil Desch, who had just gone to Sheffield to head the Department of Applied Science; Arthur Schuster, the Manchester physicist; Sir James Dobbie, the recently retired director of the Government Chemist's Laboratory; and the school inspector A. Abbott. On the tensions within the committee, and the non-publication of its report, see the comments some years later by one of the members 'AA' (A. Abbott) to Mr Pearson, 20 August 1928, ED 24/1997.
[101] University Grants Committee 1921, 11. However the idea of a five-year training for engineers ran so counter to the practice of three-year degrees that it seemed to have little future. See the discussion in ED24/1997.
[102] Clerk, 'Note on University as Distinguished from Technical Education', 5.
[103] Fleming 1930, 13. However, Fleming's many other writings rarely mentioned 'applied science' .
[104] Divall 1990, 93–95.

applications of a number of sciences'.[105] The dismissive rejection from the Board of the Faculty of Science caused an angry reaction among proponents concerned with supporting the textile industry. The university only defused this by offering a committee to investigate the establishment of a faculty offering degrees in 'technology', though it seems never to have been appointed.[106] Applied science fitted the University of London; technology did not.

Conflict over the proper balance between science and practice came into the open at the University of Sheffield in 1919. The retiring professor of metallurgy, an active consultant to local industry, was expected to be replaced by his deputy. Instead, a group of younger colleagues rebelled and insisted on an open process that ultimately appointed the distinguished outside candidate Cecil Henry Desch. He was so enamoured of science that he had been married in a positivist ceremony inspired by Auguste Comte's secular religion of humanity.[107] For Desch, any local practice needed to be rooted in pure science, which he taught with vigour. On the other hand, when he left Sheffield for the National Physical Laboratory in 1932, his successor dismissed his approach as inconsistent with both the department's identity and the needs of local industry and denounced the obsolescence of the department's equipment.[108] The hunt for balance in the treatment of metallurgy at Sheffield highlighted the alignments between the concept of applied science and the loyalties of its regional embodiment.

Local orientation did bestow local influence. A 1932 survey emphasised that these institutions' regional bases might have limited their scope but bestowed life and support.[109] Two decades later, Sheffield historian W. H. G. Armytage described the early development of the civic universities as an essential part of community infrastructure.[110] They mapped closely on to a role pioneered by the land grant colleges in the United States. Andrew Carnegie had suggested these as a model to the college about to become the University of Birmingham, and his philanthropy rendered the advice particularly compelling.[111]

---

[105] Senate Minutes, University of London, 'Position of Technological Studies in the University' (20 December 1933), para. 1067, 58 UoL ST 2/2/50, University of London, London, Senate House Library, Special Collections, quoted by kind permission of the University of London.

[106] Senate Minutes, University of London, 20 December 1933, para. 1068, 62.

[107] On Desch, see McCance 1960; and on his predecessor, Sanderson 1978.

[108] Mathers 2005, 118.   [109] Barker 1932, 90.   [110] Armytage 1953, 243–64.

[111] See 'Report of a Visit to Colleges and Universities in the United States and Canada, Made in November 1899 at the Suggestion of Mr Carnegie', 11. Inserted in the University of Birmingham Advisory Subcommittee minutes, 9 January 1900, UB/MC/B/2/6/5, Cadbury Research Library. See Ives et al. 2000.

It is striking that the two Conservative Prime Ministers of the 1930s (Stanley Baldwin and Neville Chamberlain) had both studied metallurgy at the University of Birmingham's predecessor, Mason College. The elite were, therefore, close to such regional centres, reflecting public awareness. Across the sector, there were constant appeals for local finance for new laboratories. As development occurred, departments advertised opportunities for scholarships. So, during the two decades following the armistice of 1918, science was framed by the aspirations released by institutional innovation.

The universities also taught the arts, principally to aspiring schoolteachers, but applied science bestowed their character. Its greater visual appeal was one explanation for this prominence given by the theologian H. G. G. Herklots. He blamed the 'Open Day' at which the manufacture of medicines, experiments, microscopes, and machinery took pride of place and made much more dramatic displays than literary or even abstract scientific concepts.[112] Such episodic boosts to visibility were part of a broader context of a constant drip of news about applied science from the universities. The local press reported fund-raising drives for new laboratories and courses for such practitioners of applied science as opticians.[113] New forms of validating professionals, the civic universities and science were defining one another.

Looking back to the depressed employment years of the 1930s, many critics observed a sharp swing towards the arts. Sarah Barnes, however, has analysed the figures in detail. Using Manchester's example, she has shown that the simplistic use of data can easily construct a misleading image. 'Commerce' was classified as an 'arts' subject, and the engineering degree, BSc (Eng), offered by the University, was classified as 'pure science'. Reclassifying this qualification and separating commerce reveals a largely stable distribution in which arts degrees attracted about 40 per cent of the students. There was, if anything, a drift towards technology and commerce.[114] Moreover, the significance of the interwar years' entrenchment and conflicts has come to be recognised.[115] These new institutions sought a role that would mediate between universal values and the ancient status of 'universities', on the one hand, and local service, on the other. Accordingly, the meaning of the applied science they embodied was locally defined.

---

[112] Herklots 1928, 30.
[113] For instance, in 1928 Manchester University advertised a new course for opticians emphasising 'applied science'. 'Status of Opticians: Manchester Scheme of College Courses', *The Manchester Guardian*, 26 January 1928.
[114] Barnes 1996, 284.
[115] See, for instance, Morse 1992, 177–204; Barnes 1996; Vernon 2009.

Conclusion

In Britain, the demand for a qualification system for artisans made 'technology' not just a word but a significant category of competence. It could frighten employers for its encroachment on 'private' knowledge but enthuse policymakers concerned with the capabilities of the British workforce. Technology was an alternative model of modern knowledge appropriate to new careers. Such promoters as John Donnelly and Philip Magnus deserve to be better remembered and the influence of the City and Guilds better understood.

Novel industries were growing up with demands for radically new kinds of expertise. Expanding and multiplying colleges promoted the terms that validated their approaches. Under the flags of technology and 'higher technological education', polytechnics and institutes of technology saw themselves as an alternative to the new universities. Their proponents could draw on the success of German and American industry and education to claim the superior value of their interpretation of modernity.

By contrast, the exponents of the complementary category of 'applied science' emphasised the authority of the progressive and moral nature of public science. It was associated with civilisation and breadth as well as useful knowledge. Coleridge's legacy could still be seen in the demand that universities offer more than technical training. Class distinction, of course, often underlay the differentiation. Though not mentioned explicitly, it was encoded in such hierarchies as the artisan trained in technology, and university graduates, most often associated with applied science. The two categories played complementary roles in a competition that endured until the end of the twentieth century.

Stage 2
Research in the Early Twentieth Century

4   The Dawn of the Twentieth Century

Introduction

'The world changed around 1910', reflected writer Virginia Woolf, instancing the modern cook who was 'a creature of sunshine and fresh air; in and out of the drawing-room, now to borrow *The Daily Herald*, now to ask advice about a hat'.[1] This quotation has become a classic record of social transition at the beginning of the twentieth century. In stories about twentieth-century modernity told to an increasingly educated and prosperous public, applied science proved to be a key character. Its power was demonstrated vividly at the 1909 Business Efficiency Exhibition, held at London's newly opened Olympia hall. The experience of 'hustle and applied science' so impressed a *Manchester Guardian* journalist that it seemed to go beyond an assemblage of individual gadgets doing faster by machine than had previously been done by hand and instead provided insight into a new civilisation.[2] Across the industrialised world, the trinity of social development, new products, and applied science was causing awe and bewilderment. Conversations about modern society in newspapers, parliaments, novels, on the radio, in local meetings, and across public culture gave new uses to the term.

Instead of pedagogy, the principal reference came increasingly to be 'research'. The press popularised applied science through its embodiment in huge public laboratories; the military; funding bodies supporting industry, agriculture, and medicine; industrial research; and government functions. The term's meaning was constructed and reconstructed with the organisations whose success seemed to validate its significance.

---

[1] Woolf 1928, 4–5.
[2] H.W.H., 'Business by Machinery', *The Manchester Guardian*, 25 October 1909. Copyright Guardian News & Media Ltd 2022. 'H.W.H.' was the pen-name of 'H. Wilson Harris', later the diplomatic correspondent of *The Daily News*. '"H.W.H." at Ealing', *West Middlesex Gazette*, 29 March 1924, British Newspaper Archive/British Library Board.

Therefore, the period of the hectic emergence of a host of new organisations and awareness of research is of particular interest here. This chapter introduces such issues and deals with the formative two decades before World War One. It addresses the question: How could research be discussed using the old language of applied science? Where was the continuity?

Three key themes structure the analysis: the continuing challenge of foreign powers; the growth of great institutions, particularly the National Physical Laboratory; and the attraction of supporting applied science to governments committed to maximising national efficiency but with minimal interference in the market. Although the focus will be on the few years between 1899 and the outbreak of war, the enthusiasm for research in that period needs to be set into a longer early twentieth-century context.

At the beginning of the century, the British government's annual commitment to scientific research amounted to tens of thousands of pounds sterling. In 1914, it exceeded £100,000.[3] By the end of the decade, in the years immediately after the war, it was measured in millions. Talk about applied science came to serve both as a tool for policymakers allocating these vastly greater resources and as a resource for the broader culture, making sense of a new age. Whereas in more technical contexts, the nomenclature of 'applied research' and 'industrial science' emerged, even before 1914, politicians and the general press were describing the new category with the old term 'applied science'. Therefore, its connotation shifted–from representing a branch of pedagogy to denoting a key step in developing new products.

This was not the work of just a few campaigners. The following two chapters will show how the complex bureaucratic transformations, before and during World War One, proved critical in the progressive shift in discourse about applied science. Moreover, culture was important too. In the public mind, new products came to be associated with 'research' rather than 'invention'. Recently, such historians of technology as Christine MacLeod and Bernard Carlson have reflected on the shift from

---

[3] The funds made available to the Medical Research Committee from the National Insurance Fund were £50,000 a year; see Alter 1987, 174. The Development Commission gave £50,000 a year from 1911 for agricultural research and education; see Russell 1919. In the early 1930s Julian Huxley calculated a total national investment of four to six million pounds per year in research, of which about half came from the government. A similar figure comes from adding the grants of the the Medical Research Council, the Agricultural Research Council, the Department of Scientific and Industrial Research, and the roughly equal expenditure of the armed forces. Bernal arrived at similar figures; see Bernal 1939, 63.

a belief in the inspired inventor to reliance on the organised research laboratory at the time.[4] This was particularly characteristic of the domains of chemistry, aeronautics, and electronics, which seemed to be the vanguard of an industrial transformation.

The distinctive 'civilisational' quality will be dealt with further in Chapter 6. Here, one should note that this was closely connected to talk about the consequences of advances in industrial production. Thus, the radio was said to result from, and was the driver of, applied science; it was the agent of a way of life that engendered reflection on the new scientific civilisation, and its ongoing development was symbolic of scientific progress.[5] Similarly, the conversion of coal to oil, Britain's largest civil interwar research project, could be anticipated as an achievement of the applied science of chemistry, the means of rescue of the unemployed miners, and its progress was a symbol of the contribution made by science to culture. Such accomplishments and potentialities were neither miscellaneous nor misconceived. On the contrary, their stories expressed an 'ideology' establishing research work in applied science as a marker of a new civilisation.

Research

At the beginning of the twentieth century, the activity of 'research' combined a confusingly rapid rate of growth and a bewilderingly small scale. In Britain, a country with a workforce of over 12 million, the activity occupied credentialed scientists numbered in the thousands, not the tens of thousands.[6] Its high cultural visibility was quite out of proportion to its small, if rapidly increasing, size.

Talk of the growth of the research enterprise was also in the air during the years around World War One. A visiting American chemist studying research institutions across Europe in 1907 commented, 'Even in

---

[4] Carlson 1997 has explored the American scene. Christine MacLeod has reflected on the parallel process of attributing new products to organised science rather than the inspired inventor in early twentieth-century Britain; MacLeod 2007.

[5] This combination was explicitly drawn upon by the American journalist Floyd Gibbons, working on behalf of GE, in his broadcasts entitled 'Adventures in Radio'. See, for instance, Floyd Gibbons, 'First Article', University of Maine, Raymond H. Fogler Library, Special Collections, Floyd Gibbons Papers, MS 200, B126 f016.

[6] A survey at the beginning of the twentieth century found just over 100 graduate chemists with membership of the Society of Chemical Industry, and another found that 1,500 chemists worked in British industry, but most were uncredentialed. Cardwell estimated about 300 graduate chemists. Cardwell 1972, 187–227. A modern study estimated about 100 academic physicists in Britain in 1900. See Forman et al. 1975, 12.

England there is abroad the spirit of applied science.'[7] After the war, *The Observer* newspaper told its readers about a 1922 universities meeting, where the chemist Michael Forster reportedly noted that 'research was in "everybody's mouth"' and that, consequently, young men wanted to get into it, though they knew little about what it entailed.[8] We have seen how civic universities identified with applied science in the pedagogical context. As for their scholarship, much of that was also exemplary 'applied research'.

Nonetheless, at first, the activity was symbolised by the new dedicated research laboratories in industry and in the service of the state. By 1915, more than 300 industrial laboratories had been established in the United States. Despite the shock of the Great Depression, more American chemists reported they were working in industrial research than in colleges and universities during the mid-1930s.[9] The mix of investigations conducted in the new public and private institutions defined the classification of science for a generation. Such linkages were reinforced by the glamour of famous well-funded laboratories employing hundreds of people with architecturally dramatic buildings and massive equipment.[10]

Press usage provides an indicator of changing fashions. During the late nineteenth century, from the 1860s to the 1880s, *The Scotsman* newspaper printed the term 'research' on average about 100 times per year.[11] The average in the 1900s and 1910s was five times higher, and the canter in the pace of change gave way to a gallop after the war. In the 1920s and 1930s, the average number of uses doubled again to exceed a thousand per year. No longer was science understood as a taught body of knowledge; instead, it was characterised by rapid change. New physical objects, such as radios, medicines, and dyes, gave meaning to the research from which they stemmed.

To many, at the time, it seemed evident that German and American leadership in industrial research laboratories and government research centres had to be contested. Politicians such as R. B. Haldane, scientists such as Lord Rayleigh, and industrialists such as Hugo Hirst (co-founder of Britain's General Electric Company) pointed to Germany's Physikalisch-Technische Reichsanstalt (1887), the central laboratory of BASF (1888), and the Kaiser Wilhelm Institutes (1911) or America's

---

[7] Duncan 1907, 243.
[8] 'Federal Idea in Universities. Mr Fisher on the Pooling of Resources', *The Observer*, 14 May 1922.
[9] Thackray et al. 1985, table 5.9, p. 353. The crossover was 1934.
[10] See the comments of Cardwell 1972, 15. Although the 1972 edition is cited here, the comments were also in the first edition of 1957.
[11] These figures are based on a search on 'Britishnewspaperarchive' in December 2022.

private General Electric (GE) Research Laboratory (1900), Mellon Institute (1913), and slightly later 'Bell Labs' (named 1925), and its public laboratories such as the Bureau of Standards (1901) and the Bureau of Mines (1910). Subsequently, historians have studied in detail their foundation and activities. George Wise, for instance, has shown how Willis Whitney, leading the widely respected GE laboratory in Schenectady, carefully balanced the resources allocated to different sorts of work.[12] The laboratories' varied activities typically combined short-term service projects with longer-term explorations of what Sabine Clarke has characterised as 'fundamental research' in a practical context.[13] To use a chemical metaphor, GE industrial research was not a single compound but rather a mixture, like gunpowder containing potassium nitrate, sulphur, and carbon. Indeed, one could reasonably compare the disruptive effect of the GE laboratory to gunpowder! The routine testing, the experimental work of William Coolidge, who produced thin tungsten wire – a key to the electric lightbulb's success – and Irving Langmuir's Nobel Prize–winning study of surface chemistry were all ingredients of the mixture developed in Schenectady.

Britain, too, saw its first large industrial, civil government–funded, and military research laboratories before World War One. In the chemical industry, industrial research laboratories emulating German practice can be traced back to the 1890s. The best known of these, the Wellcome Physiological Research Laboratories, were established in 1894 by the Burroughs Wellcome pharmaceutical company. Henry Dale, a force in research administration until the 1960s, started work there in 1904 and directed them from 1906. The Wellcome Chemical Research Laboratories followed on their heels.[14] If, in general, such British private institutions still had a low profile in the pre-war era, new, much better-known research-related public bodies formed a rich ecology. The public came to hear about the National Physical Laboratory (1899), the Imperial Institute (for colonial research established in 1887, nationalised in 1900) in South Kensington, the Development Commission (supporting Agricultural Research, 1909), the Science Museum (displaying research to the public, 1909), and the Medical Research Committee (funding research, 1913). Illustrated articles in the public press emphasised the physical reality of the new organisations and the applied science they promoted.

---

[12] Wise 1985; see also Reich 1987.   [13] Clarke 2010.
[14] On Dale and the Wellcome Laboratories, see Tansey 1989 and Tansey 1990. For early British industrial research in industry in general, see Sanderson 1972, 20, and Edgerton and Horrocks 1994.

Public intellectuals of the early twentieth century reflected on the organizational innovation they had witnessed. Thus in 1928, the press reported a lecture by Henry Tizard, physicist and Secretary to the government's Department of Scientific and Industrial Research (about which more in the next chapter).[15] Tizard used the occasion to reflect on the embodiment of applied science in laboratories that were not just huge but also collaborative. As we shall see in Chapter 6, the biologist Julian Huxley, grandson of Thomas, would return to this theme in broadcasts and print during the 1930s. Here, it might suffice to point to his 1936 article reflecting on the remarkable contemporary growth of institutionalised science.[16] Within a few years, J. D. Bernal had built upon Huxley's aperçu of science as a social function in his book *The Social Function of Science*. In this first internationally comparative quantitative study of scientific agencies, Bernal saw the establishments and applied science as a seamless whole.[17]

With combined budgets a factor of ten higher than earlier investment in government research, the new bodies of the early twentieth century found their role and place in the nation's life subjected to constant struggle. This dynamic involved their political sponsors, managers, the civil servants who supervised them, and scientist-leaders. As a result, the potential and scope of applied science were defined not just by individuals' philosophical expressions and aspirations but also by these organisations' responsibilities, the challenges they were expected to meet, and their widely publicised achievements. The following three sections explore the continuity of narrative between pedagogy and research, industrial policy, and the emphasis on the connection between pure and applied science.

The Continuity of Narrative

The story of 'applied science', told here, was entwined with the reply to German educational standards. The narrative of the foreign challenge only strengthened at the beginning of the twentieth century. The socialist Sidney Webb caught the mood in a 1901 pamphlet:

Moreover, the man in the street, though he knows nothing accurately, has got into his mind the comfortable conviction that Germany and the United States are outstripping us. Not merely in general education and commercial 'cuteness', but also in chemistry and electricity, engineering and business organization in the

[15] 'Value of Applied Science: Large Sums Saved to Industry Research Board', *The Manchester Guardian*, 15 February 1928.
[16] Huxley 1936, 206.   [17] Bernal 1939; see also Steiner 1989.

largest sense. Nothing would be more widely popular at the present time, certainly nothing is more calculated to promote National Efficiency, than a large policy of Government aid to the highest technical colleges and the universities.[18]

Increasingly, during the early twentieth century, the calls were for both science education and research. Pedagogy, therefore, did not drop out. Instead, the nineteenth-century rhetoric of keeping up with the Germans through education was complemented in the twentieth century by the politics of research.

American success in education and industry imposed examples on the British public stage. Thus, the thoughts of the widely respected Cornell professor of engineering Robert Thurston were studied by the promoters of the University of Birmingham, encouraged to visit the United States by benefactor Andrew Carnegie.[19] In a speech he gave in 1902, just at the time of the Birmingham visit, Thurston addressed the question of research. While not reducing the whole corpus of applied science to dependence on developments in pure science, these should be drawn upon by the engineer.[20] The stories of American colleges training vast numbers of engineers again linked the educational narrative to research achievement.

The German challenge, deeply interconnected with military competition, created the most fear. In 1906, British chemists grimly commemorated innovation in dyestuffs, and retold perhaps the most familiar story of national research humiliation.[21] The occasion was the fiftieth anniversary of the Englishman William Henry Perkin's invention of mauve, the first synthetic organic dye. Subsequently, German manufacturers had established and dominated the worldwide industry, sustained by large research laboratories staffed by qualified chemists.

It was not just a matter of dyestuffs. Scientists would frequently point to the development of modern optical instruments by Ernst Karl Abbe, Carl Zeiss, and Otto Schott at the British glass industry's expense. At the top of British politics, R. B. Haldane carefully monitored developments across the North Sea. Not only had he studied at Göttingen in his youth, but he also took an annual holiday in Germany, writing to his mother daily of the pleasure of the visits. Never himself a trade Minister, Haldane was the Liberal party's Minister of War, responsible for the army, from 1905 to

---

[18] Webb 1901, 15.
[19] The University of Birmingham, 'Report of a Visit to Colleges and Universities in the United States and Canada, Made in November 1899 at the Suggestion of Mr Carnegie', 11. Inserted in the University of Birmingham Advisory Subcommittee minutes, 9 January 1900, UB/MC/B/2/6/5, Cadbury Research Library.
[20] Thurston 1902. Thurston's attitudes to applied science are explored in Kline 1995.
[21] Travis 2006.

1912. For him, military and industrial resilience were intimately interconnected. The search for a modern means of grounding these in scientific
progress would drive him to support unprecedented institutional
development.

Debates over short-term policy were rooted in a national reassessment
and endeavours to increase national efficiency. The previous century's
campaigns for promoting applied science education provided a template
for promoting the efficiency of other aspects of national life. Formed in
1905 through the energetic leadership of the ever-present science journalist
and activist Norman Lockyer, the British Science Guild was intent on
introducing scientific expertise into the processes of government.[22]
It elected Haldane as the passionate but also smoothly diplomatic first
president. Despite their differing political allegiances, the Hegelian philosopher and Liberal Imperialist was a close friend of the Conservative leader
and theistic philosopher Arthur Balfour. Both were concerned with preparing their country for the future through strategies of continuity with the
past.[23] Each found legitimate and compelling arguments in appeals to
applied science, which justified both investments in education and research
and hope for the future. Repeatedly, these respected philosophers addressed
the tensions between the urgency of reform and the pull of tradition.

Through carefully chosen strategies, Balfour and Haldane would lead
opinion until the late 1920s. Their expectations must be understood in
terms of both their interpretation of the German experience and British
political economy. In his classic study of the so-called efficiency movement, G. R. Searle identified three particularly relevant groups: modern-
izing Conservatives, such as Balfour; Liberal Imperialists like Haldane;
and, further to the left, associates of the Fabian Society, such as H. G.
Wells and Beatrice Webb.[24] These people were closely interconnected.
As well as a close friend of Balfour, Haldane had been best man at
Beatrice and Sidney Webb's wedding.[25] Key scientists were associated
closely with the politicians. R. B. Haldane's brother was a distinguished
physiologist, and so was his nephew, J. B. S. Haldane. Lord Rayleigh, the
Nobel Prize–winning first chairman of the National Physical Laboratory
and the go-to scientist of the early twentieth century, was the brother-in-
law of Arthur Balfour (in turn the nephew of Lord Salisbury, a Prime
Minister with deep interests in science). Scientists and politicians told a
coherent set of stories of science and modernity.

---

[22] On the British Science Guild, see Hull 1994 and MacLeod 1994.
[23] On Balfour, see Root 1980 and Adams 2007. On Haldane, see Haldane 1929; Matthew
1973; Ashby and Anderson 1974; Campbell 2020.
[24] Searle 1971   [25] See Matthew 1973, 288.

Figure 4.1 The engaged philosopher: R. B. Haldane at his desk in the
War Office. *The Sphere*, 4 April 1908. Courtesy Illustrated London
News/Mary Evans Picture Library.

Ardent modernisers such as the up-and-coming Liberal Cabinet Minister Winston Churchill sought to build political interest in the policies of Wilhelmine Germany.[26] In the following years, the Liberal government indeed introduced several German-inspired measures. In April 1909, David Lloyd George, then Chancellor of the Exchequer, delivered a radical budget, including a national insurance scheme on German lines. Among other measures, Lloyd George's speech to Parliament proposed investment in 'scientific research in the interests of agriculture'.[27] During that summer, the King laid the foundation stone for Imperial College of Science and Technology, the British 'Charlottenburg', whose concept and realisation Haldane had driven. The same month, the government announced the award of independence to the Science Museum next door. The rise of the South Kensington complex, in turn, attracted attention to the subject of expertise and modern society. *The Daily Express* headline on its front page was 'What Britain Owes to Science'.[28] At a time when electricity networks were being added to the nation's physical infrastructure, an applied science network was added to its institutional infrastructure.

Science and Industrial Policy

The support of applied science research funding as a commercially unintrusive industrial policy tool was both radically modern and compatible with long-held principles. The ideal of an unobtrusive state still had a wide attraction and, to many, was lived reality. At the opening of his classic 1965 history of early twentieth-century Britain, A. J. P. Taylor recalled this distinctive quality of the pre-1914 world.[29] Even then, however, the call for efficiency, and the introduction of powerful public institutions to bring it about, were promising substantial changes to the Victorian inheritance. Nonetheless, the challenges posed by these reforms were limited compared with the much more radical call for imperial tariff reform.

Through the second half of the nineteenth century, 'Free Trade', permitting tariff-free imports, had been a talisman of the Liberal party and, indeed, of the Conservatives. As German competition grew in the

---

[26] See, for instance, Churchill's letter to the Prime Minister (Asquith), 29 December 1908, published in Churchill 1969, 466.

[27] Incorporated in David Lloyd George's statement of Revenue and Expenditure for 1909–10 in *Hansard* HC Deb. 29 April 1909, col. 494.

[28] 'What Britain Owes to Science', *The Daily Express*, 9 July 1909, British Newspaper Archive/British Library Board.

[29] Taylor 1965, 1.

1880s and growth remained weak during the so-called Great Depression of the late nineteenth century, calls for 'Fair Trade' multiplied. At first, the free traders maintained a natural dominance, but the political balance changed radically in 1903 when the Liberal-turned-Conservative Joseph Chamberlain made his move. Since the 1860s, this leading manufacturer had been promoting science education as an answer to foreign competition. Yet from 1903, he changed tack, as he sought to resolve problems of commerce and imperial solidarity with a single solution. An imperial customs union, protected by tariffs, analogous to the German Zollverein that had led to the formation of an empire, would be the answer to both. Ironically, as Frank Trentmann has argued, Chamberlain's manoeuvring actually increased support for Free Trade by polarising opinion.[30] Combatting the attack on its credo, the Liberal party was outraged by the proposal. The Conservatives, led by Balfour, were sceptical of a policy so brazenly challenging the status quo. Haldane accepted the reality of Chamberlain's two problems, commercial efficiency and imperial solidarity, and celebrated their successful raising to public debate but refused to see them as one.[31] Writing to a friend, Haldane expressed his hatred of protection and its effects, subsequently arguing in the preface of this associate's book that instead, 'the key to progress is method and the education in the widest sense of our people'.[32] Though Chamberlain suffered a disabling stroke in 1906, three elections – of 1906 and at the beginning and the end of 1910 – further polarised opinion in the country, forcing the refreshment of the free trade movement as the heart of a network of policies.

The Liberals' policy response to Chamberlain has been succinctly described by Robert Olby in terms of three initiatives: the introduction of labour exchanges to help the unemployed, national insurance to provide support for the sick, and the Development Commission, which supported agriculture.[33] The first imported a German practice; and the second, a translation of German health insurance, which justified the funding of a research organisation that later became the Medical Research Council. The third was a body supporting agricultural research. So, government support for scientific research, related to, but distanced from, the day-to-day operations of the market, lay at the heart of the Liberal response to the challenge of tariffs. It was both politically

---

[30] Trentmann 2008, 185.   [31] Matthew 1973, 225.
[32] R. B. Haldane, 'Preface' to Ashley 1904, vii–xxiii, on xxii. Also see Haldane to Percy Ashley, 20 September 1903, Hal P.5906, f. 46, National Library of Scotland, quoted in Matthew 1973, 168.
[33] Olby 1991, 512.

feasible and robust in its effects, and compatible with laissez-faire in its ideology. Just as respect for the independence of the firm set the parameters of state-supported education and examination, so the same attitude shaped the support of industry and the role of government aid for research.[34] This strategy contrasted with Chamberlain's mercantilist approach, which was politically much more challenging.

As half a century earlier, it is striking that agriculture proved to be a critical battleground. As in the previous century, tariffs were attractive to some, but for others, applied science seemed to offer farmers the beguiling vision of an alternative to protection. Accordingly, within just five years of the 1909 Development Act, the new Development Commission would fund eleven agricultural research institutes. A. D. Hall, the director of Britain's leading agricultural research laboratory, the Rothamsted Experimental Station, turned the grand gesture of politicians into a strategy.[35] In turn, his vision would lead to the establishment of the Agricultural Research Council in 1931. Both generations of funding organisation supported impressive scientific research and the creation of new strains of wheat.[36] This was but one example of how government policy and specialised new laboratories drove applied science from outside academe. Even more important was the first and most influential of the new institutions, the National Physical Laboratory (NPL).

National Physical Laboratory

The NPL's prestige, size, and budget loomed over more specialised civilian and military institutions. Experience there proved an important passage point in many careers, before World War One and for years thereafter. Typically, the trigger for its foundation was a significant German initiative that sparked what was, at first, a small British response. The Physikalisch-Technische Reichsanstalt (PTR), founded in 1886, was a major Berlin laboratory dedicated in the first place to the definition of standards essential for the new electrical and precision industries. Although the British did have a centre conducting analogous work, in the 1890s, it became known that German organisation was dwarfing the scale of the private British endeavour. So, at the end of the decade, the Prime Minister approved the extension of the laboratory enterprise managed by the Royal Society to form the slightly larger NPL, state-funded and headed by the physicist Richard Glazebrook.

---

[34] For the principles of firm autonomy, see Dobbin 1993, 21.   [35] Olby 1991.
[36] Palladino 1990; Palladino 1993; Vernon 1997; Charnley 2013.

The government created the NPL with the limited ambition to ascertain the value of natural constants of commercial value and standardise and verify measuring instruments. These could cover a wide range, so, for instance, in 1907, the laboratory began testing taximeters. The budget was, nonetheless, small. In 1904, five years after the foundation, the Executive Committee could point to an annual vote of £4,000, built upon an original capital investment of £19,000. Such sums compared miserably with the PTR investment of £200,000 by then. In the United States, the National Bureau of Standards, with a comparable remit, had benefitted from £115,000 capital investment, the committee warned.[37] With such competition, one may not be surprised that ambitions for a much grander project were manifest even during the debates around the foundation. The critical move to widen the laboratory's remit occurred as early as the opening in March 1902 by the Prince of Wales. Armed with royalty's prestige, the Prince shifted the vision from establishing constants relevant to industry to focusing on applied science more generally. 'The object of the scheme', he modestly suggested, 'is, I understand, to bring scientific knowledge to bear practically upon our everyday industrial and commercial life, to break down the barrier between theory and practice, to effect a union between science and commerce.'[38] A historian of the Laboratory, Russell Moseley, would point out how this royal interpretation served as the mandate for the NPL's much more ambitious development.[39] Nonetheless, if the Prince gave some legitimacy to ambition, the Treasury would always seek to drag the organisation and its budget back to its original purpose.

Glazebrook, and his allies in politics and the scientific establishment, had to fight an enduring battle for growth. There were no predestined places for this ambitious new laboratory in either British institutions' economic ecology or their leaders' mental maps. Although the NPL proved to be successful, it battled constantly to be close enough to the needs of commerce to be useful, and distant enough from the market for its work to be consistent with laissez-faire. The laboratory also had to complement rather than compete with the new universities. A 1905 meeting of forward-thinking scientists and politicians, convened to boost support for the Laboratory, was nervously attended by Chamberlain,

---

[37] Moseley 1978, 240. On the preceding Kew laboratory, see Macdonald 2018.
[38] 'The National Physical Laboratory: Opening by the Prince of Wales', *The Times*, 20 March 1902. Reprinted with the kind permission of His Majesty King Charles III.
[39] Moseley 1978, 236. Moseley suggested that the text may have been shaped by Lord Rayleigh, the NPL chairman.

who had championed the recently founded University of Birmingham. He feared direct competition with his pet project. Chamberlain duly suggested universities were already doing the work proposed for the new institution. Only nominally supporting the initiative, he left the meeting early. His concern was immediately noted, but not addressed by Haldane, in the chair, and Sir John Brunner, chemical magnate and strong anti-protectionist, then moved the motion to urge for more funding for the NPL. The motion passed, although this had no immediate consequences by itself.[40] More substantially responding to Chamberlain's criticism, Glazebrook subsequently published an article firmly separating the two ways science could contribute to industry: through education, on the one hand, and testing and research, on the other. The NPL was devoted just to the second.[41] Meanwhile, from the other side of the new laboratory's business, and at precisely the same time, consultant chemists objected strongly to routine testing being publicly subsidised and undermining their core business.[42]

Despite these challenges from competitors, the director realised his ambitions for a great laboratory through two funding coups. One was installing a massive testing tank for the shipping industry – funded to the tune of £20,000 by the benevolent shipowner A. F. Yarrow. Named the 'Froude Tank' after the innovator of hydrodynamic-based ship design, this was the equivalent of a large wind tunnel in the age of maritime trade, for the supreme naval power.[43] For instance, it served in experiments to explain the 1911 collision between the world's largest liner, the RMS *Olympic*, and the battlecruiser HMS *Hawke*, whose parallel courses, a hundred yards apart, should have prevented a mishap.[44] The other early expansion was towards aircraft.

The nineteenth-century industrial challenge from Germany was expanding to military matters. New weapons such as the submarine and the torpedo, rendering the navy's great capital ships vulnerable, threatened the English Channel's protective security. Further potential threats arose from the invention of the Zeppelin airship and the aeroplane. Just five years after the Wright brothers' first flight, in 1908, H. G. Wells published *The War in the Air*. In this novel, world destruction follows from a war fought initially by airships, superseded by much more destructive aeroplanes. By December the same year, a subcommittee of

---

[40] 'The National Physical Laboratory', *The Times*, 5 July 1905.
[41] Glazebrook 1905, 245–46.   [42] Moseley 1978, 243.
[43] Matsumoto 2006, 121–24.
[44] See the reports in 'Notes' 1912, 322; 'Summary of the News', *Yorkshire Post*, 5 April 1912, British Newspaper Archive/British Library Board.

the Committee of Imperial Defence, including R. B. Haldane, now War Minister, held hearings on the airborne threat. At that moment, airships had greater capability, but the prospects for developing aircraft with their high speed, if currently low carrying capacity, were unknown.[45] In July of the following year, within months of these sittings, Louis Blériot crossed the Channel in a flying machine. Mr Selfridge's new London department store displayed his plane as a modern wonder to be celebrated. However, Blériot's achievement also served as a warning. In reporting a 1913 address on the strategic significance of airpower by Major Frederick Sykes of the Army's Royal Flying Corps, *The Daily Express* summarised the threat: 'Britain is no longer an island.'[46] In the era of anticipated war, aeroplanes seemed to have the potential to be a powerful disruptive force.

Rather than capitulating to public demands for the immediate pur-chase of numerous aircraft, Haldane turned to his conception of applied science. It was not even yet clear why planes remained airborne. In the wake of the Boer War a few years earlier, he had sat on a committee with Lord Rayleigh dealing with the improvement of high explosives.[47] The experience of this scientist-led process had been formative for Haldane, who soon drew upon it when planning the organisation of aeronautical research.[48] At Göttingen, his alma mater, the distinguished professor of mathematics Felix Klein had founded the Motorluftschiff Gesellschaft, dedicated to aeronautical research, and established a testbed for models in 1907. Haldane thus turned to Lord Rayleigh to lead a British 'Advisory Committee for Aeronautics' and, in August 1909, put his argument to Parliament. Although the aeroplane would one day have great military potential, that was not yet so, and he declared: 'even if the British Army had 200 aeroplanes of the best present construction we should not be one bit further on than we are at the present moment. That being so, obviously there is a great deal of scientific investigation to do.'[49]

---

[45] 'Report and Proceedings of a Sub-Committee of the Committee of Imperial Defence on Aerial Navigation,' 28 January 1909, CAB 16/7, TNA. See Gollin 1981.

[46] This expression was published as the moral of an address at the Royal United Services Institution (see Sykes 1913) by *The Daily Express*. See 'Matters of Moment. Mastery of the Air', *The Daily Express*, 27 February 1913, British Newspaper Archive/British Library Board.

[47] Hugh Fletcher Moulton, revision of Dora M. Silberrad, 'Pro Patria Artibusque: A Record of a Scientist's Work for His Country, Compiled from Existing Documents', Silberrad Archive, the Science Museum, folder A49, 10; Sharp 2003; Brock 2008, 448–49.

[48] Haldane 1929, 282–83; see also 164–65.

[49] See R. B. Haldane's statement on the debate in 'Naval and Military Aeronautics', *Hansard* HC Deb. 2 August 1909, cols. 1565–66.

In the current year, Haldane explained, the government would be spending £78,000 on aeronautical development, including machines, sheds, and science practised at NPL, more than France and, though less than Germany, not considerably less. The current economic value is perhaps £70 million. By March 1914, the Committee had spent £33,000 on aeronautics alone.[50] Haldane's Committee led the design of a plane, the BE2, with which the country would begin the war. The work also sustained widespread social awareness. In 1910, the popular *Illustrated London News* offered readers a full-page illustrated spread about the Committee's activities at the NPL (see Figure 4.2).[51]

Beyond the Froude tank and the Committee on Aeronautics, the NPL raised considerable sums from routine instrument testing. These it used to support longer-term programmes, particularly on materials for which it acquired a name. The capital investment in the Laboratory grew from an initial expectation in 1900 of £14,000 to the £156,000 spent by 1914.[52] This sum was comparable with expenditure in the first decade of its German counterpart. By 1911, NPL's earnings from work done exceeded £17,000, and its total revenue was £30,000, as high as that of its German counterpart.[53] The next year's annual inspection-day exhibition featured plans for a new experimental driving track, enabling a better design of tyres for motor cars to drive on any surface, using American testing equipment. Also on display was the 'Eda calculating machine', which 'anticipates the future too'.[54] This analogue computer, transferred from the India Office, reportedly substituted 'brass for brain' and was used to predict tides in Indian ports.[55] The disappointment of a

---

[50] Glazebrook 1917, 10. The Committee would be the model for the 1915 American formation of the National Advisory Committee for Aeronautics (NACA), itself the basis for NASA.

[51] 'The British Air-Office: A Place of Wind-Towers, Whirling-Tables, and Gales Made to Order' 1910.

[52] For investment, see Glazebrook 1933, 18 and 35.

[53] 'National Physical Laboratory: Annual Inspection', *The Times*, 16 March 1912. The operating budget of the PTR in 1913 was 670,000 RM, equivalent to £33,000 (Cahan 2004, 196) though revenue from fees at NPL was considerably larger than that of its German equivalent, which earned 115,000 RM (about £6000) from fees in 1913 (Cahan 2004, 216). On the other hand, Glazebrook himself suggested that the cluster of Berlin institutions, which together were equivalent to the NPL, had a total budget of £70,000 in 1913. Glazebrook 1917, 10. Glazebrook claimed the equivalent budget of Washington's National Bureau of Standards was £100,000.

[54] 'Road Endurance Tests: The National Laboratory Experiments', *The Yorkshire Post*, 16 March 1912, British Newspaper Archive/British Library Board. On Eda, see 'The National Physical Laboratory', *The Manchester Guardian*, 16 March 1912.

[55] 'Tide Prediction', *The Graphic*, 8 May 1909, British Newspaper Archive/British Library Board. The machine drew on an earlier design by Lord Kelvin to be seen now at the Science Museum in London.

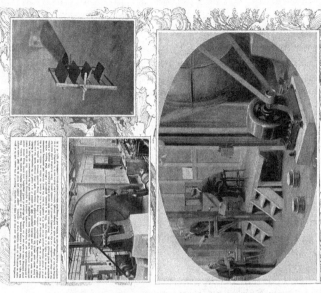

Figure 4.2 The NPL envisaged: 'The British Air-Office: A Place of Wind-Towers, Whirling-Tables, and Gales Made to Order', *The Illustrated London News*, 2 April 1910. Courtesy Illustrated London News/Mary Evans Picture Library.

Manchester Guardian journalist indicated current expectations of the NPL as the home of applied science. Of the category's three branches most familiar to the public, two, cinematography and 'wireless', were not in evidence. However, it had to be admitted that the third, aerodynamics, was well represented.[56]

NPL was a favourite of *The Illustrated London News* even before the First World War. Its diverse scenes and enormous scale made it particularly photogenic, and images of it were appropriate to the new illustrated magazines. After the war, with a broader range of institutions, the magazine's avid readers would see dramatic pictures of 'Lightning in a Laboratory' from the high-voltage laboratory of Metropolitan Vickers or detailed images of the making of oil from coal at Billingham.[57] Nonetheless, the NPL kept its high profile and exotic appeal to the press even then. Scientists too were impressed. Writing in 1936, Julian Huxley described his admiration for the immense scale of the institution.[58] The industrial and military outcome of such large establishments during the early twentieth century, public and private, ranging from stable aircraft to radar and the first oil from bituminous coal, would become objects of national pride.

The Pure Science Connection

In an address during World War One, as new weapons such as poison gas, aeroplanes, and submarines were being developed, deployed, and countered, the Royal Society's president, J. J. Thomson, argued: 'Applied science may lead to reform in our industry, it is to pure science we must look for revolution.'[59] Here was an expression of faith that would resonate throughout the twentieth century. Its implications were, however, less certain than its bravura.

Indeed, 'applied science' sometimes carried the connotation of empiricism and short-termism, as distinct from 'pure' science. On other occasions, it was distinguished from its pure cousin by management, context, and intention, rather than by content. These two different connotations were often conflated, for the idea of 'pure' science was not clearly expressed either. Indeed, the new institutions of applied science in Britain would promote the vision of a close alliance between 'pure' and

---

[56] 'National Physical Laboratory', *The Manchester Guardian*, 16 March 1912.
[57] '"Lightning" in a Laboratory: Man-Made "Thunder-Storms"', *The Illustrated London News*, 15 March 1930, British Newspaper Archive/British Library Board. For Billingham, see Figure 6.1 in this volume.
[58] See Huxley 1936, 206.   [59] Thomson 1917, 95.

'applied'. In a frequently cited World War One speech entitled 'The National Physical Laboratory: Its Work and Aims', Walter Rosenhain, the highly respected founder/head of metallurgy at NPL, suggested pure science, applied science, and implementation in industrial plants represented three 'phases' in the application of science.[60] To understand his meaning, we need to appreciate Rosenhain's professional work on the 'phases' of alloys.[61] He was concerned with the structural diversity within a metal. Typically, in a given alloy, different homogenous phases coexisted. So, within the scientific community, there were different phases. Applied scientists were developing the knowledge and theories generated by pure science but made them useful by translating them to the complex world of real manufacturing. Sir William Bragg (who also concerned himself with studies of the crystallographic structure of iron) deployed Rosenhain's model of science. He used it, for instance, in his much-lauded 1928 Presidential Address to the British Association, and it appears in the title of the catalogue to the display organised by the Royal Society at the 1924 British Empire Exhibition.[62]

For Rosenhain, the special role of the NPL was to serve 'especially applied science'.[63] However, he emphasised the close relations required for pure and applied science:

We thus see that on one side the worker in applied science finds the field of his activities not only in the most intimate contact with, but actually overlapping that of the worker in pure science. This contact and overlapping, however, bears good fruit on both sides, since the experience of practical problems frequently serves to open up whole regions of new phenomena to the purely scientific investigator, while on the other hand the stimulus of the most advanced scientific thought prevents the worker in applied science from losing touch with the true spirit of his science and its latest development.[64]

In Rosenhain's model, there was no necessary unidirectional flow of information. Still, applied science dealing with complex systems followed on from pure science, which employed simplified models of nature. This vision could be accommodated within the long-standing interpretation of a chasm between pure science and industrial application.

[60] Rosenhain 1915–16.
[61] Rosenhain's definition of a 'phase' of 1914: 'For this purpose the constituents of an alloy are classified as "phases," any constituent which is spacially distinct from the surrounding region being termed a "phase". Each phase within itself must be regarded as completely homogeneous and of one and the same chemical composition throughout.' See Rosenhain 1914, 119.
[62] Bragg 1928, 363. Bragg was then director of the Davy-Faraday Research Laboratory at the Royal Institution, and the Royal Institution archive holds the many laudatory messages he received from different quarters. Also see Royal Society 1925.
[63] Rosenhain 1915–16, 228. See also Kelly 1976.   [64] Rosenhain 1915–16, 219–20.

To Rosenhain, Glazebrook, and an increasing number of their successors, applied science was no competitor to pure science but rather its validation. The account of Michael Faraday's reply to the Prime Minister's question about the utility of induction – that it would be taxed one day – seems to have little grounds in history.[65] It has, though, served as a foundation myth, repeatedly cited as the allegorical justification of a belief in the essential role of pure science as the initiator of a revolution in applied science. The 1924/25 annual report of the Department of Scientific and Industrial Research expressed the credo:

But the time-lag between the new discovery in the laboratory and its commercial exploitation has sometimes been very great. Two generations passed before Faraday's prophecy that his electro-magnetic experiments might interest a Chancellor of the Exchequer began to be fulfilled, and although the lag tends to diminish as applied science is consciously and systematically cultivated, it must always be considerable in new investigations of a complex kind and of large industrial significance.[66]

The specific sense of a time lag between pure-science discovery and applied-science application would have continuing importance. Meanwhile, the appeal of a model in which applied science grew from pure science reflected industrial research's growth. Writers linked accounts of the rise of such institutions closely to stories about the potential and the limitations of science, and descriptions of innovative breakthroughs, particularly in the aircraft, electrical, and chemical industries.[67] Cultural critics have often pointed to a snobbish scorn of pure science for its applied cousin. At one level, this reflected the inherited Kantian epistemic hierarchy; at another, it represented the defence of their own self-esteem by poorly paid academics and their clannish tendency to downplay other professional tribes. However, whatever feelings of snobbery professors experienced, it was not proper to express them publicly. For example, the pages of *The Times* and *Nature* would be enlivened by irate complaints of the slights, followed by angry objections by accused miscreants that they had been misunderstood.[68] In 1903, a letter to *The Times* from a distinguished civil engineer complained that Oxford's

[65] Cohen 1987.
[66] [DSIR] 1925, 'Report of the Committee of Council', pp. 1–16, on 2–3.
[67] Richard Gregory, protégé of, and successor to, Norman Lockyer, was perhaps the foremost documenter of the presumed progression from pure science to technical progress.
[68] See the debate over the attitudes to pure and applied science implied by the policies and payscales of DSIR in the letter page of *Nature* in 1920: Church 1920; Soddy 1920; Williamson 1920a; Williamson 1920b.

professor of astronomy had 'branded those who applied science to some purpose of ordinary use, and of profit, as suffering from "vulgarity of mind"'. The professor apologised immediately and explained that his point was merely 'historical' and made 'half-seriously' referring to the views of the American physicist Henry Rowland.[69] He had fallen for the temptation to express himself because, beyond philosophical debates, applied science was acquiring massive physical reality dominating all other manifestations of science.

New organisations for deploying science, initially founded with short-term ends but with little philosophy, evolved quickly within the constraints of military challenge and political discord. Repeatedly, charismatic leaders, seeking to justify the allocation of resources, converted politically acceptable boundaries and desired institutional roles into philosophical realities – even if their origins had been contingent and their acceptability resulted from tactical considerations. Such American trailblazers as Willis Whitney at GE or Kenneth Mees, responsible for research at Kodak, shared their visions and established them as truths through public speeches and the press. They described the organisation of research in the service of industry and often defined applied science at the same time, giving administrative form and scientific category both cognitive and practical legitimacy. In his article differentiating NPL from a university, Glazebrook explained, 'It is the business of such a laboratory to solve conundrums for the good of industry; it is the business of the University to advance knowledge for the benefit of all, and to train its students so as to aid in the advance.'[70] The role of the material institution and the understanding of the abstract category it promoted were intimately intertwined.

In the United States, J. J. Carty, vice-president of the American Telephone and Telegraph Company (AT&T) and shortly to found the prestigious 'Bell Labs', gave the 1916 Presidential Address to the American Institute of Electrical Engineers on 'The Relation of Pure Science to Industrial Research'. He began with the now thirty-six-year-old extract from Thomas Huxley's speech at the opening of Mason College denouncing the use of applied science and then progressed to the dependence of the applied upon the pure.[71] A year later, Cambridge-based scientists concerned about applied science's wartime prominence

---

[69] The insult was attributed to a Royal Institution discourse. See Frederick Bramwell, 'Vulgarity of Mind', *The Times*, 9 June 1903; and the reply, Herbert Hall Turner, 'Vulgarity of Mind', *The Times*, 18 June 1903. Although Turner's apology was published a week after the criticism by Bramwell was published, its contents make plain it was written the same day.
[70] Glazebrook 1905, 245.   [71] Carty 1916.

published a book entitled *Science and the Nation*, echoing Carty's argument. They, too, began with the familiar quotation from Thomas Huxley's speech. Also focusing on a research context, the editor's preface to *Science and the Nation* replaced Huxley's denial of applied science's existence with an indictment of its widespread visibility. It explained that the authors wished to 'enable the reader to grasp in its true perspective the relation of pure science to applied science, "the worker in pure science discovers; his fellow in applied science utilizes"'.[72] The expression was passionate, though its meaning was belied both by the work conducted by AT&T and by the contributions of several of the authors to the Cambridge volume itself.[73]

The arguments of *Science and the Nation* were elegantly answered by the NPL's Richard Glazebrook in a pamphlet on the central role of the University of Cambridge.[74] Formerly himself a leader of Cambridge's Cavendish Laboratory, he paid respectful obeisance to Faraday as the saint of pure science, but then promoted a three-stage account of science's contribution to industry: the man of science in his laboratory, the laboratory of industrial research developing new processes, and finally the works laboratory controlling the quality of raw materials. He could prove the outstanding importance of applied science research by citing exemplary German models: the industrial research laboratories introduced by the chemical companies, the PTR, and the Kaiser Wilhelm institutes. Glazebrook praised the enemy research laboratory of the German company BASF and America's GE facility as the models of the class to which the National Physical Laboratory belonged.

Some industry leaders, such as the director of the Mellon Institute in Pittsburgh, also emphasised that their institutions were temples of applied science.[75] Kodak's English-born laboratory director, Kenneth Mees, was particularly influential in his home country.[76] His 1922 manual of industrial research was widely read, and his experience frequently cited in Britain. Mees had a clear belief in his enterprise and emphasised the interdependence of pure and applied science, in any case differing only in their objectives.[77] However, the distance of time should

---

[72] Seward 1917, vi. Even the unattributed quotation indicated the silent but close American rhetorical links. The quotation had previously been used in a 1898 presidential address to the Geological Society of America: J. J. Stevenson, 'Presidential Address', *Bulletin of the Geological Society of America* 10 (1899): 83–98, see 97, and there may have been other uses on either side of the Atlantic.
[73] Gooday 2012.   [74] Glazebrook 1917.
[75] Weidlein and Hamor 1936, 12. See also Kline 1995.
[76] On Mees, see Sturchio 2020. See also Shapin 2010, which deals at length with Mees' approach to research management.
[77] Mees 1920, 3.

not bestow concordance between these charismatic leaders. At Westinghouse, another American company with global influence, Peter Nutting, a protégé of Mees, presided over a laboratory torn between the pursuit of fundamental science and engineering culture, which he subsumed as 'applied science'.[78] Charles Cross, consultant chemist and developer of rayon, on the other hand, made a distinction from industrial science: 'Industrial science is concerned both with the particular objectives of applied science and the general philosophy of Industry.'[79] Others, such as Glazebrook, emphasised the role of the industrial laboratory as mediating between science and application. Meanwhile, the press addressing the lay public converted all these diverse concepts into 'applied science'.

Interest in the institutions translated into enthusiasm in the public arena with which advertisers and successful writers could engage. This awe could also be tinged with fear. In 1912, less than two years before the outbreak of war, full-page advertisements from Colt's Calculators of Manchester encouraged would-be purchasers with the example of the German and French admiralties and the thought that 'modern warfare is an affair of mathematics and applied science'.[80] Frank Froest, London's chief detective, who had planned the capture of wife-murderer Dr Crippen using a pioneering radio message to a ship, gave more peaceable affirmations of the power of science. On taking retirement in 1912, he published a series of detective novels, demonstrating how different the current organised scientific reality was from the solitary genius of the fanciful Sherlock Holmes novels.[81] A review in *The Spectator* magazine explained that Froest showed how even a leading detective was part of a team with 'the latest resources of applied science at its disposal; the control of telephones and telegraphs; the swiftest means of transport; and, when occasion requires, special facilities for extracting information from crooks and criminals'.[82] The future poet laureate John Masefield began his 1911 novel *The Street of Today* with a conversation about 'Mendelism', then recently celebrated as the basis of a modern genetics. A *Scotsman* newspaper review brought out the

---

[78] Kline and Lassman 2005. For Nutting's use of applied science, see Nutting 1919. For a broader review of conceptions of research, see Godin and Schauz 2016.

[79] Cross 1923–25, 1.

[80] See the Colt's advertisement in *The Manchester Guardian*, 4 October 1912.

[81] On his retirement a newspaper described Froest as 'the exact antithesis of Sherlock Holmes'. See 'The Gentleman from Scotland Yard', *Pall Mall Gazette*, 21 September 1912, British Newspaper Archive/British Library Board.

[82] 'The Grell Mystery', 1913. Quotation courtesy of *The Spectator*.

meaning of the whole book, suggesting that the hero was 'out to teach people how to apply science to life'.[83]

The new civilisation offered hope to many, but even before World War One, it had negative connotations for some. In his study of popular magazines, Peter Broks has shown how, with the beginning of the new century, uncritical adulation of industry's latest products declined, and the tone of treatment changed. The press was now criticising what had been seen as modern miracles as challenges to jobs, the environment, and health.[84] During the war, when it came, discussion of science would be intimately connected with the short-term issues of weapons, battles, munitions, and institutions. High explosives, aeroplanes, and poison gas came to be intimately associated with applied science, and recasting it to be other than an agent of human destruction would be an enduring challenge.

Conclusion

In the talk across the public realm, applied science as research seemed less well defined than in the bureaucratic discussions of a tool for allocating money. Nonetheless, it did develop strong associations and connotations. The press described and illustrated extensive new laboratories in Britain and overseas. Three qualities of this new research-centred model provided continuity with the earlier focus on pedagogy. First, the research narrative continued such traditions of nineteenth-century rhetoric as the national need for applied science in the face of international competition and the requirement to fight the old-fashioned rule-of-thumb methods in British industry. Just as before, more of it seemed to be required. Second, the support of applied research within a laissez-faire industrial policy inherited from the nineteenth century served well its new proponents, such as Haldane and Balfour. It was ready-made for an industrial policy: it addressed familiar problems and did not challenge British tradition. Like its teaching, its investigation also respected the freedom of British industry from government intrusion. Science offered public knowledge, while company secrets remained safe. The third continuity lay in the relationship to pure science. The concept emphasised scientists' continuing relevance and justified increased funding. Whereas, in the nineteenth century, the promise of a new cohort of

---

[83] 'Current Literature: *The Street of Today*', *The Scotsman*, 27 March 1911, British Newspaper Archive/British Library Board. For the book itself, see Masefield 1911.

[84] Broks 1996, 117. For the ambivalent reflections of intellectuals, see Wilson 2010.

scientifically trained managers in industry promised leadership in a new industrial revolution, now successive governments placed their hopes in scientific research–based disruptive innovation. Applied science also served the nation's military against an enemy with analogous methods and ambitions. All these trends were strongly encouraged by the experience of World War One.

5   From the Magi to Industrial Function

Introduction

In the struggle to sustain the nation's economy and society accompanying World War One, the concept of 'applied science' was widely deployed and further enriched. Its meaning in the public sphere came to be embodied by well-known new public and private bodies. Therefore, we must now ask how it matured through wartime and post-war administrative developments and the debates over research amongst the military services, civilian agencies, and private industry. This chapter deals with the entrenchment of hopes through a series of new establishments that would collectively endure until late in the twentieth century. More than anywhere else in this book, it deals with the details of specific research projects as illustrations of the working of individual institutions. Through their thinking over funding priorities, new bodies often formulated and promoted their own conceptions of applied science. They both responded to public opinion and helped shape widely shared understanding.

Traumatic to the lives of young people and of old empires, World War One facilitated tectonic change within Britain's politics. From 1915, British attitudes to science were shaped by public expectation that soon – when peace came – the country would face a harsher world of commercial competition. As early as June 1916, halfway through the conflict, a meeting was held among the allies in Paris to negotiate industrial relationships in the hoped-for post-war era; this was followed by internal British re-evaluation under the rubric of 'reconstruction'.[1] When the war did ultimately end, there was a brief period of widespread optimism for the economic future. The height of those hopes was evoked by H. G. Wells a few years later, who recalled, 'Productivity at the touch of the

---

[1] See Johnson 1969; Cline 2014. On education and the hopes of reconstruction, see Jenkins 1973; Sherington 1974.

new spirit of collective organisation was to leap up like a man who has sat on a wasp.'[2]

The nature and causes of industrial change during the interwar years were complex. However, research had acquired even greater prominence during World War One as a tool of both industrial and military policies. The Germans' military success and their early use of such new weapons as the submarine, gas warfare, and fighter aircraft packed a persuasive punch more powerful than campaigners' speeches.[3] Even if post-war optimism soon ended and budget cuts quickly sabotaged the initial ambitions of government-funded institutions, the multiplication of bodies associated with applied science seemed a dramatic and note-worthy outcome of the war. Meanwhile, fear of Germany did not disappear with the silencing of guns. By 1921 American journalist Will Irwin was reporting the use of a new term, 'the next war', in America and across Europe.[4] So, although, until 1932, the British government held formally to the principle that there would be no significant conflict within the next decade, there was constant support for research directly to build on recent experience and sustain the development of new weapons. From 1934, the British began to rearm, if at first to deter rather than to fight a war.[5]

The outputs of scientific research were multiplying at a rate that seemed dizzying to both scientists and the public. A 1936 analysis suggested that in a single recent year, the world's chemists had published almost as many research papers as during the entire nineteenth century. The editor of *Chemical Abstracts* pointed to the four-fold growth in chemistry over the thirty years to 1939.[6] Besides, industrial companies were conducting much more unpublished research. Philosopher and mathematician Alfred North Whitehead, reflecting on the emergence of an aeroplane-flying society, concluded: 'The greatest invention of the nineteenth century was the invention of the method of invention... One element in the new method is just the discovery of how to set about bridging the gap between the scientific ideas, and the ultimate product.'[7] In Whitehead's formulation, the process, not just the contents, was a defining characteristic of modernity.

It was, of course, not entirely clear or universally agreed how best to bridge the gap between scientific ideas and the ultimate product. Historians of naval warfare have examined, for instance, the real differences over addressing the challenge of German submarines, which

---

[2] This was expressed in Wells' novel *The World of William Clissold*. See Wells 1926, 241.
[3] Katzir 2017.   [4] Irwin 1921, 1.   [5] Dunbabin 1975.
[6] Thomas 1936; Crane 1944.   [7] Whitehead 1926, 120 and 121.

threatened the merchant marine during World War One. Agricultural historians have explored tensions between, on the one hand, those organisations that emphasised close engagement with farming practice, such as the National Institute of Research in Dairying, and, on the other, those that aspired to draw on coherent programmes of fundamental science to make impressive progress. The development of new strains of wheat applying modern genetics made obvious and dramatic progress – but their impact on farmers' practice was more doubtful.[8]

In recent years, Jonathan Harwood has described the varying negotiation in different places between the competing pressures of academic and external expectations on researchers.[9] Without prejudging the outcome, our approach, looking back, must take account of the actors' lack of a consensus over what should be the desirable balance. Developing a new system was a creative process in which aspirations for research programmes, institutions, national industrial and military policies, and applied science were constructed together.

The Military

Of all the legacies left to science by World War One, financially, the greatest was the military's expenditure. The intention here is not to emphasise its significant scale, which has been done well by Edgerton and others.[10] Here, the concern is to explore its classification. In July 1925, the new MP and future Prime Minister, Captain Harold Macmillan, queried reports suggesting that the British government was spending £4 million annually on research and development. He asked in Parliament what this sum comprised. Although Macmillan was assured that it was impossible to separate pure and applied research, behind the scenes, civil servants were investing thought and investigation into the question.[11]

In 1928, a well-known botanist, Lawrence Balls, published a method of discriminating between science's categories. Its relevance to military research was explored in an internally circulated memorandum by Frank

---

[8] See, for instance, Palladino 1990; Vernon 1997; Charnley 2013.
[9] See, for instance, Harwood 2005; Harwood 2006; and Harwood 2010.
[10] See, for instance, Edgerton 2006.
[11] See the written answer on 'Scientific Research' reprinted in *Hansard* HC Deb. 27 July 1925. See also 'Research Expenditure Memorandum by the Treasury', CR I 5, Committee of Civil Research, Research Co-ordination Sub-Committee, CAB 58/ 107, TNA.

Smith, the Navy's director of scientific research.[12] Smith distinguished between kinds of investigation according to scientists' freedom across three dimensions: method, aim, and subject. A pure scientist was free on each of these. By contrast, 'applied research' was characterised by freedom of method, but, here, both aim and subject were constrained. In 'fundamental research', the scientist could choose the goal, but the subject was specified. Smith found that most military research counted as 'applied' according to his system. Strikingly, he shared Rosenhain's view of these categories as forms of work rather than epistemic hierarchy.[13] Indeed, his model complemented Rosenhain's, integrating the phase model into a more general classification.

Historians have analysed the military research of the time, conducting particularly careful histories of aeronautical and naval research styles.[14] They have looked in detail at debates over organisation, at the relationships between science and practice in the investigation of aircraft stability and lift, and at submarine detection, telemechanics (remote operation of ships and planes), and gyroscopes. Each portrays a process of discovering politically possible paths between initiation and implementation within the complex constraints of public interest, scientific knowledge and ambition, engineers' and practitioners' skills, and technical limitations. They show how, between 1914 and the late 1920s, military figures, politicians, elite scientists, and campaigning journalists argued over the specific remit of government research establishments concerned with applied science. Through such debates, scientists fought out the distinction of their work from 'pure science', on the one hand, and development and manufacturing, on the other. Here, interest lies not in those individual results but rather in the range of strategies made possible by the bodies characterising applied science.

The detailed studies of aeronautical and naval research enable us to trace the enrichment of the wartime Baconian belief in science's potency and the potential of its most brilliant stars to change the world. Historians have documented the problematic relations between the services with their traditions of achieving and managing innovation and the

---

[12] Balls 1926; F. E. Smith, 'The Interpretation of the Words Research, Experiment, Test, Trial, Development and Standardisation', 4 January 1927, I(R) 20, Committee of Civil Research, Research Co-ordination Sub-Committee, CAB 58/107, TNA. Later, Smith would become Secretary of the government's Department of Scientific and Industrial Research.

[13] The concept of science as work has been explored in recent years by Jon Agar. See Agar 2012.

[14] See MacLeod and Andrews 1971; Hackmann 1984; Soubiran 2002; Leggett 2016.

academic scientists drafted to achieve breakthroughs.[15] Yet, as it emerged in the 1920s, the system was sufficiently robust for the principal concern to be its fiscal generosity.

Aeronautics

The previous chapter recalled how, on the heels of the Wright brothers' first flight, the Aeronautical Research Committee was commissioning research at the National Physical Laboratory (NPL). Planes were designed either by private companies or by the government-run Royal Aircraft Factory at Farnborough.[16] Potent they might be, but what flying machines might best do was unclear. At first, armies used them principally for reconnaissance. The BE2 plane, developed at Farnborough, was exceptionally stable but neither manoeuvrable nor fast. When, in the summer of 1915, German Fokker fighters, equipped with synchronised machine guns firing through their propellors, entered the skies over the Western Front, the BE2 proved vulnerable, and numerous British pilots died. The consequent angry press campaign linked enemy machines' superiority to the state's repression of private enterprise. Although an urgent enquiry dismissed many of the criticisms, the government removed the head of the Royal Aircraft Factory and prohibited his organisation from building fighting machines. When the independent branch of the military, the 'Royal Air Force', with the same initials, was formed in 1918, the Farnborough 'Factory' was retitled the 'Royal Aircraft Establishment' (RAE).

The line between Farnborough and manufacture was drawn, therefore, by fiat from above. The establishment's relationship with pure science was resolved less directly. The NPL studied models in wind tunnels, while Farnborough tested full-sized aircraft in flight. In this process, the engineer Leonard Bairstow, who led the NPL aerodynamics section, played an essential part. In a detailed study of his work, Takehiko Hashimoto has convincingly argued that Bairstow played an important mediating role between mathematicians and engineers, two otherwise separate tribes, and that Bairstow enabled the successful development of practical approaches to the design of stable aeroplanes.[17] By contrast, according to David Bloor, in the analysis of lift – why planes

---

[15] See, for instance, Gusewelle 1977. For a review of the parallel experiences of the combatant countries, see Agar 2012, 89–117.
[16] For first-hand descriptions of research at the early NPL and Farnborough, see 'The Centenary Journal' 1966. I am grateful to Andrew Nahum for guiding me to this source.
[17] Hashimoto 2007. See also Hashimoto 2000.

stayed in the air – Cambridge mathematicians had an overwhelming and, it turned out, detrimental authority.[18] Instead, it fell to the more practically minded mathematician and physicist Hermann Glauert and his colleagues at Farnborough to incorporate their German competitor Ludwig Prandtl's wartime achievements into British aeronautical culture. Through the 1920s, RAE research maintained this orientation. With the assistance of Ben Lockspeiser (one day to be the first president of CERN), A. A. Griffith led studies of elasticity, cracking of metals, and creeping of fracture, which would subsequently make possible the first jet engine. Farnborough was productive, and its annual budget of £80,000 was significant compared with many industrial research laboratories, though much smaller than that of the still-dominant NPL.[19] It would have a complex, if productive, relationship with private industry that Hermione Giffard has documented.[20] The roles of Bairstow and Glauert as 'translators' and their establishments were models of their kind in the complex network of institutions of the 1920s. They helped define the possibilities and the limitations of applied science.

The Navy

Aircraft were new weapons strongly associated with science from their early years. The Royal Navy, for the past century and a half the world's greatest maritime force, had a very different culture. It also went through a deeply divisive debate about the proper place of applied research and its relationship to pure science and practice. Early in the war, a profoundly unpopular research model was imposed from above, and the navy achieved a long-term resolution only after hostilities had ended. Fortunately for historians, the negotiations were recorded in words so

---

[18] Bloor 2011.
[19] The total expenditure on research by the Air Ministry in 1926 was estimated as £206,000, or about a million American dollars, roughly equivalent to the budget of NPL. Director of Scientific Research [H. E. Wimperis], 'Air Ministry Research Organisation', Committee of Civil Research, Research Co-ordination Subcommittee IR. (R) 18, T161/290, TNA. See also H. E. Wimperis, 'Research Expenditure', Committee of Civil Research, Research Co-ordination Subcommittee C.R. (R) 7, CAB 58/107, TNA, in which Wimperis distinguished between expenditure on research and the headline sum of £1,373,000 for Air Ministry 'Experimental and Research Services'. He pointed out that 'it is incorrect to speak of this financial provision being for Scientific Research, since the construction of experimental types of aircraft is included: and, as will be seen from the items, a considerable sum is included for inspection and stores' (annual expenditure on electrical stores alone amounted to £91,000).
[20] Giffard 2020. She too points out the sensitivity shown in Britain to the boundaries of the public and private sectors.

straightforward that, a century later, the reader can feel like an observer watching the drama unfold. Thus, these discussions are worth following in some detail.

In the pre-war years, innovators, including the engineer Charles Parsons and the charismatic admiral Jackie Fisher, had already brought about a revolution in ship design, bringing in the new technologies of the steam turbine, oil firing, and the massive dreadnought battleships.[21] In small pockets of activity around the country, enthusiasts carried out investigations and forced through many novelties. Meanwhile, the culture had changed less than the machines, and naval practice generally drew on long empirical experience rather than new scientific findings. There were also severe failures. Little preparation had, for instance, been made for protection against submarines. Within months of the war starting, a single German submarine sank three obsolete British cruisers on a single day. The press decried German leadership in deploying new weapons and the lack of coordination of Britain's scientific response. The government had to respond.[22]

In 1915, Arthur Balfour, now responsible for the Admiralty, handed the problem to Jackie Fisher, who had recently resigned from operational command of the navy, having disagreed with Churchill, its political head. Determined to achieve 'Big Conceptions and Quick Decisions', under the title Board of Invention and Research (BIR), Fisher assembled a group of scientific luminaries known for their genius.[23] The small Central Committee, chaired by Fisher himself, was dominated by experts without professional naval experience, including J. J. Thomson, discoverer of the electron; physicist and president of the Royal Society, the industrial chemist George Beilby; and the ship-designer Sir Charles Parsons, inventor of the marine turbine. In Fisher's words, the twelve-person consulting panel members were 'as famous as the Magi'.[24] Again, predominantly not naval, the panel counted both the recent Nobel Prize–winning physicist William Henry Bragg, known for his work on finding the structure of crystals using X-rays, and physicist Ernest Rutherford, remembered for identifying the atomic nucleus. Although the new arrangement ran with the public mood, sceptics felt it was constructed to get Fisher back into power in the navy and ran against tradition. The BIR had two responsibilities. One was to identify any worthwhile

---

[21] See Leggett 2015; Hackmann 1988.

[22] See Balfour to Haldane, 12 June 1915, ff. 184–85, Balfour papers 49724, the British Library Board.

[23] On this group and its limitations, see Gusewelle 1977; also Macleod 1971 and Leggett 2016.

[24] On the comparison with the Magi, see Bacon 1929, vol. 2, 286.

prospects amidst the many inventions suggested by the public, and in its short life, it addressed about 40,000 suggestions. The other was to conduct its own research.

Running parallel to the navy's existing research system, the BIR, led by its eminent central committee, was accountable not to the navy itself but to the government's new Minister, Arthur Balfour. It comprised special-ised sections, each served by a subcommittee made up exclusively of distinguished scientists and engineers. Their titles indicate their vast range and comprehensive ambitions: I, airships and general aeronautics; II, submarines and wireless telegraphy; III, naval construction; IV, anti-aircraft equipment; V, ordnance and ammunition; and VI, armament of aircraft, bombs, and bombsights. The section with the highest budget, with distinguished leadership and prominent in the public memory, was Section II, responsible for finding submarines and radiotelegraphy. The advisors to this section were people whose names would recur in the history of wartime science. They included William Henry Bragg, Ernest Rutherford, Richard Glazebrook of the NPL, distinguished chemist and engineer Richard Threlfall, and the leading electrical engineers Charles Merz and William Duddell. By the end of June 1917, their section alone had dealt with more than 14,000 suggested inventions.[25]

The elite scientists were convinced that hydrophones, which could pick up a submarine engine's sound underwater, were the key to detecting the feared U-boats. In September 1915, academic physicists were dispatched to an existing centre of investigation in Hawkcraig, on the north bank of the estuary of the Forth (Firth of Forth), opposite the city of Edinburgh. They encountered Commander Cyril Ryan, already hard at work, reporting directly to the Admiralty but with an approach very different to that of the physicists. A member of the new team was shocked to discover that he had no idea of who Rutherford was, but nonetheless Ryan was a talented inventor.[26]

Petty squabbling between the competing groups, expressing different models of research, went up the chain of command as far as Balfour, the Minister. Though Bragg himself was dispatched to direct research at Hawkcraig, the relationship problems remained unresolved, and soon the physicists from the BIR moved to Harwich on the east coast of England. The Admiralty team under Ryan at Hawkcraig and the BIR scientists at Harwich each offered their designs. Neither group, however, could devise hydrophones with military significance, and, ultimately, the

---

[25] [Sothern Holland chair], 'Report on the Present Organisation of the Board of Inventions and Research', Appendix 3, ADM 116/1430, TNA. Hereafter Sothern Holland Report.
[26] Wood 1965, 203.

navy met the submarine threat through a system of convoys.[27] Detection
using active 'sonar', known in Britain as ASDIC, was experimentally
developed during the war, based on work by the great French physicist
Paul Langevin. However, it was not yet operational at the armistice.[28]

The tensions between the navy and BIR teams proved momentous for
research organisation, even if the devices the parties developed were not
themselves particularly significant in the field of battle. By 1917, relationships had become so bad that the Admiralty established an enquiry
under the senior engineer and civil servant Sir Alfred Sothern Holland.
His report began with a damning condemnation of the traditional trial-
and-error approach of the navy and its lack of systematic research and
sympathy for science. It 'possessed no research department of its own
and was satisfied to carry on its business on old-fashioned "rule of
thumb" lines'.[29] As in the earlier pedagogical debates, so in the research
context, 'rule of thumb' was seen as the antithesis of the desirable
modern way.

Sothern Holland used the hydrophone example to contrast what was
happening in practice with an ideal research process. That would have
identified and resolved fundamental physical problems leading to the
development of a prototype that would then progress to a device
'redesigned for service use by a practical engineer' and be tested under
service conditions.[30] Such a sequence had been carried out 'in the case of
the aeroplane':

The other method is that of 'trial and error,' which may in some cases lead very
rapidly to a result, but practically never to the best possible result, and more often
to failure. A little bit of both is the usual inventor's process, and there is danger at
scoffing at such a method when fostered by a touch of genius. In fact, we consider
that with the hydrophone in its present stage of development more has been
achieved by the 'trial and error' process, as the scientific method has not yet had
time to permit results to be reaped.[31]

The report recommended that the Admiralty replace both the existing
BIR and the many small research centres with a well-funded central
research institution run by a director of research and a naval superintendent. Equipment alone would require £360,000. Accordingly, the
scientist would be in a hegemonic position, and, furthermore, leading
scientists would have direct access to decision-makers in the navy.[32]

---

[27] Waters 1980.
[28] Sonar worked by emitting a sound wave and then picking up echoes from, for example,
submarines using piezo-electric detectors. See Hackmann 1984; Katzir 2012.
[29] Sothern Holland Report, 7.   [30] Sothern Holland Report, 6.
[31] Sothern Holland Report, 6.   [32] Soubiran 2002, 69. See also Hamilton 2014.

Records of the confidential responses reveal the ambitions and prejudices of both sailors and scientists. Admiralty rejoinders to the Sothern Holland report were scathing. Lord Jellicoe, chief of the naval staff, was adamant that small groups under the control of heads of naval departments would be the only organisation that could work:

It is not possible for the scientist ever to obtain a close acquaintance with Service requirements and that close knowledge of what is practicable at sea which is possessed by the Naval Officer. There is also a danger of scientists devoting their attention to the purely scientific aspect as opposed to those which have a definite Service application.[33]

With a well-known opposition to convoys, the navy was, though, losing debates over how to fight the submarine challenge at the time. Thus, the contribution of William McLellan, a well-known electrical engineer, proved to be influential. His approach would create research groups combining a sailor, a theoretical scientist, and a practical engineer.[34] So, the response was not to reject research informed by science but to question its accountability. As the Secretary to the Navy said in his comments on the report, 'It must be made clear that the DRE [Director of Research Establishments] is the assistant and advisor of the technical departments and not in any case their dictator.'[35] Here was a vision of science as embedded within a client body, not separate from it. What that would mean was yet to be decided. Yet the Sothern Holland report had been a seminal moment for British science.

In January 1918, a new directorate of Experiment and Research was inserted into the BIR under the leadership of William McLellan as deputy director and his business partner Charles Merz as director. With the ending of the war, reorganisation and learning continued. At the end of 1920, the physicist Sir Frank Smith, formerly at the NPL, was appointed 'director of scientific research'. He pulled together the 'Admiralty Research Laboratory' (ARL, at first called the Naval Research Institute), established next to NPL in 1921. Smith himself classified naval research as generally applied science.[36] Looking back a quarter of a century, a later long-standing assistant director of research at the Admiralty recalled that the laboratory of the 1920s had

---

[33] Admiral J. R. Jellicoe, 17 October 1917, 'Remarks of CNS', Summary of Opinions', Sothern Holland Report.
[34] William McLellan, ADM 116/1430, TNA.
[35] Testimony of Sir O. Murray in 'Report of Sir Sothern Holland's Committee, Summary of Remarks of (1) Sir. O. Murray (2) Controller (3) Third Sea Lord', ADM 116/1430.
[36] See Smith, 'The Interpretation of the Words Research, Experiment, Test, Trial, Development and Standardisation'. For new roles, Wood 1965; Soubiran 2002.

been dominated by 'donnish types' inclined to dismiss the 'application of research' as work for 'artisans'.[37] But perhaps he was also expressing the irritation of a leader committed to ever-further improvement.

The navy maintained other more specialised laboratories, for instance, the Signal School, where it concentrated 'wireless telegraphy', but the ARL held a wide brief, taking a more fundamental tack.[38] It would not design or build the devices needed to aim the long-range guns that had come into use. Nonetheless, it could develop and patent designs, test, and recommend commercial products for naval use. NPL profoundly influenced that laboratory as it did the RAE. Both had grown out of traumatic political experiences, learning about the connections with pure science of which they needed to boast, but also the practicality they were required to demonstrate. Like other large organisations such as America's GE, the Admiralty used a mix of strategies under its envelope of applied science.

For completeness, the army too should be mentioned. It had its invention committee, the 'Munitions Inventions Department', and addressed significant challenges by deploying elite scientists, including the location of enemy guns and the reply to gas warfare.[39] In 1916, the army opened the research centre at Porton Down in Salisbury Plain to develop Britain's own chemical weapon capabilities. This centre would survive World War One and be the basis for World War Two, the Cold War, and ongoing military research on chemical warfare into the twenty-first century. In 1929, the military research groupings supported by the army, navy and air force would collectively employ about 400 scientists, slightly more than half the total working directly for the government.[40] The scientists had taken command, and applied science had shifted from Commander Ryan's empiricism to the application of fundamental science of the Braggs and the translation activities of the ARL.

The discussions among policymakers in wartime also reached a much broader audience. In 1916, R. A. Gregory, *Nature* editor and H. G. Wells' close friend, published *Discovery, or The Spirit and Service of Science*. This widely read book emphasised the close links between scientific discovery and the development of successful weapons. Without

---

[37] Buckingham 1952, 101.
[38] For the details of the post-war work at ARL, I am indebted to Soubiran 2002 and express my gratitude to Sébastien Soubiran for a copy of his doctoral thesis.
[39] See Pattison 1983; MacLeod 2000; Van der Kloot 2005. On Porton Down and the role of chemists, see MacLeod 1993; Carter 2000.
[40] HM Treasury 1930, 40–44. Edgerton long ago pointed out that more scientists were employed by the military than civilian wings of government during the interwar years. See Edgerton 2006, 117–22.

either evidence or accuracy, Gregory interpreted the Wright brothers' building of the first aircraft as drawing upon the scientific analysis of their bitter rival, the Smithsonian astronomer Samuel Langley.[41] More generally, Gregory portrayed pure science's cultural nourishment as an essential complement to students and researchers in applied science. The wartime engagement of academics associated with fundamental questions, whether J. J. Thomson, Rutherford, or Bragg, also emphasised the interdependence of pure and applied science. The military experience now offered the promise of peacetime breakthroughs.

MRC and DSIR

In civilian life, medical research mirrored the victory of the elite scientists in the Admiralty. There, physiologist Walter Morley Fletcher led the Medical Research Committee from its launch just before World War One. He was a bureaucrat who was also an evangelist for medicine's transformation by science, particularly laboratory science. Immediately after the war, Fletcher, with the title of 'Secretary', took the organisation to its Medical Research Council (MRC) status with ambitions to cover all sectors of biomedicine. Meanwhile, the government established a Ministry of Health for the first time. Each body had its own research ambitions, and in 1924, they reached an agreement on the partition of responsibilities.[42] The MRC would address experimental clinical and laboratory research, whereas the Ministry would be responsible for aetiology, 'field' enquiries, and 'applied' or health-related research.

Successive secretaries of the Council argued that medicine and biology did not readily fit a template formed by chemistry and physics. They invested heavily in both fundamental science and health-oriented work, such as the use of vitamins. The MRC contended, therefore, that 'pure' and 'applied' were categories that were inapplicable to its work.[43] Fighting for a singularly integrated science model, it vehemently rejected the use of patents and any subservience to the parallel government Health Ministry. The approach of the MRC expressed a philosophy of an independent science and dependent 'applications'.[44]

---

[41] Gregory 1916, 290. On the relationship between the work of Langley and the Wright brothers, see Crouch 2002.
[42] 'Relations between the Ministry of Health and the Medical Research Council', 12 February 1924, MH 123/498, TNA. See Austoker 1989, 25–26.
[43] For more on the relations between 'pure' and 'applied' in the early years of MRC, see Bud 2008.
[44] See, for example, the testimony of the MRC Secretary Walter Morley Fletcher to the Patents Committee, published in Board of Trade 1931, paras. 3306–408, pp. 195–205.

Most other science (except agricultural research) came to be funded through the new Department of Scientific and Industrial Research (DSIR). With roots planted in 1915, this was envisaged as an organisation that would strengthen British industry in peacetime. It was the outcome of decades of thinking about national efficiency before the war and became a flagship of reconstruction after it. More perhaps than any other body, it gave meaning in the public sphere to 'applied science' in British research during the twentieth century. Thus, when *The Manchester Guardian* newspaper compared the annual report of 1935 to a 'gargantuan pie of applied science', it was not just telling the reader about the DSIR but also using the institution to illustrate applied science.[45] The body's importance to promoting and understanding both term and concept justifies detailed attention here.

The proposal for the organisation was drafted by an eminent doctor and protégé of Haldane, Christopher Addison. Although little remembered today, he would be a major actor in many of the defining events of British research and a close ally of wartime leader David Lloyd George. Working in the Board of Education during the early part of the war, he drafted the strategic proposals for a new body supporting training and research.

In his political views, Addison was a typical Edwardian-Liberal proponent of applied science institutions, convinced of the importance of efficiency, education, and research in combating protectionism. From the beginning of his tenure in the Board of Education, he committed thoughts on science to his diary. As early as December 1914, he had reflected on the need to 'link up science and trade more than we have done'; otherwise, 'we shall meet with a strong Protectionist cry after the war is over to keep out the German competition'.[46] A vigorous campaign for technical education was waged in 1914 as the war settled into a stalemate fought with new weapons.[47] So it was with Haldane's support and within the Board of Education, and not perhaps the more obvious Board of Trade, that much broader arrangements for wartime and postwar science emerged. The title of the paper submitted to the Cabinet indicated its roots in pre-war Liberal party rhetoric and ambitions: 'Scheme to Provide Advanced Scientific, Technical and Commercial Education and for the Encouragement of Research to Promote

---

[45] 'Many Inventions', *The Manchester Guardian*, 30 December 1935.
[46] C. Addison, Diary, 9–10 December 1914, MS Addison dep. d. l, Papers of Christopher Addison, University of Oxford, Oxford, Bodleian Library, Special Collections. Quoted by kind permission of the Bodleian Library. See also Daglish 1998.
[47] Daglish 1998.

National Efficiency'.[48] However, before Ministers could implement this document, the Liberal government of Asquith fell, to be replaced by an administration in coalition with the Conservatives. The educational scheme's author, Addison, moved to the Ministry of Munitions in the new regime. Without him, plans were redrafted to reflect wartime urgency and express an evolving concept of applied science. They lost the primacy of pedagogy, and the final title of the White Paper, which appeared in July 1915, was 'Scheme for the Organisation and Development of Scientific and Industrial Research'. The new document shifted the funding emphasis to research. It was firm in the importance of the 'development of scientific and industrial research as will place us in a position to expand and strengthen our industries and to compete successfully with the most highly organised of our rivals.'[49] Accordingly, the government created an entirely novel national organisation, an advisory committee on scientific and industrial research, which became part of a new Department of Scientific and Industrial Research a year later.

Despite the change in emphasis, the continuity with previous moves to improve pure and applied science education was striking. The founding administrative leaders were educational administrators: the first chairman, William McCormick, had been head of the Carnegie Trust, which funded Scottish universities. He was also a member of the committee that led up to the University Grants Committee (UGC), of which he was the first chairman. The Secretary, effectively the chief, Frank Heath, had worked with McCormick on university funding.

An overseeing committee of eminent scientists supervised this alliance of senior civil servants. The historian Christine Shinn has pointed out that not only was this Advisory Committee of DSIR headed by the same man as the UGC, Sir William McCormick, but the two organisations worked on similar principles. There were parallels between the management of research associations by DSIR and of universities by the Committee.[50] Both 'parents' offered matching funding to their clients and practiced academic visitations. So, just as the other body promoted the idea of the university, DSIR established a national, if flexible, research system. There were also clear parallels with the management of NPL (soon taken over by DSIR). Lord Rayleigh, who chaired its Supervisory Committee, sat on the DSIR's Advisory Council, which also

---

[48] Addison to President of the Council, 20 March 1915, 'Summary of Proposed Grants', ED 24/1581, TNA. This was put to the Cabinet in April 1915 as 'Proposals for a National Scheme of Advanced Instruction and Research in Science Technology, and Commerce' (ED 24/1581). There is a huge literature on the foundation of DSIR. See, for instance, MacLeod and Andrews 1970; Varcoe 1970; Hull 1999, 461–81.
[49] Board of Education 1915, 1.   [50] Shinn 1986, 205–11.

included three industrial chemists, two engineers, and two physicists.[51] Among the other distinguished members was the chemist and industrialist George Beilby, chairman of governors of the Royal Technical College at Glasgow, offering the university's 'applied science' degrees. His interest was fuel economy, and he would lead the Fuel Research Board established by DSIR shortly after its formation. We have already encountered Beilby and another DSIR Advisory Council member, Richard Threlfall, in the van of the BIR at the time. This leadership engendered distinctions between a class of elite scientists who were broadly happy that they had influence, a younger group frustrated that they had neither influence nor power, and civil servants who could get things done.[52]

Industrialists were considered, but not included, an omission particularly striking considering the institution's justification as industrial therapy. Roy MacLeod and Kay Andrews, writing in 1970, shortly after the demise of DSIR, saw in its foundation the separation of teaching from research and of research from industry.[53] As a result the organisation sometimes had the air of talking about industry from the outside rather than from an industrial perspective. Nonetheless, as a key part of the new scientific infrastructure, it quickly achieved visibility, which had often eluded science.[54] For instance, the department received the energetic coverage of the Leeds-based *Yorkshire Post*, with its long-standing interest in the civic universities in its vicinity. In the two decades 1920–1939 the newspaper cited DSIR 338 times.[55] Arthur Balfour's Privy Council introduction to its 1925 DSIR annual report quoted Prime Minister Stanley Baldwin claiming that if science could not 'save' industry, it could provide 'resilience'.[56] Three decades later, Armytage enthusiastically described the beneficial impact of DSIR in promoting cooperation, delegating to universities, and encouraging departments.[57]

From its launch, the department was self-conscious about its ambit. Because it had both scientific and industrial stakeholders, DSIR paid respect to the language and concerns of each. As the first annual report appeared in September 1916, elite scientists were planning their book on the importance of pure science. At the same time, the new body sought to influence and encourage industry, using its annual reports for such

---

[51] MacLeod and Andrews 1970, 40.   [52] This thesis was developed by Hull 1999.
[53] MacLeod and Andrews 1970.
[54] Roy Macleod has pointed out how much the Germans envied the new institution and then copied it. See Macleod 2009a.
[55] www.britishnewspaperarchive.co.uk, accessed April 2023, British Newspaper Archive/ British Library Board.
[56] DSIR 1925, 'Report of the Committee of Council', 1–16, on 2–3.
[57] Armytage 1953, 253.

outreach. Today, these offer the historian a valuable opportunity to overhear the voices of a previous century balancing competing demands from its stakeholders. Each report contained an introductory section by the civil servants responsible under the signature of the government Minister responsible; descriptions of the various categories of work by institutional beneficiaries concluded the main text. Between these two sections was an overview, officially signed off by the scientifically distinguished Advisory Council. This articulated the values and ambitions of the department's civil servants and advisors reaching out to the key stakeholders. Therefore, a variety of voices contributed to these reports.

Each year these documents emphasised the importance of a balance between the categories of science. In the first annual report of 1916, having accepted that 'properly speaking' there was no difference between the subcategories of science, the Advisory Council pointed out that 'they have to deal with the practical business world in whose eyes a real distinction seems to exist between pure and applied science.'[58] The Council defined pure science as having 'no other aim than the creation of new knowledge', a process that could not be externally organised. 'On the other hand, it is necessary for the modern State to organise research, including those simpler types of research which we may call investigation, into problems which directly affect the well-being of large sections of its people.'[59] In turn, the latter were concerned either with applied science or with an aspect of pure science with 'a specific end in view'. Sabine Clarke has studied the use of this last category that the DSIR wished to identify as 'fundamental'.[60] Undoubtedly, this category was the organisation's favourite during the early 1920s. However, she points out that its use was not consistent across the period and faded in the 1930s.

In evaluating the use of technical terms, we need to be aware of the vernacular context. While amongst the newspapers covered by the digitised 'Britishnewspaperarchive', *The Scotsman* deployed the term 'fundamental research' the most, even it counted less than a couple of dozen usages in the whole of the 1920s. By contrast, it used each of 'pure' and 'applied' science considerably more than a hundred times during the same period.[61] Contemporaries did reflect on the options. The businessman Henry Greg suggested to a meeting of the Textile Institute in 1916,

---

[58] [DSIR] 1916, 'Report of the Advisory Council', 5–44, on 10.
[59] [DSIR] 1919, 'Report of Advisory Council', 12–78, on 13.   [60] Clarke 2010.
[61] According to 'Britishnewspaperarchive' (accessed December 2022), in the 1920s *The Scotsman* used 'fundamental research' 21 times, 'applied science' 115 times, and 'pure science' 140 times, British Newspaper Archive/British Library Board.

'The terms pure research and applied research were useful and worth sticking to, and they fell in with ideas already made familiar by the terms Pure and Applied Science.'[62] Two decades later, *The Times* similarly supported the general convenience of these established categories, despite the theoretical benefits of such new terminology as basic research and concerns about distinguishing between pure and applied.[63]

In 1926, the annual report of the department's Advisory Council for the past year incorporated thinking about the entire first decade of its existence. For the first time, it reflected the input of a policy subcommittee established in January 1925.[64] The report described the overall policy as 'to build up gradually organisations to form a link between pure scientific research on the one hand, and industrial applications on the other'.[65] Its analysis completely avoided the use of 'fundamental research' and, as if it had never used the term, laid out its list of categories of research. The report of the Advisory Council reflected, 'The terms pure science, applied science, industrial research, "ad hoc" investigations, are frequently used; but not, we suspect, always, with the same meaning.'[66]

The Advisory Council report sought clarity and public acceptance by dividing its treatment between 'The State of Pure Scientific Research' and 'The State of Industrial Research'. The latter it introduced by reflecting, 'Applied scientific research has two main objects – improvement of existing and introduction of new industrial methods and products. Both are necessary for the industrial health of the country.'[67] So, here, DSIR equated industrial research to applied scientific research.

---

[62] Greg 1916, 244.   [63] 'Science', *The Times*, 1 January 1936.

[64] For information on the 'Policy Committee', see 'Final Report of the Policy Committee', A.C. Minute 57, appended to Minutes of the Meeting of the Advisory Council, 9 February 1927, DSIR 1/5, TNA. The annual report was finalised and discussed in the absence of both the Administrative Chairman since DSIR's foundation (William McCormick) and the Permanent Secretary (Frank Heath). The acting chair was the politician, chemist, and industrialist (G.) Christopher Clayton, a PhD who was a board member of the United Alkali Company (and when Imperial Chemical Industries [ICI] was formed in December 1926, a board member of ICI) and MP for Widnes. Henry Tizard was Acting Secretary for part of that autumn before becoming Permanent Secretary the following year. For the process of finalising the report, see the Advisory Council minutes for 6 October 1926 and 1 December 1926 (both chaired by Clayton), DSIR 1/5.

[65] [DSIR] 1927, 'Report of the Advisory Council: Research on a National Basis', 24–25, on 25.

[66] [DSIR] 1927, 'Report of the Advisory Council: The State of Pure Scientific Research', 18–19, on 18.

[67] [DSIR] 1927, 'Report of Advisory Council: The State of Industrial Research', 22–24, on 22.

It incorporated 'Research on a National Basis' as a further subheading within that category. The paper, therefore, subsumed all the work funded by DSIR under the headings of pure and applied research. It emphasised the challenge repeatedly cited through the annual reports: 'we have to hold the balance between scientific and industrial research, and in times when national economy is urgent to give help where help is needed'.[68] Hence, when the Advisory Council reported on food investigation in June 1926, calling for substantial investment, it addressed the balance of 'applied research' and the 'purely scientific research' on which the former was 'based'.[69]

A few years later, as the country recovered from the Depression, the Advisory Council reflected on the transformation in life wrought by the application of science over the previous quarter-century. Those advances had come about, 'in the main, not as the result of new and revolutionary scientific discovery, but by the steady and organised application to the problems of industry of discoveries already made'.[70] The message from the various voices of DSIR was that industrial progress flowed from a mix of longer-term and applied scientific research. Although the lines between 'pure' science', 'fundamental research', and 'applied science' could be problematic, and, certainly, the relationships were not all one-way, this cluster of related activities was considered key to future British prosperity. It was a model that fitted the government's wish to demon-strate an industrial policy that did not interfere with the market, and the ambitions of scientists who wished to show they were useful in whichever part of the system they worked.

The new department had adopted three distinct roles. First, it sup-ported the training of post-graduate students across the country, the academic research projects on which they were engaged, and the investi-gations of a limited number of academic investigators, in the interwar years typically totalling a few dozen. Second, it ran its own research laboratories, and it also funded problem-driven projects in a variety of institutions on topics, which the 1926 report summarised as 'Research on a National Basis'. Third, it oversaw and co-funded collaborative indus-trial research for specific industries. Of these, the first received the least funding and visibility, and the third won the most press attention. It was also the most closely associated with 'applied science'. This term was,

---

[68] [DSIR] 1927. 'Report of Advisory Council: The State of Pure Scientific Research', 18–19, on 18.
[69] 'Report of the Advisory Council on Food Investigation', A. C. Minute 121, 3 June 1926, DSIR 1/5.
[70] DSIR 1935, 'Report of Advisory Council: Science and Industry', 11–17, on 14.

however, often associated by the press with news about DSIR in general, and it was one that the organisation itself took very seriously.

In 1927, Arthur Balfour introduced a DSIR survey of industrial research with a widely reported worry that 'In the fundamental discoveries of pure science this country, I believe, has taken its full share. About applied science, I am not so sure.'[71] The occasion was the tenth anniversary of the DSIR programme supporting cooperative industrial research. In the immediate aftermath of World War One, the government allocated a million pounds to a new network of industry-specific 'research associations', equivalent to trade associations. They were to jointly support specialised research laboratories initially co-funded by the government on a pound-for-pound basis. The associations were quickly attractive, with 2,500 companies signed up by 1920.[72] When the million pounds ran out in 1934, new arrangements were made, and, by then, British industries had created thirty-four associations.[73] In retrospect, they came to seem underfunded by government and industry alike and, for many, a failed experiment.[74] Yet, at the time, they were a substantial part of the industrial policy of governments committed to economic liberalism, which outlived World War Two to be an important part of the post-war settlement, surviving into the 1980s.[75]

By 1937, industrial support for the research associations had more than doubled since the early 1920s and was almost twice government-derived income.[76] Initially, according to its annual report, DSIR had intended that the principal concern of the research associations would be 'fundamental research'.[77] The expressed intention was to leave applied science to individual companies, and the role here of the association was to encourage rather than replace commercial enterprise. However, as the 1920s rolled on, demand for the associations to conduct more specific work became clear. The laboratories' annual reports were full of reflections on the need for balance between the requirements of more general, further from the market, 'fundamental' research and special, closer to market, investigations. The 1926 report of the Advisory Council showed that the desired balance differed according to industries' histories and the

---

[71] DSIR 1927a, Arthur Balfour, 'Preface', iii. On 27 April 1927 many newspapers carried verbatim reports of Balfour's 'Preface'.
[72] Varcoe 1981, 442.   [73] Varcoe 1981, 440.   [74] See, for instance, Mowery 1986.
[75] Allen 1948, 3. See also Tomlinson 1994; Foreman-Peck and Hannah 1999. On industrial policy in the first sixty years of the twentieth century, see Grove 1962.
[76] DSIR 1938, 'Growth in Financial Resources of Research Associations since 1920', in 'Report of the Advisory Council', 9–21, facing p. 9.
[77] [DSIR] 1920, 'Report of the Advisory Council: Relation of the Work of Research Associations to Works Practice', 38–39, on 38.

scientific competence of member firms. Moreover, as we have seen, the use of language changed and applied science came to be a legitimate description of the work of the laboratories. By 1927 when the ten-year review was published, Balfour could follow his lament, 'about applied science I am not so sure', by the reassurance 'But the efforts recorded in this volume show that a real advance has been made.'[78] The widely publicised report described the practical benefits to industry and evoked the old battle hymn of applied science. Too often, other countries' industrialists seemed to be ahead. Nevertheless, there was hope, and with more support from industry, it could become a major factor in the recovery of prosperity.[79]

An example of such a centre was the Cotton Industry Research Laboratory in a great house in Didsbury, south Manchester (the Shirley Institute), opened in 1922. In the autumn of 1916, the Textile Institute, the trade body, met to discuss the formation of a research laboratory. This would support the industry, which, after agriculture, was the largest in the country. The association was founded three years later, immediately after the million-pound fund was available, and soon reached a budget of £40,000 per year.[80] Typically for the sector, this sum more than doubled by 1935/36, and the Shirley Institute employed 270 staff.[81] The chairman reminded the annual meeting in 1925 that the ambition was 'To try to understand the chemical and physical changes produced during the manufacture and so to establish, gradually, a broad roadway along which future advances may be made'. However, he warned that experience had quickly challenged such rhetorical flourishes: 'We are finding that this road making business is a slow and laborious process.'[82] The encounter with such difficulties was not for lack of scientific talent. The director of the Shirley Institute, A. W. Crossley, was a former professor of chemistry at King's College London and had established the gas warfare research laboratory at Porton Down during

[78] DSIR 1927, Arthur Balfour, 'Preface', iii.   [79] DSIR 1927, 43.
[80] The initial meeting was reported in 'Textiles and Science: Important Discussion', *The Manchester Guardian*, 14 October 1916, and 'Problems in the Textile Industry. Scientific Research. Co-operation of Manufacturers and Universities', *The Yorkshire Post*, 14 October 1916, British Newspaper Archive/British Library Board. See also British Cotton Industry Research Association, Report of the Fifth Annual General Meeting, 30 September 1924, 6, Papers of the Cotton Industry Research Association, Manchester Central Library, Manchester, Archives and Local History, GB 127.M801. This archive holds the other annual reports cited below. On the institute, see Black 1949.
[81] DSIR 1937, 'British Cotton Industry Research Association', 116–18, on 118.
[82] Kenneth Lee, 'Chairman's Report', British Cotton Industry Research Association, 'Report of the Sixth Annual General Meeting', 15 September 1925, 1, Annual Report, quoted with kind permission of BTTG/Manchester Central Library.

'the war'. At first, there were four departments – physics, chemistry, colloid chemistry and physics, and botany – each headed by the holder of a DSc degree and staffed by half a dozen graduates.[83] Indeed, the director had to address criticism that the establishment had drawn 'a heavy veil of technical language over our publications'.[84] It was more dedicated to the term 'fundamental science' than any other such body.

Faced with the lack of relevant studies in the past or currently, by academe, the 1926 DSIR survey accepted that the Cotton Industry Research Association had been 'forced to devote much of its time and effort to purely scientific research'.[85] The Advisory Council supported this response and pointed to investigations on preventing mildew and the quality of cotton as an electrical insulator as examples of the Institute's more immediate industrial results. Shortly after, in 1933, the association reported it was bridging the gap with industry and had conducted 250 special investigations for members in the past year.[86] It also ran active subcommittees dealing with bleaching and finishing, spinning and sizing. The Research Association needed to reassure its industrial members that it was responding to commercial needs.

If the associations were criticised by industry for their lack of practical relevance, they were also subject to assault by the scientific community. In 1923, a critical article in *The Times* suggesting that the DSIR planned to terminate the scheme infuriated Kenneth Lee, Cotton Industry Research Association chairman.[87] He learned the author of the piece was Chalmers Mitchell, the Secretary of the Zoological Society of London, whose significance within science he discussed with the head of another association.[88] However, the challenge was headed off, and the scheme survived.

Although many of these research associations had low budgets and limited scientific impact, they did give science and engineering innovation a local presence. They were visible to a degree disproportionately

[83] 'The British Cotton Industry Research Association' (1923): file 1/1/2, Box 1, Papers of the Cotton Industry Research Association.
[84] British Cotton Industry Research Association, 'Report of the Fifth Annual General Meeting', 25 September 1923, 8, Annual Report, quoted with kind permission of BTTG/Manchester Central Library.
[85] [DSIR] 1927, 'Report of the Advisory Council: Research for Particular Industries', 30–43, see 41.
[86] DSIR. 1933, 'The British Cotton Industry Research Association', 103–4, on 104.
[87] 'The Progress of Science. Industrial Research Grants', *The Times*, 9 October 1923. See Lee 1923.
[88] R. S. Hutton to A. W. Crossley, 3 November 1923, Box 1, Papers of the Cotton Industry Research Association, GB 127.M801.

greater than in-house corporate research laboratories. News reports about the research associations varied strongly by region. In Sheffield, news about the Cutlery Trades Research Association was closely associated with the university's department of applied science. *The Falkirk Herald*, whose locality housed a British Cast Iron Research Association laboratory, specially reported its news. Such coverage reflected writing on local events, the celebration of recent developments, and advertising for jobs. It showed how DSIR had evolved from its World War One origin in the improvement of national efficiency to having a place in local life.

By the mid-1930s, success seemed within touching distance. As a whole, the country had experienced the worst of the Depression much less intensely and more briefly than the United States, and some regions had recovered rapidly.[89] Industrial research was booming. Research associations were particularly associated with industries comprising many smaller companies. In the newer cartelised branches of manufacturing, including chemicals, such dominant corporations as Imperial Chemical Industries (ICI) conducted large amounts of research without DSIR intervention.[90]

The upbeat report of the DSIR for the year 1935/36 summarised recent achievements, pointing to the 'historian of the future' who would find the previous five years marking a turning point in the country's industrial outlook.[91] The report's implication that British industry was exploiting applied science was drawn out by *The Times*.[92] The summary in *The Yorkshire Post* focused more upon the benefit of research outcomes to individual readers and local manufacturing.[93] Whether or not the research associations were an effective method of promoting innovation, so far as these newspapers were concerned, DSIR had promoted applied science, British industry had engaged, and life had improved. As shown by newspaper and published reports of the DSIR and NPL, on the one hand, and the minutes of their advisory committee meetings, on the other, scepticism was to be heard generally in private.

---

[89] Whereas the GDP of the United States fell 27 per cent between 1929 and 1933, overall, in Britain, GDP fell 6 per cent between 1929 and 1931 and then rose. See Crafts and Fearon 2013, chapter 1.

[90] For British industrial research before World War Two, see Edgerton and Horrocks 1994.

[91] DSIR 1937, 'Report of the Advisory Council', 11–21, on 15.

[92] 'Science and Industry', *The Times*, 8 February 1937.

[93] 'Improving Standard of Living. Scientific Research into Industry. Unshrinkable Wool', *The Yorkshire Post*, 8 February 1937, British Newspaper Archive/British Library Board.

The Research Boards and Committees

The first annual report of the stand-alone Advisory Council that preceded the formation of the DSIR itself remarked that 'research in pure science should be as much our care as research in applied science'.[94] In the annual statement of 1930, announcing that Ernest Rutherford was taking the Advisory Council's chairmanship, the department emphasised how its grant support for student and assistant work fulfilled that mandate.[95] These grants, however, came to constitute only 5 per cent of the department's budget.

The balance, the remaining 95 per cent, was invested in supporting its research institutes (above all, the NPL) and work commissioned by about twenty coordinating committees and research boards addressed to 'national purposes'. In 1926 the Committee of the Privy Council, which was the responsible body for DSIR, perhaps concerned about any politically dangerous 'misreading' of the Advisory Council's ten-year report and the range of activities of industrial associations that followed immediately, emphasised the correct reading. It also explained the breadth of work of the boards working in the national interest:

The Advisory Council point out that in the field of applied research the duty of the Department is rather to stimulate than to replace private enterprise, save in certain directions where the needs of Government itself or the direct interests of large sections of the community justify the expenditure of public funds on a large scale over long periods for the solution of industrial problems. Outstanding examples are the study of our fuel resources and of the preservation and transport of food-stuffs.[96]

One can get a strong feeling of the meaning of the expectation of practical benefits the funding of science might bring by reading through the lists of investigations on the following: fuel, food investigation, forest products, building research, chemistry, engineering and physics, Geological Survey, illumination, the Severn barrage, stresses in railroad bridges, adhesives, dental investigations, British Museum laboratory, fabrics, current meters, gas cylinders, minor metals, springs, and behaviour of materials at high temperatures. It seems the category of national interest expressed engagement with a host of industrial and governmental concerns.

Three boards with the innocuous names of Chemistry, Physics and Engineering were dedicated to relations between military and civilian

---

[94] [DSIR] 1916, 'Report of the Advisory Council: Contact with Industry', 14–16, on 15.
[95] DSIR 1931, 'Report of the Advisory Council: Introductory Remarks', 11–14, on 13.
[96] [DSIR] 1927. 'Report of Committee of Council', 1–12, on 2.

problems. For example, one of the Physics Board's earliest projects was supporting work into the new Bakelite plastic required by the Air Ministry. DSIR noted that the NPL was working closely with the Electrical & Allied Trades Association. However, when it came to elasticity work at the Air Ministry, which involved fatigue in ductile materials, a general theory of rupture, and a new form of altimeter, there were questions of who should pay, where it should be done, and, ultimately, the classification of the research:

The representative of the Air Ministry present at the meeting maintained the view that the work of the Elasticity Section was pre-eminently of general scientific interest and was of the nature of pure rather than applied research, and he stated that the Director of Research therefore considered that the Royal Aircraft Establishment, Farnborough was not a suitable place to carry out the investigations. The view of the Board, however, after discussion and perusal of a report on the work of the Section submitted by the Air Ministry was that, although the work was of general scientific interest, its applications were primarily of importance to aeronautical engineering, they were of opinion that the policy of separating such work from the Royal Aircraft Establishment was to be deprecated, and that it was clearly of advantage for the investigators to be in close touch with the practical developments of aeronautics.[97]

Ultimately, the Physics Board reimbursed Farnborough for what the annual reports of 1923 and 1924 identified as applied research on elasticity.[98] The Physics Board also supported work on X-rays. In 1915 the father and son team of William Henry (W. H.) and William Lawrence (W. L.) Bragg had won the Nobel Prize for Physics for their work exploring crystals. More immediately and on a much larger scale, the Woolwich arsenal had explored the potential for studying flaws in castings and the nature of enemy shells. Now, these teams came together with Major Kaye of the National Physical Laboratory (then experienced in using X-rays to detect faults in wooden laminates used in aircraft, and famous for his tables of physical constants assembled with T. H. Laby from 1911). They wrote a memorandum on the need for X-ray research, including the need for national development of X-ray tubes and the work of the Military Research Establishment at Woolwich. The first of its benefits would be examining the interior of otherwise opaque structures, such as fuses or aircraft components, and flaws in metalwork. Only second did the document discuss the use in examining crystal structures that, if perfected, could illuminate the 'characteristics and conditions of

---

[97] Ll. S. Lloyd to the Secretary, Air Ministry, 28 December 1922, in file 'P.R.B. Elasticity Research, General Papers', DSIR 36/4004, TNA.

[98] [DSIR] 1923, 'Physics Co-ordinating Research Board', 70–71, on 71; DSIR 1924, 'Physics Co-ordinating Research Board', 64–66, on 65.

all solid materials' and, third, medical applications.[99] A year later, W. H. Bragg wrote a supplementary memorandum on using X-rays for studying crystals.[100] Among recent results, he highlighted the crystalline structure of cordite, a military explosive. This bid resulted in a grant of £2,000 from the department for the following year. In the longer term, it made possible the success of a programme of X-ray crystallography, which, a generation later, with his son and colleague William Lawrence Bragg among the leaders, would lead to the decoding of structures of proteins and DNA.

The combination of practical interests and the need for fundamental understanding also provided a precedent for Peter Kapitza at Cambridge's Cavendish Laboratory. He was working on a battery able to be discharged very fast to produce intense magnetic fields.[101] Initially, the Physics Board paid Kapitza and his assistant's salaries and an almost equal sum for equipment. It also received a special report on the companies interested in exploiting his work, the patent situation, and government-reserved rights.[102] DSIR then transferred responsibility for supervising Rutherford and Bragg's projects to the central Advisory Council, which spent about £8,000 on the work of its two elite members in 1925/26.[103] This subvention amounted to more than one-fifth of the entire sum expended on individual grants to 'students and workers'. Some of those grants served to fund research in universities, while DSIR's own laboratories conducted further research. In total, the military-connected Physics, Chemistry, Engineering, and Radio Boards spent the budget of a large industrial laboratory, about £60,000 in 1925/26.

A sensitivity to the categories of 'pure', 'fundamental', and 'applied' was evident across the boards' reports, particularly in the early 1920s. Thus in 1923/24, the Food Investigation Board distinguished between

---

[99] See W. H. Bragg, 'Memorandum', November 1920, Physics Research Board, DSIR; 'Memorandum on the Development of X-Ray Work', considered at meeting of the Physics Research Board, March 1921; also see W. H. Bragg, G. W. Kaye, and V. E. Pullin, 'Report to the Physics Co-ordinating Research Board by the Sub-Committee on X-Ray Research', also considered March 1921, all in DSIR 36/3979, TNA.

[100] W. H. Bragg, 'Memorandum by Sir Wm. Bragg on the Analysis of Crystals by X-Rays', 12 July 1922, DSIR 36/3979, TNA.

[101] 'Draft Report of the Board to the Advisory Council on Researches Undertaken by Dr. Kapitza under Sir Ernest Rutherford's Supervision', DSIR 36/3974, TNA.

[102] 'Statement Prepared for the Information of the Board Regarding the Steps Taken for the Exploitation of Dr. Kapitza's Inventions Arising from His Investigations on the Production of Intense Magnetic Fields', DSIR 36/3974. Two companies, Metropolitan Vickers and Messrs Watson and Son, were interested in taking licences on this work, which DSIR was at pains to point out was Crown property.

[103] [DSIR] 1927, Appendix IV – Finance, p. 123.

'fundamental work' and 'applied investigations' it funded.[104] For example, it described 'continuation of work upon the use of ice containing an antiseptic' as 'partly fundamental and partly applied'. Applied investigations included 'combined gas storage and refrigeration trials', the diseases of apples in storage, and 'a study of the preservative action of carbon monoxide and carbon dioxide'.[105] The gas storage work led to new means of transporting both meat and fruit from Australia. The research on the preservation of apples was so successful that DSIR Secretary Henry Tizard could write humorously to Hardy, Superintendent of the Food Investigation Board, 'Annually we serve up brown-hearted apples as dessert to assist the digestion of our Parliamentary Estimates.'[106] Sally Horrocks has emphasised how the Board cited its benefits by highlighting the utility of its research, even though it did little to overcome the problems of implementation.[107] Reflecting on the preference of its director for fundamental over applied research, Eric Hutchinson has pointed out the scientifically unexplored territory in which this biochemically oriented group operated.[108]

The Adhesives Research Committee explained its funding of research, principally driven by the problems of wooden aeroplanes. Towards the war's end, so many planes were being built and the casein used in such short supply that manufacturers feared a national glue shortage. This threat required research towards a more rational use of the scarce resource, a search for new raw materials, and a deeper understanding of adhesion.[109] Research at the Royal Aircraft Establishment was focused 'on the methods of testing glued joints', an explicitly practical issue.[110] But a project also funded under the same budget, supervised by Professor Schryver of Imperial College, a leading expert on plant proteins, was described as covering the costs of 'pure research on the chemistry of glues and gelatin' and led to seven papers on amino acid residues in gelatin.[111] The annual report described research on the nature of adhesion at Bristol University as 'fundamental'. It explained

---

[104] On the distinctions made by the Food Investigation Board, see Horrocks 1993, chapter 6. See also Hutchinson 1972.
[105] Advisory Council for Scientific and Industrial Research, Minutes of the Meeting of Council, held on Wednesday 7th November 1923, Appendix, 8–10, DSIR 1/4, TNA.
[106] Tizard to Hardy, 21 January 1926, DSIR 6/2, TNA. For the context, see Hutchinson 1972, 45–47.
[107] Horrocks 1993, 246–47.   [108] Hutchinson 1972, 50.
[109] DSIR. 1922, 'Introduction', 2–5.
[110] Advisory Council for Scientific and Industrial Research, Minutes of the Meeting of Council, held on Wednesday 7th November 1923, Appendix, 23, DSIR 1/3, TNA.
[111] Advisory Council for Scientific and Industrial Research, Minutes of the Meeting of Council, held on Wednesday 7th November 1923, Appendix, 23, DSIR 1/3, TNA.

the range of descriptions by the dual purposes of the Committee, 'not only to fill a serious want in industrial science by founding a sound technology of glue manufacture, but also to elucidate the fundamental nature of adhesion'.[112] This work led to a temporary exhibition on adhesives and their application at the newly rebuilt Science Museum in South Kensington.[113]

DSIR took over the NPL, always the most significant call on its non-industrial funds.[114] In 1936/37, its allocation to the Laboratory of £244,081 was almost ten times greater than the department's expend-iture on research grants to individual 'workers and students'.[115] During the late 1920s, NPL's roughly 150 scientists constituted about half of DSIR's directly employed scientifically qualified workforce and one-fifth of all such professionals employed by the government.[116] This central-isation required a delicate balancing act between its broader ambitions, the practical requirements of standards-setting, and the scientific com-munity's expectations. A study in 1930 showed that NPL devoted about two-thirds of its staff expenditure to research and the rest to testing.[117] During the early 1930s, NPL was testing almost half a million clinical thermometers annually. Beyond thermometers and taximeters, it con-ducted a further 30,000 tests for a thousand companies in a single year.

The utility and balance of research were praised by several general board members, such as the physicist and electrical engineer Alexander Russell and the chemist and head of research at a major railway company, Harold Hartley.[118] Similarly, the senior physicist Arthur Eddington was complimentary: according to the minutes of the NPL board meeting in 1935 he found the Laboratory doing an excellent job both in 'purifying applied science' and in 'applying pure science'.[119] But, other members, such as Lord Rutherford, worried that the balance was in

---

[112] [DSIR] 1921, 'Adhesives Research Committee', 56–57, on 56.
[113] 'News and Views' 1936.   [114] Hutchinson 1969.
[115] DSIR 1937a, Appendix V – Summary of the Expenditure of the Department during the year ended 31st March, 1937, 168.
[116] The DSIR employed 316 'scientific staff'. H.M. Treasury 1930, 40. For NPL employees, see Mortimer and Ellis 1980, 5. Keith 1982 is a most useful treatment of NPL, and of DSIR in general, in this period.
[117] 'N.P.L. Staff for 1930–31. Proportion of Staff Engaged on Routine work', in Committee of Civil Research, 'Research Co-ordination Sub-Committee: Departmental Comments and Correspondence on First Report', DEFE 15/44, TNA.
[118] Alexander Russell, 'Report of Proceedings at the Annual Meeting of the General Board', NPL, 28 March 1924, 16, DSIR 10/2, TNA. Remarks of Harold Hartley, 'Report of Proceedings at the Annual Meeting of the General Board', NPL, 10 May 1935, 13, DSIR 10/3 TNA.
[119] Arthur Eddington, 'Report of Proceedings at the Annual Meeting of the General Board,' NPL, 10 May 1935, 6–7.

constant danger.[120] From outside, the young J. D. Bernal complained of an over-emphasis on the routine.[121] Certainly, whether the balance was indeed admirable was a constant topic for the DSIR Advisory Council. From within the institution, Rosenhain, the head of metallurgy, had identified its research as 'applied science' in 1916. At the end of his NPL career, in a 1930 article on aircraft, he reflected on a triangle of needs: intercommunications between scientists and their counterparts 'on the industrial side' aware of practical problems, the 'actual applied research', and investigations free from 'consideration of immediate practical use'.[122] The result was indeed widely appreciated: in 1924, the *Yorkshire Post* reflected that 'there is scarcely one important industry in the country that has not called on the laboratory for assistance'.[123] Glazebrook himself saw his institution as a great central industrial resource.[124]

At the same time, we need to recognise the anxieties about effectively implementing this model. British industry did not seem to be drawing sufficiently on the work of the NPL. Blame for this apparent paradox was batted to and fro between scientists and industry. NPL board member and Oxford University chemist Alfred Egerton placed responsibility on the 'jingle jangle of modern civilisation', amidst which the opportunities were overlooked, and research ignored.[125] A Board of Trade report criticised the lack of a 'receiving mechanism' in companies researchers had endeavoured to inform.[126] For its part, industry complained that, too often, the investigation by NPL stopped short of being easily implementable in practice. Metrovick research director Arthur Fleming, reflecting on the paucity of commercial benefits flowing from DSIR-funded research, suggested that the government should support the building of pilot plants. He called for a new organisation to fund commercial development spanning the gap between DSIR research and industry.[127] In the years after 1945, this anxiety about the mechanism

---

[120] Lord Rutherford, 'Report of the Proceedings at the Annual Meeting of the General Board', NPL, 30 March 1926, 10–11, DSIR 10/3.
[121] Bernal 1939, 43. See also Moseley 1980.   [122] Rosenhain 1930, 647.
[123] 'National Physical Laboratory: Novel Processes Described', *The Yorkshire Post*, 25 June 1924.
[124] Glazebrook 1918.
[125] Alfred Egerton, 'Report of the Proceedings at the Meeting of the General Board', NPL, 26 April 1932, 10, DSIR 10/3.
[126] Board of Trade 1929, 217.
[127] Fleming 1933. See also Fleming 1932 and Moseley 1976, 243. Fleming's campaign for what he saw approximately as a British counterpart to America's Mellon Institute was waged intensively and seems nearly to have been successful. The proposal was rejected after an enquiry whose report was circulated to Cabinet. See 'Report of Committee on New Industrial Development', Economic Advisory Council, CAB 24/232/29, TNA.

by which applied science could benefit the economy, and indeed the adequacy of the category itself, would unfold much further. Among DSIR's earliest actions was the planning of a Fuel Research Station in Greenwich, East London. Opened in 1920, this was the first of the centres explicitly established under the department's auspices. It was also the recipient of substantial amounts of money. Keith has pointed out that in the early 1920s, according to some measures, its net income from the government exceeded that of the National Physical Laboratory.[128] Generically, DSIR described fuel research as 'applied research'.[129] Founded during World War One, the Station's mandates were to evaluate Britain's coal reserves and to find ways of processing coal to minimise smoke and convert it into more valuable products such as otherwise imported petrol. Later it added the development of alternative fuels through fermentation.[130]

Smoke was a significant challenge to life in Britain's cities; it caused chronic bronchitis (labelled overseas as the 'English disease'), dirt, and the choking smoke-enhanced fogs known as smogs.[131] Heating coal at high temperature in the absence of air (carbonisation) yielded four products.[132] These were coal gas for heating and lighting; ammoniacal liquor, which was useful for making fertilisers; an oily mixture referred to as coal tar, which could be distilled to produce a motor fuel, then called 'benzole'; and coke, which was itself a smokeless fuel. By reducing the reaction temperature, a manufacturer could optimise smokeless fuel production, and produce 'creosote'. This oil, widely used as a wood preservative, contained a different mix of hydrocarbons to petroleum but could also be converted to petrol. From 1923 until World War Two's outbreak in 1939, the laboratory would also put considerable effort into hydrogenating coal to oil. This huge enterprise will be explored at greater length below, because it became the signature applied science enterprise of the interwar years.

The boundary between commercial and public knowledge was a constant challenge for the research associations, the research boards, and the National Physical Laboratory. For example, in the 1924/25 report on the Fuel Research Board, the Advisory Council recommended that it was in the national interest to publish the results of tests on promising full-scale

---

[128] Keith 1982, 125.
[129] DSIR 1932. 'Report of the Advisory Council: Fuel Research', 17–19, on 18.
[130] On the early work of the Fuel Research Board (the parent body of the Fuel Research Station), see [DSIR] 1917a, 'Report of the Advisory Council: Fuel Research Board', 17–19, and the Fuel Research Board's biennial reports beginning with [DSIR] 1917b. The hydrogenation work has been described by Stranges 1985, while the overall work of the Board was reviewed by Keith 1982, 117–26.
[131] See, for instance, Mosley 2016; Thorsheim 2017.
[132] See Gordon 1934 for a summary of the products of both high- and low-temperature coal carbonisation and their uses.

low-temperature carbonisation plants requested by private owners and not to charge for the tests. On the other hand, in 1929, the Council, contemplating building a new tank at the National Physical Laboratory, pointed out that it did not charge overheads for individual tests and proposed to charge the shipping industry half the cost of the tank.[133] Applied science was occupying a blurred space between public and private knowledge.

During World War One, the exigencies of conflict had led to unprecedented government intervention in the economy and industrial regulation. In general, that system was quickly dismantled after peace returned, in the name of 'normalisation'.[134] However, the work of DSIR, its research associations, and other moves of successive governments to underpin research stood out as major exceptions. Indeed, during the 1920s, we see debate about support for industry through science in Britain, with shifting positions through the decade. In practice, the industrial associations found they had to conduct considerable work directed toward the complexity of individual contexts. By 1926 the rhetorical importance of 'fundamental research' had waned somewhat, and the DSIR Advisory Council described the department's work for industry and the national interest as 'applied scientific research'. Here was a conception of applied science as the application of pure science. As important as its usage in internal discussions, this was the interpretation in the forum of public debate.

Haldane's Principle

Two philosopher-politicians reflected upon and affected the development of the new bodies during the immediate post-war recovery and the 1920s: Arthur Balfour and R. B. Haldane. Individually, their impact upon applied science was considerable, and since they had different political affiliations, one or the other was in authority through a succession of elections. Thus, Haldane served as Lord Chancellor in the short-lived Labour government of 1924. Between 1919 and 1922 and again from 1925 to 1929, Balfour was 'Lord President of the Privy Council', gently overseeing many civilian research organisations, including DSIR. Responsibility for these was consciously distanced from the hierarchies of the client departments, thanks to Haldane.[135] He had argued for such

---

[133] DSIR 1925, 'Report of the Advisory Council: Fuel Research', 20–21, on 21; DSIR 1929, 'Report of the Advisory Council: William Froude National Tank', 28–29, on 29.
[134] Tawney 1943.
[135] Although Haldane was out of power by the time the DSIR was formed, Addison had felt strongly supported by him during the writing of his proposal, for instance, at lunch on 30 March 1915; see Addison 1939, 71. His subsequent formulation of the MRC and Agricultural Research Council demonstrates Addison's own commitment to the Haldane Principle. The Machinery of Government report uses DSIR as the prototype of research organisations whose oversight would require the new Minister.

independence when he chaired the Committee on the Machinery of Government in 1918/19. The 'principle' he articulated then would be invoked and silently modified over the next fifty years.

Haldane recommended that there should be a special research bureau in addition to the research closely connected with the government's day-to-day activities (such as managing disease outbreaks in particular localities – a special interest of the Ministry of Health). In a personally written working paper, a copy of which he shared with Prime Minister David Lloyd George, Haldane made a pitch for someone much resembling himself at the helm. He argued that the Lord Privy Seal, currently in charge of DSIR, was too busy, urging that instead:

The Head of this Department ought to be essentially a trained thinker. It might well be possible to select for the work an individual sufficiently detached from Party politics to go on in one or two successive Administrations and so to bring about the continuity of work and Knowledge which has proved itself in this war of vital importance in the Committee of Imperial Defence.[136]

Haldane emphasised that 'research and general inquiry, if kept separate from administration, can be centralised without seriously interfering with devolution of local administrative services'. There was no implication here that this meant focusing on freedom from government control or on pure science. Indeed, Haldane gave the example of the Fuel Research Board, addressing 'the systematic investigation of the highly practical scientific problems which had to be solved before the maximum of bye-products [sic] can be obtained from coal'. In addition to the DSIR, he envisaged support of other areas, including medical and, in future, perhaps aspects of agricultural research.[137] Introducing and underlying this approach was what Haldane emphatically called, in his working paper, his first 'principle'. This was 'the necessity of systematic reflection

[136] The working paper is to be found in the Lord Passfield papers, 13/3, Paper 16, 'Machinery of Government Committee, Memorandum by Lord Haldane', 11th January 1918, ff. 781–87, see 4, London School of Economics Library, Archives and Special Collections; and the separately typed copy sent to Lloyd George, Papers of Lloyd George, LG/F/74/12, House of Lords Library. I am grateful for the kind permission of the LSE and of Mr Richard Haldane to publish the quotations from it. See also Hume 1958. The final diluted report was published as Ministry of Reconstruction 1918. The 'principle' expressed here by Haldane was rather different from the entirely 'hands-off' rendition cited around 1960. See Edgerton 2009.

[137] 'Clearly the organisation might develop so as to include research in medicine, the diseases of animals, fishes, and plants, in agriculture, botany, and forestry, in problems connected with surveys (including questions, of geodesy, geology, meteorology, and oceanography), in the preparation and use of statistics, and possibly in matters of education and of historical and political science.' Ministry of Reconstruction 1918, 34, para 70.

and inquiry as preliminaries of action, to an extent which shall be much greater than has been the case in the past'.

Haldane concluded that government-funded research should fall under two heads, one directed tactically by operating departments, the other controlled centrally, on a longer horizon. Accordingly, applied science was not just the servant of policy but also its antecedent. There was nothing unusual about this. In America's GE, the global model for industrial research at the time, Willis Whitney took considerable trouble to distance his research department from the company's day-to-day affairs. The Philips company's 'Nat. Lab', established in 1914, also self-consciously balanced the demands of existing product divisions and the corporate need for new products.[138]

Fortunately for Haldane, during the immediate aftermath of the war, Addison, his protégé, was the President of the Local Government Board responsible for health. In an otherwise unlikely self-denying act, Addison suggested that the newly named Medical Research Council (MRC) should be independent of the new Ministry of Health. Scientific bodies should not be 'in any way under the direct control of Ministers responsible for the administration of health matters'.[139] A decade later, when Addison was appointed Minister of Agriculture, he pushed through the 1931 formation of a new Agricultural Research Council (ARC), which also had an arm's-length relationship to its Ministry.[140] Both councils were strong proponents of a model of unified science and did not wish to see their operations restricted by bureaucratic categories. Yet when Walter Morley Fletcher needed to justify the MRC affiliation with the Lord Privy Seal, rather than the Minister of Health, he argued that its projects 'can be conveniently associated with sister research organisations in the other fields of applied science'.[141] Haldane's principle of corralling longer-term research, separately from executive departments, had created the need to emphasise inclusion within that envelope.

Governments implemented Haldane's approach to the management of research in another way too. When Arthur Balfour became Lord Privy Seal in the new Conservative government of 1925, he developed a plan, formulated by Haldane under the previous administration, for a body modelled on the Committee of Imperial Defence. The new Committee

---

[138] De Vries 2005.
[139] Local Government Board 1919, para. 11, p. 3, Wellcome Collection. See also Thomson 1973.
[140] DeJager 1993.
[141] Walter Morley Fletcher, 'Constitutional Position of the Medical Research Council', Committee of Civil Research, Research Co-ordination Sub-Committee, C.R. (R.) 26, n.d. c. 1926, TNA.

of Civil Research (CCR) sought better development of the civilian economy by improving the use of research results and improved departmental
coordination.[142] A 1925 report on East Africa's economy prioritised
scientific research and the applications of science to support colonial
economic development.[143] Former German support for the Amani
Research Institute in the newly mandated territory of Tanganyika (now
part of Tanzania) had, moreover, set a standard for the British to match.
Lasting five years, the CCR provided the forum for discussing the linkages between fundamental and applied research to industry, at home and
across the empire. Its activities marked a brief attempt formally to
manage British and imperial science as a single whole.

The empire's need for research coordination had been proclaimed as
part of its development, even before World War One. Though initially
established as a private organisation, the Imperial Institute had soon been
taken over by the government, first by the Board of Trade, and then
(1906) by the Colonial Office. As in so many other domains, the war
transformed attitudes towards the empire, whose benefits to its members
needed increasingly to be evident.[144] Certainly, the imperial context was
not frequently mentioned in newspaper articles using the phrase 'applied
science', beyond the statements that it could facilitate the exploitation of
the empire's resources and that staff trained in applied science would
serve development well.[145] However, from the mid-1920s, 'colonial
science', defined by Worboys, its historian, as 'British scientific effort
for the Colonial Empire' and seen principally as applied science, would
begin to thrive.[146]

---

[142] For the Committee of Civil Research, see MacLeod and Andrews 1969. The authors
suggest that the Committee represented the fulfilment of Haldane's vision of central
research department; see p. 704. Andrews 1975, 211, gives evidence from the Beatrice
Webb diaries of Haldane's belief that his suggestion was the realisation of the
Machinery of Government investigation. For the purpose of the department in
Fletcher's eyes see Macleod and Andrews 1969, 689.

[143] Secretary of State for the Colonies, 1925, 80–93.

[144] On the early development of the Imperial Institute, see Worboys 1990. More generally
on this period, Hodge 2007.

[145] For the invocation of the wonders of empire, see, for instance, John Buchan, 'Palestine
Immigration: Mr Buchan urges Increase of Quota', *The Scotsman*, 23 April 1932,
British Newspaper Archive/British Library Board; on the importance for the Empire,
see the report of the Imperial Conference, Subcommittee on Research, chaired by
Balfour, 'Imperial Research: Lord Balfour Commends Applied Science', *The
Yorkshire Post*, 23 November 1926, British Newspaper Archive/British Library Board.

[146] Worboys 1979, 400. I am grateful to Michael Worboys for permission to quote his
thesis. I am using the term 'colonial science' here as Worboys defined and characterised
it (p. 15). Note, however, the low base: in 1922 the annual grant for the Colonial
Research Committee was reduced to £1,000. See Malmsten 1977.

The eighteen subcommittees of the Committee of Civil Research studied an array of problems ranging from the supply of radium for medical purposes to research in the colonies. Amongst these, it was responsible for coordinating and accounting for the increasingly significant government investment in research across all fields within Britain. Though the Committee itself did not have access to research funds, it was complemented by a separate funding organisation, 'The Empire Marketing Board', established to promote empire trade without imposing tariffs.[147] During an albeit short life (1926–33), £1.8 million, about 65 per cent of the Board's funds, were devoted to grants for research, largely on applied biology. While from a strictly British perspective, agriculture was already a minor part of the economy, from an imperial point of view, agricultural products were very important and held great prospects as raw materials. Consequently, British chemists and biologists returned time and again to the possibility of making chemicals using imperial agricultural products – what their American counterparts were calling 'chemurgy', and they called 'biotechnology'.[148] Enthusiasts included the British chemists William Pope and Harold Hartley and biologist Lancelot Hogben. Hartley concluded a much-cited high-profile lecture in 1937 by conjuring up a dream of designing plant strains specified to produce exactly the products required by industry.[149]

Concerned with next steps as well as far horizons, and attempting to promote imperial development, the Empire Marketing Board, for instance, funded work on the parasites of insect pests. Worboys has suggested the Board envisaged its role as mirroring – for applied biology across the British empire – what DSIR was doing for the applied physical sciences in Britain.[150] Thus scholars have shown that such disciplines as tropical medicine, entomology, botany, mycology, ecology, tropical agriculture, and forestry were deeply affected. Although major imperial institutes were developed in Tanganyika (Amani) and Trinidad (Imperial College of Agriculture), at least at first, much of the Board's research funding went to such English bodies as Oxford's Imperial Forestry Institute and Cambridge's Low Temperature Research Station.[151]

---

[147] Worboys 1979, 256–94. See also Macleod and Andrews 1969; Self 1994; Atkins 2005.
[148] See Bud 1993, 46 and 76–78; Uekötter 2021.
[149] Hartley 1937, 172. For the similar ambitions of the geneticist Rowland Biffen, see Charnley 2013.
[150] Worboys 1979, 257. On applied biology, see Kraft 2004. This paper explores how the term 'applied biology' replaced 'economic biology'. The principal subjects discussed in *The Annals of Applied Biology* (founded as *Journal of Economic Biology* in 1905) between 1914 and 1938 are given in Marsh 1953.
[151] Anker 2001; Atkins 2005; Hodge 2011, 15.

It was, naturally, no trivial matter to transfer advice thousands of miles across the empire to specific complex natural and human ecologies. Despite the power vested in the imperial authorities and the authority of science, indigenous peoples, colonial officers, and 'experts' alike were often aware that they needed to draw on many other kinds of knowledge to make use of the imported findings.[152] In any case, applied science had been specifically conceived to complement local knowledge. Even within Britain, its acceptance depended on a recognised distance from local conditions and expertise.

In 1933, responding to worldwide protectionism, a thoroughgoing tariff system was finally imposed against non-imperial goods, and, accordingly, the status of science funding changed. The Committee of Civil Research had already gone, and now, the Empire Marketing Board closed. The potentially powerful combination of the Committee and the Board lasted but four years and lacked the coherence to make a broader impact.[153] Nonetheless, until a more comprehensive review in 1940, other agencies, such as the Colonial Development Fund (set up in 1929) and the Carnegie Trust, filled only a part of the niche left behind.[154] The Empire Marketing Board had proved to be a model, with an influence that stretched decades into the future, demonstrating innovative marketing techniques, sophisticated filmmaking, statistics collection, and presentation and support for applied science.[155]

Although the raising of imperial ambitions did not transform the connotations of the term 'applied science', the interwar funding of imperial science did reinforce the marriage with agriculture. Both the scientists and their colonial officer masters came to prefer the term 'fundamental research' to describe work that they considered less constrained than 'applied research'.[156] However, support for imperial science boosted both the funding and the profile of applied biology within Britain, and ensured that 'applied science' referred to more than physics and chemistry.

Coal Case Study

How did industrial institutions and expectations fit together? We have seen how DSIR and others equated industrial research with applied

---

[152] Hodge 2007; Tilley 2011. See also Chambers and Gillespie 2000.
[153] See Macleod and Andrews 1969.
[154] On the Colonial Development Fund, see Abbott 1970. Also see Drummond 1971 and Malmsten 1977. Scientific research received about £60,000 per year through this avenue between 1929 and 1940. See also Malmsten 1977.
[155] See 'Empire Marketing Board', www.colonialfilm.org.uk/production-company/empire-marketing-board (accessed December 2020).
[156] Clarke 2010.

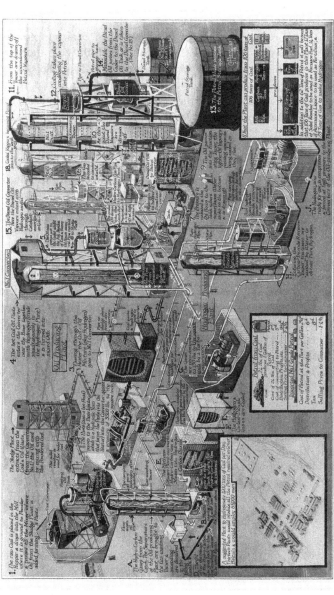

Figure 5.1 Applied science triumphant: Billingham coal oil plant. Generating petrol from coal using the hydrogenation process. Illustration by George Horace Davis in *The Illustrated London News*, 21 November 1931. Courtesy Illustrated London News/Mary Evans Picture Library.

science.[157] In the years after World War One, leading industrialists believed that research would be the source of disruptive change, of which they could be the controllers. It would provide safety-assuring answers to those concerned with the commercial and social challenges of innovations already transforming civilisation. The future of coal was among the greatest of such issues. Producing the national fuel, the coal industry still employed more than a million workers, concentrated in a few areas, and had a special meaning in the country. But, in a downward spiral since its peak in 1913, the industry found that export markets were disappearing, and new competitors were emerging. Oil was making remorseless inroads into the demand for the fuelling of ships and soon even trains. Coal use was increasingly efficient in industry and the home, and, between 1913 and 1933, consumption per head fell by almost a third.[158] The industry's decline entailed deep suffering, particularly in areas where the jobs of more than half the male workforce were in the mines, with endemic unemployment rates of more than 20 per cent. The fate of the industry was a national disgrace and a political challenge.[159] In 1929, when the DSIR Secretary, Henry Tizard, wanted to point to the opportunities for science to transform old industries, the first example he took was the use of coal.[160]

Typically, in 1925, the government turned to a Royal Commission to investigate the industry's structure and prospects.[161] Mining itself was an old-fashioned business and could be better mechanised; however, without new markets, such measures would lead to further unemployment. So, recommendations of new forms of utilisation and research were high up in the Commission's report, and, after all, coal offered the chemist an entire harvest of chemicals. The chemicals produced from the coal tar of high-temperature carbonisation were legion. The semi-coke produced by low-temperature carbonisation was a smokeless fuel that also yielded a satisfying and entertaining flickering flame in the family hearth, which was the destination of about one-quarter of Britain's coal consumption. Three somewhat different corporations that explored the potential of research in the coal industry illustrate how their activities defined both applied science and themselves. These were the Low Temperature Carbonisation Company, ICI, and Powell Duffryn.

---

[157] Horrocks has looked in detail at how major firms in the British food industry conducted research before World War Two; see Horrocks 1993. See also the tensions within a single company explored by Divall 2006.

[158] Political and Economic Planning 1936, 6.   [159] Supple 1989, 6; and Boyns 1987.

[160] Tizard 1929.   [161] Royal Commission on the Coal Industry 1925, vol. 1 *Report*.

By the 1920s, businesses were exploring over a hundred processes for the low-temperature carbonisation of coal.[162] One firm, the 'Low Temperature Carbonisation Company Ltd', drawing on pre-war work with a smokeless fuel branded as 'Coalite', showed particular signs of success. Though the enterprise ran into financial difficulties in 1926, it was revived by the energetic engineer and former airman Colonel Whiston Bristow.[163] He led a remarkable recovery by developing new assemblies of retorts and a portfolio of coal tar by-products, including creosote for wood-preserving, flotation agents for the mining industry, a powerful disinfectant, coal-petrol for cars and planes, and coal-diesel for trucks.[164] Early in the 1930s, the organisation started to sell its fuel to the Royal Air Force and supplied several squadrons. The octane rating of ninety was higher than the sixty-five then characteristic of the petroleum-sourced product. By 1937, a few hundred service stations for cars around the country had started to sell Low Temperature Carbonisation's petrol (refined by the Carless Capel company, which had previously introduced the word 'petrol') under the brand 'Carless-Coalene'.[165] Meanwhile, Coalite, the smokeless fuel, was being sold for government buildings under an initiative to encourage both new technologies of coal utilisation and smoke abatement. The company exhibited at the London exhibition associated with the 1928 Public Health Conference. Such niche markets as government buildings and the air force were just the low-competition, high-profile launch vehicles that, two generations later, the American business analyst Clayton Christensen came to see as crucial for the beginnings of disruptive technologies.[166] Emphasising the costs of research and development and the need for shareholders to show patience, Colonel Bristow's speech to the 1939 annual meeting

---

[162] See Slosson 1928. For the specifically UK processes, see Brownlie 1927.
[163] See the long biography of Bristow: Moore 1943. Also 'The History and Growth of the Solid Smokeless Fuel Market with Particular Emphasis on the Coalite Company's Developments in This Field', presented at Coal Industry Society: West Midlands Branch, 1 February 1972, in Box J106/A4, Papers of Doncaster Coalite Ltd, Collection D7350/UL, Derbyshire Record Office, Matlock.
[164] See 'Company Meetings: Low Temperature Carbonisation Ltd', *The Manchester Guardian*, 19 July 1939. The range of chemicals sold by the company is discussed in Travis 2013.
[165] 'Low Temperature Carbonisation Ltd', *Western Mail & South Wales News*, 22 December 1936, British Newspaper Archive/British Library Board; 'Company Meetings: Low Temperature Carbonisation Limited. Great Increase in "Coalite Sales"', *The Manchester Guardian*, 23 June 1937; W. A. Bristow, 'Smokeless Fuel from Coal: Oil Production at the Barugh Works', *Financial Times Fuel and Power Supplement*, 5 December 1932, 11–12.
[166] Christensen 1997.

exemplified pleas to investors in evaluating a start-up that would become familiar in the post-war era.[167] Initially, the development of 'Coalite' was a matter principally of engineering rather than chemistry, and certainly not applied science in the sense of DSIR. However, as the range of products widened during the 1930s, applied chemistry became more critical. In 1938 the company recruited an ICI chemist, Wilfrid Devonshire Spencer, to head a research laboratory established in London.[168] The next year the company opened its refinery, making a variety of oil products from the coal tar it was producing.[169] An alternative way to obtain liquid fuel from coal was to transform it chemically by adding hydrogen. In 1913, Friedrich Bergius at the University of Hanover developed a high-pressure process for converting the soft brown coal, known as lignite, although a plant began operating only after the war. Not long after, DSIR's Fuel Research Station launched work in Britain, building on German experience and exploring the new process's potential.[170] The initiative led to interwar Britain's most extensive and best-known industrial research project. Its rationale has been studied from the perspectives of the military and a few key research institutions, while its meaning to the coal industry has been neglected. However, it was on account of its promise to rescue this crumbling pillar of the British economy that the quintessential applied science project attracted interest from the press and civil society.

In 1927, the recently formed Imperial Chemical Industries (ICI), chaired by lawyer and politician Alfred Mond, also started research on synthetic oil. The company's historian has emphasised the combination of corporate and public interests behind the decision.[171] The conversion of coal to oil would solve ICI's practical problems of a surplus of hydrogen and over-dependence on a saturated fertiliser market, increase demand for coal, and, in the eyes of political supporters, assure fuel supplies for the navy, reduce unemployment, and ensure political stability. Mond himself died unexpectedly in 1930, but ICI pursued

---

[167] 'Company Meetings: Low Temperature Carbonisation Ltd', *The Manchester Guardian*, 19 July 1939.
[168] 'Low Temperature Carbonisation Limited: Colonel Whiston A. Bristow's Speech', *The Manchester Guardian*, 4 July 1938.
[169] 'Coal Oil Refinery: New Plant Opened at Bolsover by Secretary for Mines', *The Derbyshire Times*, 19 May 1939, British Newspaper Archive/British Library Board.
[170] This work has been thoroughly studied by Anthony Stranges. See Stranges 1985.
[171] Reader 1977, 234–37. Mond himself was very involved in trying to assure political stability through collaboration between business and unions. See McDonald and Gospel 1973. For political support for coal-oil, see the seventy-nine-page *Labour's Plan for Oil from Coal* (Labour Party 1938).

its research on bituminous coal conversion, leaping ahead of German developments.

The ICI coal-oil plant, producing more than 100,000 tons a year of petrol, opened in October 1935 to international publicity.[172] The *New York Times* published several articles about the new plant. It was the first commercial factory anywhere to hydrogenate bituminous coal into oil.[173] Typically, the press across the land translated the hugely complex engineering achievement into an achievement of applied science.[174] In Glasgow, the *Sunday Post* ran an article, 'Oil from Coal by a Chemist', explaining that the process involved 'higher chemistry and applied science'. The report attributed 'the whole secret of the process' to the use of the right amount of heat and the catalysts' constituents. The *Aberdeen Press and Journal* headlined their report of the Billingham plant's opening 'Fairy Tale of Science Comes True'.[175]

These developments were not unconnected. ICI soon found it more convenient to process the liquid creosote produced alongside 'Coalite' by Low Temperature Carbonisation Ltd, than raw coal (and they could shut down the plant quickly in the event of German bombing).[176] Therefore, the Coalite and the ICI processes could work interdependently: producing both advanced smokeless fuel for the hearth and high-octane petrol for the fighter. In the event, Billingham would be the only factory of its kind to be built in Britain.[177] At the time, though, it was seen by many as the forerunner of several more.

As the coal-oil plant was rising, unemployment was still punishingly high. Particularly intractable was the plight of the South Wales valleys.

---

[172] The *New York Times* published two articles about the opening at Billingham, one an editorial, the other a photoessay emphasising that this heroic achievement was sustained only by a substantial subsidy. See 'Gasoline from Coal', *New York Times*, 17 October 1935; Waldemar Kaempffert, 'This Week in Science: Gasoline Made from Coal', *New York Times*, 10 November 1935.

[173] Most German plants hydrogenated brown coal, which is chemically different from the bituminous coal generally exploited in England. The first German plant to convert bituminous coal at Gelsenhammer was opened a year after Billingham. For a good technical description, see Kirk 1998.

[174] See, for instance, 'Oil from Coal', *Hartlepool Northern Daily Mail*, 16 October 1935, British Newspaper Archive/British Library Board.

[175] 'Oil from Coal', *The Sunday Post*, 4 August 1935, British Newspaper Archive/British Library Board; 'Oil from Coal and Water: Fairy Tale of Science Comes True', *Aberdeen Press and Journal*, 16 October 1935, British Newspaper Archive/British Library Board.

[176] The particular suitability of the crude tar produced by low-temperature carbonisation for hydrogenation was widely discussed, for instance, in 'Fuel Research in Great Britain' 1936.

[177] During World War Two a plant using much of Billingham's design was built by ICI at Heysham on the west coast, but it started with a different raw material, imported gas-oil extracted from petroleum, and not coal-sourced creosote. See Edgerton 2011, 181–93.

Not only were they primarily dependent on coal, but much of their coal reserves were in the form of 'anthracite', burning at a high temperature and especially suitable for steam-raising. This traditional export market had collapsed due to competition from hydroelectricity, petroleum, and foreign suppliers. So, when the National Industrial Development Council of Wales and Monmouthshire was formed in 1934, it too intensively explored the potential of a combination of low-temperature carbonisation and hydrogenation. On further examination, though, this approach seemed uneconomic for the Welsh coalfields.[178]

Wales was also the home of Britain's largest coal company, Powell Duffryn. By the mid-1930s, this combine employed 37,600 workers, managed ninety-three pits, and conducted a distinctive research programme, albeit one which historians have hitherto overlooked. In mid-1932, meetings with ICI explored the option of developing a plant that would convert 200 tons of Welsh coal daily to oil. Deciding to go its own way, Powell Duffryn recruited as its research manager the Cambridge PhD manager of the ICI plant itself.[179] Dr Idris Jones would establish a new laboratory with, by 1936, four research chemists, five apprentices, a secretary, and a sales liaison manager.[180]

A blizzard of research reports documented advances on two fronts. One was that of urgent and practical questions. The laboratory developed a process to reduce waste, binding small, unsaleable pieces of smokeless anthracite with bituminous coal.[181] Applied science was also trumpeted in an ambitious return to the conversion of coal to oil. This time, the main emphasis of Powell Duffryn was on the Fischer-Tropsch synthetic process for oil production, also developed in Germany.[182] Steam was passed over hot coke to produce a mixture of carbon monoxide and hydrogen, which were then catalytically combined to make oil. It seemed more appropriate for South Wales anthracite than

---

[178] 'Report and Findings of Special Coal Research Committee Set Up under the Auspices of the National Industrial Development Council of Wales and Monmouthshire by the Commission for the Special Areas (England and Wales) October 1935/December 1939', Papers of National Industrial Development Council of Wales and Monmouthshire, DIDC/7, Glamorgan Archives. See Counsel 1987.

[179] Sir Ben Bowen Thomas, 'Obituary. Dr Idris Jones. Research at the Coal Board', *The Times*, 10 July 1971.

[180] Idris Jones, 'Powell Duffryn Associated Collieries Ltd. Research Department, Report for the Three Months Ended 31st March 1936', Coal 14/622, TNA.

[181] Mark Williams, 'Phurnacite Plant, Abercwmboi', www.aberdareonline.co.uk/history/phurnacite-plant-abercwmboi (accessed August 2023). This work resulted in the products Phurnod and Phurnacite.

[182] See Powell Duffryn Associated Collieries Limited, 'Documents Submitted to the Committee on Imperial Defence, Subcommittee on Oil from Coal', Coal 14/982, TNA.

the hydrogenation approach and was a very sophisticated process, but senior management had to consider the practical and the scientific together. Stephenson Kent, the director most associated with research, justified the investigation to his management committee in terms of otherwise wasted coal that might now become usable.[183]

For some years, the company lobbied for government aid for a South Wales coal oil plant. Its endeavours had a high public profile and even laid the basis for a Hammond Innes thriller about coal oil, *All Roads Lead to Friday*.[184] This fictional story, published in 1939, involving victory over industrial espionage and ending with the triumphant opening of a plant, may have offered a rosier future for coal-oil in Britain than Powell Duffryn could effect. However, it dealt realistically with the uncertainty over what was known about a process and by whom. Ultimately, rather than new coal processing methods, World War Two temporarily solved the Welsh valleys' employment problems.

The growing fears of future conflict in the 1930s raised the spectre of sunken tankers, dead sailors, and a petrol shortage. So, in 1938, the government's Committee of Imperial Defence convened a committee under Lord Falmouth to investigate the strategic importance of Britain's oil dependence. Powell Duffryn submitted a lengthy submission on their work on the Fischer-Tropsch process.[185] The unfortunate conclusion was that a few hydrogenation or synthesis plants would not be economically competitive with oil. And because they would be vulnerable to bombing, neither would they be militarily beneficial. Nevertheless, realising the dream of coal hydrogenation, with its promise of employment for miners, would continue to be a powerful scientific story.

In 1938 Glasgow hosted a world-scale British Empire exhibition. The DSIR took responsibility for the coal exhibition, one of three (together with iron and steel, and shipbuilding) that would show the public the place of scientific knowledge and research in the country's life.[186] At the exhibit's heart, on opposite sides of a central case, the committee positioned large models of the Billingham coal hydrogenation plant, more than two metres high and almost three metres long, and of the Fuel

---

[183] Powell Duffryn, Central Committee, Minutes of Meeting 25 January 1934, 3, in DPD/ 2/1/3/5, Glamorgan Archives, Cardiff.

[184] Innes 1939.

[185] Powell Duffryn Associated Collieries Ltd, 'Documents Submitted to the Committee on Imperial Defence, Subcommittee on Oil from Coal', COAL 14/982, TNA.

[186] Empire Exhibition, Scotland 1938, 111. Responsibility for the coal exhibit was taken by a small subcommittee (seven members) chaired by Sir Frank Heath, formerly the Secretary of DSIR. The committee included a nominee of the ICI chairman, Lord McGowan.

Research Station's Fischer-Tropsch synthesis plant.[187] The DSIR was sharing with the public the dream of rescuing the coal industry with applied science.

This chapter has dealt with the research of three kinds of coal-related enterprises, each seeking salvation through applied science. The Low Temperature Carbonisation Company was still a relatively small corporation whose future relied on disrupting the current use of coal, driven by innovative engineering and marketing. ICI was an entirely different type of organisation. The company was a national champion and research leader needing to promote the national interest as well as its own. The development of hydrogenation enabled it to do both with a single high-profile, high-prestige project of strategic significance.

The third of the companies, Powell Duffryn, was a major player in the disastrously declining coal industry. For this company, applied science promised means of improving its day-to-day business and a way of providing hope to staff, shareholders, and management alike that better times were around the corner. That the prospects animating these efforts did not themselves transpire did not detract from their inspirational power at the time.

Radio Case Study

Coal was on the way down. Radio was on the way up, and it was as representative of modernity in the early twentieth century as the internet would be a hundred years later. In 1927 the British Broadcasting Corporation (BBC), formerly a private company, was re-established as a public service, with a mandate to serve the nation and to air programmes addressing the public demand for worthy content. Meanwhile, the number of radio sets doubled every five or six years, increasing from 2.5 million in 1924 to 11.8 million in 1935.[188] For British commerce, government, and the DSIR, three areas of importance quickly appeared: linking the empire, navigation, and public broadcasting. Underlying the challenge of all three was the problem of improving

---

[187] See the file labelled Department of Scientific and Industrial Research, BT 60/47/9, TNA. The hydrogenation plant model, which was almost three metres long and two metres high, is described in a letter from ICI to O. F. Brown of the Exhibition, 25 October 1937, in this file. See also in this file 'Glasgow Empire Exhibition, Coal Hall in Government Pavilion', appended to Minutes of the Second Meeting of the DSIR 'Glasgow Empire Exhibition Committee', 26 January 1938; and 'Glasgow Empire Exhibition 1938, Coal Subcommittee', 3 December 1937, 4. The Coal Hall is treated in Department of Overseas Trade, 1938, 24–43; for illustrations, see pages facing pp. 32 and 33. The key is on p. 42.

[188] Anduaga 2009, 112.

Figure 5.2 Applied science for the modern public sphere: the first
transmitter station used by the BBC. '2LO' was built by Marconi in
1922. Rebuilt by the Science Museum in 1954. © Science Museum
Group.

the newly developed thermionic valves, also known as vacuum tubes, that
underpinned the design of interwar radio. Addressed in substantial private research laboratories and at the DSIR-sponsored Radio Research
Station (RRS), the valves proved to be the basis of the increasingly
diverse electronic devices (see Figure 5.2).

Understanding the strange properties of radio waves, whose messages
could quickly degrade but also traverse vast distances, involved leading
physicists and promised practical implications. Thus in 1924, the DSIR
Advisory Council approved an additional expenditure of £1,000 by the
board to study transmission from rotating loop aerials in support of its
approved work on directional wireless.[189] Edward Appleton, a
Cambridge demonstrator who became professor at KCL in 1924,
explored the properties of the atmosphere's charged layer, which
explained the possibility of trans-oceanic transmission. Its technical

---

[189] Minutes of the Advisory Council for Scientific and Industrial Research, 1923–24,
2 July 1924, 2, DSIR 1/4.

name, 'ionosphere', was due to Robert Watson Watt, then a young physicist at the Meteorological Office, founder-director of the RRS in 1927, and later famous for his work on radar.

Research, practice, and education were intimately linked, and the KCL department became a vital nursery for future radio researchers. Himself taking on the leadership of the whole of DSIR in 1939, Appleton's later publications show how he had embraced the language of pure and applied science.[190] As Jeff Hughes demonstrated, the context was the enthusiasm of young physicists, for whom radio was not just an exemplar of science; it fed back into the practice of useful research.[191] Training in, and the experience of fiddling with, radios had been one of the unexpected benefits of serving in World War One. Whatever the job to which former soldiers returned, these men provided a cohort of enthusiastic and knowledgeable customers for new devices. By the mid-1920s, thirty periodicals addressed the enthusiasts' interests.[192]

The politics of limited government intervention were an important influence on the character of British applied science. But this was shaped, too, by the nature of privately funded activities. Like the great national laboratories, the industrial research centres that grew up from the early twentieth century addressed competing priorities. Beyond direct economic benefits, they needed to contribute to the status of large combines, justifying their size and oligopolistic place in the economy.[193] Movements between private and public research reinforced similarities between the sectors. Thus, we have seen how Crossley moved from Porton Down to the textile industry's Shirley Institute. Similarly, C. C. Paterson, the first director of the substantial laboratories of Britain's General Electric Company (GEC), launched just after World War One, had previously led the first electrical work at the National Physical Laboratory. Several scientists and technicians came with him to GEC, which was aspiring to reach global standards from a standing start.[194] They were now making electric light bulbs, which, albeit simpler, shared production challenges with radio valves. Later, Paterson would reflect on the demands faced by his group in post–World War One Britain. The company had addressed the need for better production control and had to develop new materials for 'gettering', or maintaining the vacuum within bulbs.[195] Gas-filled lamps were another challenge, and the expertise that made these possible was related closely to that

---

[190] See, for instance, Appleton 1945 and Appleton 1959.   [191] Hughes 2018.
[192] Anduaga 2009, 107.   [193] Niblett 1980.   [194] Clayton and Algar 1988.
[195] Paterson 1945, 25. Robert le Rossignol, who had previously worked in Germany on light bulbs and earlier on the Haber process, worked there. See Sheppard 2017.

deployed in designing the valves needed in the new broadcasting era. By 1925, GEC had supplied the components for the world's most powerful transmitter, the BBC's 'national station' at Daventry. The other electrical combine in the country, Associated Electrical Industries, was formed in 1928. Producing light bulbs and radio valves under the 'Mazda' brand, in the late 1920s, it was spending over £100,000 a year on its research.[196] Under the leadership of Arthur Fleming, its Metropolitan Vickers Laboratory had a budget of £71,000, comparable with the RAE budget.[197] The laboratory's expertise in high voltage, and its 350,000-volt transformer, were brought by John Cockcroft to Cambridge's Cavendish Laboratory. There, with Ernest Walton, he built the world's first accelerator to 'split the atom' in 1932.[198] The tangible outputs were associated with the symbol value of industrial research. Christopher Niblett, a historian of the laboratory, points out that these were linked by Fleming's 'conception of the value of industrial science, especially of its value to the image of a progressive company'.[199] The visible research endeavour had more benefits than its knowledge outputs. It could assure workers, government, the public, and shareholders that corporations could overcome great problems and realise high ambitions. A visible research process could enable a company to capitalise expectations of a prosperous future.

Conclusion: Applied Science Institutions

The global conflict of World War One, its industrial demands, and the anticipation of its aftermath amplified British determination to build an applied-science infrastructure. The commitment to organisational innovation during and immediately after the war invested considerable expectations in the category. Criticism of weakness in the past reflected high ambitions for the future. Government policy expressed a widespread concern that Britain's institutions had been inadequate in promoting applied science. Now it was put to work to rescue the coal industry and the regions that depended upon it, prepare for the next

[196] A. P. M. Fleming, 'Recommendations on the Organisation of Research within the Company. Made to the Instructions Given in Sir Philip Nash's Memorandum, January 23rd 1929', NAEST 70 Box 3, Papers of Arthur Fleming, IET Archives, Institution of Engineering and Technology. AEI brought together the Metropolitan Vickers and British Thomson Houston companies.
[197] Niblett 1980, 43.   [198] Hughes 1998.
[199] Niblett 1980, 42. I am grateful to Christopher Niblett for permission to quote from his doctoral thesis.

war, renew the textile companies, and support such growing new industries as radio and chemicals. The denomination 'applied science' emphasised utility, but many proponents saw it drawing on pure science more closely than ever. Conversely, people identifying with pure and 'fundamental' science cited its relevance to application to demonstrate that they, too, were engaged in an economically worthwhile process. It is striking how rarely that journalists, politicians, and businesspeople publicly doubted the adequacy of science as a royal pathway to prosperity and the solution to national problems. Certainly, there were grumbles and doubts about the adequacy of DSIR, and, in Chapter 3, we have seen the rise of the category of technology. Still, even Arthur Fleming, the standard-bearer for modern engineering and the proponent of a new organisation standing between DSIR and industry, saw the roots of engineering progress in applied science. This interpretation justified an important place for applied science in government industrial policy. While DSIR committees rarely used such terms in their internal discussions, scientists and administrators used them for discourse with politicians and the press.

Long-established services, such as the Royal Navy, and old industries, such as coal, could justify hopes for rejuvenation. Even then – as the Sothern Holland report, the naval response, and the outcome highlighted – success would depend upon the integration within institutions, the synthesis of science and practical expertise, and the links to implementation. The isolated experience of neither the brilliant academic nor the inspired inventor alone would be sufficient. Wartime experience led to a grasp, if not of successful solutions, at least of the problems of linking research with practitioners. Industry, too, came to invest considerable hopes in the outcome of success.

Whereas during the war, J. J. Thomson had suggested that only pure science could offer a revolution, such reservations were not universal. In research on coal, diverse organisations, ranging from medium-sized to exceptionally large, sought to benefit from different mixes of applied science. The three defining issues of the years before World War One – the continuity of narrative, the centrality to national industrial policy, and the relationship of applied to pure science – played out further during the interwar years through new agencies. These were not just incidental features but essential to understanding the nature and working of the appeal of applied science.

Many of these bodies, and the ambitions they expressed, were clung to for decades thereafter. Nonetheless, later improvers often disparaged as too parsimonious the early nurture of the new system by all interwar governments. In the aftermath of the subsequent world war, scientists criticised the innovations introduced earlier in the century as so starved of resources that their implications had not been allowed to unfold. Later, this book will explore the attempts to build upon and later to dismiss hopes and concepts that had formerly seemed so real.

Historians have unduly neglected the DSIR research boards. Institutional entrenchment through their funding, and the activities of great laboratories, did not just bring further resources to an existing category. On the contrary, the projects they supported, and their very existence, shaped and gave substance to 'applied science' as a research category. It had become a reality, apparently independent of the political economy through which it had come to life.

6   'Western Civilisation' and Applied Science

Introduction

While the scientists and politicians were translating professional aspirations into bureaucratic language, new institutions and exciting gadgets gave vivid reality to a changing landscape of public culture. Usage across society interacted with the language of officials as public, political, and bureaucratic discussions of applied science intertwined. In the public sphere, the concept served multiple purposes, making sense of both modern times in general and the role of science. Such talk also came to be intimately connected with an intense discussion of 'modern civilisation'. Therefore, this chapter asks how applied science was treated in the public realm during the inter-war years.

Radio, itself made possible by research, and turned into a national forum by the BBC, would be a particularly effective and appropriate medium for such considerations. Its programmes celebrated and fretted over science, together with modernity, rationalisation, and modern warfare.[1] Lay interpretations also animated a cast of new characters and public figures. These natives of the coming era of broadcasting, such as Julian Huxley, Patrick Blackett, Solly Zuckerman, and John Desmond Bernal, would be influential in policy during the post–World War Two years. More immediately, politicians, industrialists, and campaigning engineers and scientists would mediate between popular aspirations and concerns, on the one hand, and policy decisions, on the other.

Today's readers of early twentieth-century newspapers may find striking the range of subjects in which 'applied science' made an appearance. It might not be surprising to see an evocation in discussions of new devices, education, industry, agriculture, medicine, and of the cognate concept, pure science. Perhaps more arresting were the opportunities for discussing the more general civilizational consequences provided by cookery, unemployment, humanity, the future of the church, Britain

---

[1] Beer 1996.

166

herself, and reports from India, China, Italy, and France. Widespread interest was reflected in the emergence of the specialist science journalist employed by a few newspapers. J. G. Crowther at *The Manchester Guardian* was followed in the trades union–sponsored *Daily Herald*, where entrepreneurial sports reporter Peter Richie Calder suggested the new role to his editor.[2] Numerous enthusiast magazines sprang up, and the interest in research on such topics as radio communication was not restricted to just a few specialists.[3] Editorials of popular national newspapers, such as *The Daily Mirror, The Daily Herald*, and *The Daily Mail*, whose readers numbered about a million, reflected on 'breakthroughs'.[4]

This interest does not mean laypeople understood 'research' in the way that professionals wanted. Fortunately, one voice from the time reported how he saw the newspapers dealing with it. Handwritten notes in the archive of the biologist Julian Huxley, grandson of T. H. Huxley, dating probably from about 1930, reflected:

The word research becomes familiar to most readers of newspapers in connection with the Medical Research Council, with war researches and such subjects as research on aeroplanes, in gas or engines, the detection of submarines or with industrial research otherwise work on applied chemistry or physics. They are vaguely aware of agricultural research stations, of fisheries research, of research in plant and animal breeding. They have probably heard the mystic words pure research & know that, at the older and less practical universities people are still to be found to devote their time to research on such curious subjects as ancient religion, the distribution of fossil reptiles or the deciphering of codices & papyri.[5]

This interpretation represented research as divided, in the public sphere, between applied science and scholarship in 'curious subjects'. Note how this formulation was strongly associated with individual institutions and projects: the military, large companies, agriculture, and the Medical Research Council. In the previous chapter, we saw how a chemist complained that the 'jingle jangle of modern civilisation' was interfering with the application of discoveries. To many other people, that 'jingle jangle' was itself the consequence of applied science.

---

[2] On Crowther, see Hill-Andrews 2015. [3] Bowler 2009.
[4] Seymour-Ure 1975, 236. The top-selling papers were *News Chronicle* (until 1930 two separate papers), *The Daily Mail, The Daily Mirror, The Daily Express*, and *The Daily Herald*. Their relative positions fluctuated. In 1920 *The Daily Mail* was by far the best-seller; by 1939 *The Daily Express* was in the lead. On the distinctive qualities of the interwar British press, see Chalaby 1996.
[5] Julian Huxley 'Research', Box 60: 8, Papers of Julian Huxley, Rice University, Special Collections and Archives, Woodson Research Center. Quotation by Julian Huxley reprinted by permission of Peters Fraser & Dunlop (www.petersfraserdunlop.com) on behalf of the Estate of Julian Huxley.

Civilisation

The bewildering technical changes experienced by citizens during the years between the world wars were summarised in a 1950 retrospect by a respected economist, R. F. Sayers. He particularly pointed to the widening presence of cars, aeroplanes, radios, electricity, plastics, and artificial fibres, and the omnipresent mechanisation.[6] Indeed, the making of the radio set and its rapidly improving technical performance, the new broadcasters such as the BBC, and the act of listening with hundreds of thousands of others collectively were transforming life. Meanwhile, access to mains electricity was widening dramatically. Generically, such changes were put down to 'applied science'. In his path-breaking 1922 novel *Ulysses*, James Joyce devoted several pages to the preference of modernist Leopold Bloom for applied over pure science. Bloom adduced this tendency to his appreciation 'of inventions now common but once-revolutionary, for example, the aeronautic parachute, the reflecting telescope, the spiral corkscrew, the safety pin, the mineral water siphon, the canal lock with winch and sluice, the suction pump'.[7] The term seemed to describe an ample space of contemporary experience.

The press and advertisers used the close linkage between 'applied science' and 'civilisation' to evoke modernity internationally. Articles appearing in *The Manchester Guardian* and *The Times* as early as 1912 rebranded Italy as a cradle of applied science.[8] In the mid-1920s, publishers promoted the book *France and the French* as dealing with such contemporary issues as applied science and social welfare.[9] Significantly, such an association related not just to western countries. In May 1909, there was interest in the foundation of Hong Kong University, which included applied science amongst its ambitions. Similarly, the British government actively massaged the country's international image through such institutions as the Empire Marketing Board. The board's architect, former Minister Walter Elliot MP, boasted of a British display that featured the work of entomologists, grass researchers, and cold storage men rather than 'beefeaters' (the traditionally uniformed guards of the Tower of London) or the ship of the sixteenth-century hero Sir Francis Drake.[10]

---

[6] Sayers 1950, 276.
[7] Joyce 1932, 682. On Joyce, modernism, and science, see Shiach 2018.
[8] 'Hustled History. Italy's Triumphal Progress. New Nursery for Applied Science', *The Manchester Guardian*, 17 December 1912; the same article appeared on the same day in *The Times*.
[9] Advertisement for Sisley Huddlestone, *France and the French*, in *The Manchester Guardian*, 11 March 1926.
[10] Elliot 1931, 740.

Correspondingly, negative evaluations of modernity as a whole often blamed applied science. In 1928 the *Church Times* published a sermon by an Oxford tutor worried that 'we' must pass on a religious message to India to complement 'applied science'.[11] It was just three years later that an address on the 'the world situation today' warned a conference on stewardship and church finance of the looming threat of a global atheistic civilisation underpinned by applied science.[12]

In the first forty years of the twentieth century, among thinkers about science, H. G. Wells was the most influential of (in his own words) 'those people who stir that remarkable salad which is called "public opinion"'.[13] Both friends and enemies associated his name and what was referred to as 'Wellsianism' with hopes raised by applied science, though he seems to have been a sparing user of the term.[14] It occurs only twice in his 900-page *Wealth, Work and Happiness* of 1931, and he clearly saw the category as just one component of the innovative process.[15] It could, nonetheless, present a social challenge. Wells' earlier fiction had expressed deeply ambivalent feelings towards the linkage of science and progress. In *Tono Bungay* (1908), he described the village of his hero's childhood in ominous terms: 'The hand of change rests on it all, unfelt, unseen; resting for a while, as it were half reluctantly, before it grips and ends the thing for ever.'[16] There was only one solution to the challenges of international governance in an increasingly dangerous world. Wells proposed rule by a 'new samurai' of scientific and technical experts.[17]

Meanwhile, Wells argued, the present day offered comforts and opportunities to the many that even the few could not have dreamed of, even 200 years earlier. Though his writing after World War One moved away from science fiction, it continued to be both astonishingly popular and steeped in science. His *Outline of History*, published in the war's aftermath,

---

[11] A. E. J. Rawlinson, 'The Mission of the Religious Life to India', *Church Times*, 13 January 1928. For analogous worries by another missionary about the anti-Christian 'Free Thought Movement' in China, see 'United Free Synod of Lothian: Dr Webster on Changes in China', *The Scotsman*, 11 October 1922, British Newspaper Archive/British Library Board.

[12] 'World Conference: Stewardship and Church Finance', *The Scotsman*, 23 June 1931, British Newspaper Archive/British Library Board.

[13] Rayleigh 1916, 23.

[14] On the 'Wellsians', see Crossley 2011. During the 1920s, Wells' name was co-cited with 'applied science' in as many as twelve articles in the 'Britishnewspaperarchive' (consulted October 2020).

[15] Wells 1932, 149 and 623. In *The World of William Clissold* (Wells 1926, 167–68) Wells reflected on the complementary roles of science and the 'innovating spirit' in the process of innovation. For a general reflection on Wells' attitude to science, see Haynes 1980.

[16] Wells 1911, 9.

[17] Wells first explored this idea in his *A Modern Utopia*, published in 1905.

found two million purchasers worldwide within a dozen years.[18] Despite earlier doubts, this book was a panegyric in favour of modern, technical, and scientific progress, emphasising material advance.

Educated women had distinctive, if long frustrated, hopes for the 'new civilisation'. Change accelerated during World War One when the need for skilled workers provided unprecedented opportunities for female chemists.[19] During the war, at Imperial College, the chemist Martha Whiteley worked on poison gas, and was subsequently honoured with an 'OBE'.[20] Opportunities outlasted the war. Dorothy Jordan Lloyd directed the British Leather Manufacturers Research Association for almost two decades from 1927. Acclaimed writer Vera Brittain pointed out in *Women's Work in Modern England* that old traditions and precedents did not need to be overcome in such areas as astronomy and aviation.[21] Reflecting in a *Daily Mail* article on the openings becoming available to young women such as her infant daughter, she saw opening up in applied science some of the most significant opportunities for women in the modern world.[22] In time, her daughter would mature into the celebrated politician Shirley Williams and one day supervise her country's science funding. Shirley's contemporary, Margaret Roberts (later Thatcher), would choose chemistry and industrial research before taking to politics. Nonetheless, in many cases, the roles open to women were subservient to male colleagues.[23]

Brittain was not alone in identifying the congruence of modernity, applied science, and a better world for women. Her friend, the journalist Storm Jameson, was also writing against the conservative enemies of all these things.[24] Two of the first three students approved for a PhD course under the Board of Studies in History of Science at the University of London were women – Florence Yeldham, who worked on medieval mathematical methods, and Dorothy Turner, who studied the history of science teaching.[25] Dr Turner published her research in a book suggesting that readers see science as a cultural activity, not as a fixed body of

---

[18] Wells 1920. This readership cited by Skelton (Skelton 2001) is based on a claim by Wells himself; see also Ross 2002.
[19] See Horrocks 2000.   [20] Nicholson and Nicholson 2012.   [21] Brittain 1928, 58.
[22] Vera Brittain, 'Let Your Daughter Invent Her Own Career', *The Daily Mail*, 26 July 1932.
[23] Horrocks 2000. For an international view based on the Kodak company, see Mercelis 2022.
[24] Storm Jameson, 'Get the Best Out of Life', *The Daily Mail*, 4 April 1931. On the association of young women with modernity and work in the popular press, see Bingham 2004, 47–83.
[25] Copies of the doctoral dissertations of each of these early graduates are held by the library of the Science Museum.

Women at Work in One of the Chemistry Laboratories

Formerly only two laboratories were available for the Chemistry Department, and so great was the over-crowding that the College was forced to put up rough army sheds in the front quadrangle to cope with the ever-increasing number of students

Figure 6.1 Applied science for the worker of the future: 'Women at Work in One of the Chemistry Laboratories' at Bedford College, University of London. Illustrated in *The Sphere*, 21 January 1922. Courtesy Illustrated London News/Mary Evans Picture Library.

knowledge. Subsequently, she amplified her thoughts in a more avowedly polemical history of science, *The Book of Scientific Discovery: How Science Has Enabled Human Welfare*. Dr Turner was struck by the new experience of leisure made possible by applied science.[26] However, much more negative evaluations frequently contrasted with this optimistic reflection.

American historian Charles Beard introduced his collective volume entitled *Whither Mankind: A Panorama of Modern Civilization* of 1928 with a striking observation about discourse in the public sphere: 'Anxiety about the values and future of civilization is real. It has crept out of the cloister and appears in the forum and marketplace.'[27] The observation of Beard, and others, was that the conversation was extending far

---

[26] Turner 1933, 254.
[27] Beard 1928, 7. Similar sentiments were expressed by the philosopher John Katz. See the beginning of Katz 1938, 1. For a modern reflection on the mood then, see Stuart 2008.

beyond small elites.[28] Indeed, today's reader may be startled by the frequency with which the general British press used the phrase 'modern civilisation' in the interwar years. Even a local organ, the *Aberdeen Press and Journal*, published 332 items using it between 1919 and 1939.[29] The close association of applied science and modernity justifies paying attention to this 'civilisation' talk.

Across the country, the proposal that 'modern civilisation is a failure' was a popular topic for groups whose discussions would then be covered by the local press. Thus in 1927, a church-hall debate on this motion was reported in detail in the major municipal newspaper *The Hull Daily Mail*.[30] A scientifically trained church supervisor expressed the central issue well during the Hull debate: in the recent war, the barbaric use of science had caused untold shameful suffering. In Berwick on England's Scottish border, where unemployment was a major issue, the local paper documented a debate on a similar motion two years later. Showing this was not just a matter of indirectly reflecting on industrial crisis, the topic was debated too by the Baldock Wesley Guild in Buckinghamshire, a county with minimal unemployment.[31] Beyond such local organisations, the issue also attracted elite public intellectuals. Thus, the national left-leaning Fabian Society had organised a series of six lectures in 1923 under the title 'Is Civilisation Decaying?'[32]

In his contribution to the Fabian Society series, Bertrand Russell spoke on 'the effect of science on social institutions'. He argued that hitherto only physical sciences had made an impact: through their effect on the imagination and through what he called physical causes, 'especially machinery'. Now, he believed, was the time to apply other sciences. There was hope from the application of birth control, eugenics, and

---

[28] For the concerns of intellectuals, see, for instance, Schleifer 2000 and Hughes 1967. See also Overy 2009, 26–64.

[29] Count made courtesy of www.Britishnewspaperarchive.co.uk (accessed July 2022), British Newspaper Archive/British Library Board.

[30] 'Hedon: Wesley Guild Debate', *The Hull Daily Mail*, 1 March 1927, British Newspaper Archive/British Library Board.

[31] For the Baldock Wesley Guild debate, see 'Baldock', *The Biggleswade Chronicle*, 14 December 1934, British Newspaper Archive/British Library Board. On the perception of unemployment in the area at the time, see 'Instruction Courses Not Needed in the County', *Biggleswade Chronicle*, 18 May 1934, British Newspaper Archive/British Library Board. For the Berwick debate, see 'Coldstream-Berwick Interdebate on "Is Modern Civilisation a Failure"', *The Berwickshire News*, 19 February 1929, British Newspaper Archive/British Library Board. On the significance of unemployment issue in this area at the time, see 'Reception of Premier's Speech at Berwick', *The Berwick Advertiser*, 31 January 1929, British Newspaper Archive/British Library Board .

[32] Fabian Society, 1923.

psychology, which would allow 'state control over emotions' using hormones. Russell concluded, 'at present, all that gives men power to realise collective passions is bad, and science leads to hell. Is there any hope of improvement in our collective passions? If so, science may become a boon.'[33] Such expressions urging the biological improvement of the population stock in the cause of civilisation were popular. It was an ambition that chimed well with the movement for national efficiency; as Prime Minister David Lloyd George warned, 'you cannot maintain an A1 empire with a C3 population'.[34] It might have looked as if 'national eugenics', then seen as 'the study of agencies under social control that may improve or impair the racial qualities of future generations either physically or mentally', was set to become an ideal applied science in Britain.[35]

Eugenics

At the beginning of the century, two scientifically prominent laboratories were established at University College London (UCL) under the charismatic Karl Pearson, the Biometric Laboratory and the Galton Eugenic Laboratory.[36] These were followed shortly after by the outward-looking 'Eugenics Education Society'.[37] Looking back, Richard Soloway has argued that members saw themselves as 'the vanguard of a propagandist movement that could bridge the gap between the new science of heredity and popular assumptions and beliefs about what made individuals, and, more important, classes different from each other'.[38] However, the two categories of 'eugenics' and 'applied science' did not overlap substantially in the public sphere. For example, in newspapers of the first forty

---

[33] Fabian Society 1923, 4. Quoted by permission of the Bertrand Russell Peace Foundation/Fabian Society. Russell would develop his ideas further in *Icarus, or The Future of Science*, published the following year. On this book and Russell's use of it to answer *Daedalus*, a reflection on science by J. B. S. Haldane, see Rubin 2005. I am grateful to Rick Stapleton and his colleagues at the Russell archive for pointing out the link between the Fabian Society contribution and *Icarus*. It should also be emphasised that Russell's views on eugenics subsequently changed. I am grateful to Tony Simpson of the Bertrand Russell Peace Foundation for pointing out this change of heart.

[34] 'The War and After', *The Times*, 13 September 1918. Although here Lloyd George was referring specifically to the need for public health, the writer and journalist Basil Barham worried his comments expressed a widespread fixation on eugenics at the expense of industry. Basil Barham, 'Industrial Efficiency', *The Globe*, 26 November 1918, British Newspaper Archive/British Library.

[35] Galton 1909, 81.

[36] On the relationship between these two laboratories, see Magnello 1999a and Magnello 1999b.

[37] Farrall 2019. The word 'eugenics' had been coined by Pearson's mentor Francis Galton in 1865.

[38] Soloway 1995, 37.

years of the twentieth century, less than 1 per cent of articles including either term also mentioned the other.[39]

At least partly, the explanation must lie in profound divisions within British eugenics and its lack of an institutional brand.[40] Albeit a polemical firebrand, in his research and teaching Pearson largely concentrated on the statistical underpinning for his biometric interpretation of heredity. He engaged too in much-publicised disputes with the Society – with which he would not collaborate. Pearson himself would not join its Council, dismissing the Society's work as propaganda. Moreover, he was angry with the Society's adoption of the competing Mendelian genetic approach to explain the inheritance of mental defects.[41]

After Karl Pearson's retirement in 1933, two men replaced him, Egon, his son and also a leading statistician, and Ronald A. Fisher, the supporter of eugenics and statistician at the flagship of British agricultural research, the Rothamsted Experimental Station. Fisher took over Karl Pearson's journal *Annals of Eugenics* and endeavoured to supersede the old battles between biometricians and geneticists with a new emphasis on genetics and mathematical statistics.[42] Having fallen out of the public eye in the 1920s, the Galton Laboratory's visibility rose again but not nearly to the level it had achieved before World War One. The 'Britishnewspaperarchive' shows that whereas there were mentions of the laboratory in twenty newspaper articles in its corpus during the single year 1911, only eighteen cited it in the entire decade of the 1930s.[43] So, although some of the work published by Fisher during his decade at UCL proved to be of the highest scientific significance, it did not directly contribute to the public conception of either the Galton Laboratory or eugenics. Again, despite frequent discussion of the other biological context of agricultural research, itself frequently discussed as 'applied science', any cross-over with eugenics was also minimal.[44] Despite Fisher's long association with Rothamsted,

---

[39] The 'Britishnewspaperarchive' (for 1900 to 1939 consulted December 2022) shows 16,862 articles citing 'applied science' and 16,200 mentioning eugenics. Yet only 98 (scarcely more than one in 200 of either set) mentioned both. By contrast, 1,753 (roughly one in ten) of the applied science articles mentioned chemistry.

[40] The divisions within eugenics in Britain have been studied through distinguished scholarship. See Freeden 1979; Mazumdar 1992; Kevles 1995; Soloway 1995; Bland and Hall 2010. I am grateful to Daniel Kevles for the opportunity to discuss this section with him.

[41] Kevles 1995, 104–5.   [42] 'Foreword' 1934.

[43] 'Britishnewspaperarchive.co.uk' (consulted December 2022).

[44] The 'Britishnewspaperarchive' corpus (consulted December 2022) shows 3,200 articles mentioning both 'agriculture' and 'applied science' in the period 1900–1939, constituting about a fifth of all applied science articles. Some 538 articles (about 3 per cent of the eugenics articles and an infinitesimally small proportion of the more than 2 million articles mentioning 'agriculture') mentioned both 'agriculture' and 'eugenics'.

out of the 119 articles in the 'Britishnewspaperarchive' explicitly referencing that laboratory across the 1930s, hardly any also mentioned eugenics.[45]

Geneticists had their complaints against the Eugenics Society too. They became alienated from the social campaigns of the Society, in which they were reduced to a small minority by social scientists, other academics and medical people, and politicians.[46] Several scientific leaders of public interest in biology, such as Lancelot Hogben, scorned popular visions of undesirable traits' genetic inheritability.[47] Many lay people also proved sceptical, and in the 1930s British lawmakers failed to support the Society's petering-out campaign to legalise 'voluntary sterilisation' of people dismissed as 'mental defectives'. Indeed, German practice from Hitler's ascendency in 1933 brought odium to the very concept of eugenics.

The leadership of the Society did seek to move from promoting classical eugenics focused upon heredity to supporting a diffuse reform movement concerned with nurture too. Without either a closely associated high-profile laboratory or scientific speciality, though, eugenics looked very different from such established applied sciences as agriculture. Therefore, while eugenics was widely discussed, it did not substantially impact the concept of 'applied science' in the years before World War Two.

Rationalisation

A single word did, however, come to represent the civilisation of applied science: 'rationalisation'. This was a framework for interpreting industrial change, but its implications were even broader. It fuelled faith in science, organisation, and planning. Alfred Mond claimed to have

---

[45] Some 119 articles mentioned both 'Rothamsted' and 'laboratory' but only three mentioned 'Rothamsted' and 'eugenics'.
[46] Farrall 2019 reported on the membership of the Eugenics Education Society (which became the Eugenics Society in 1926). Out of fifty-seven members who had entries in the *Oxford Dictionary of National Biography*, ten were academics in the biological sciences. Nine were social scientists and four were politicians, including two who served as Prime Minister, Arthur Balfour and Neville Chamberlain. The outnumbering of biologists and their lack of influence was an issue. See the frustrated letter to *The Times* from the director of the Potato Virus Institute complaining about the sterilisation proposal of the Eugenics Society. Retcliffe N. Salaman, 'Mental Health: The Question of Heredity', *The Times*, 23 September 1930.
[47] Perhaps the most notable supporter of eugenics among geneticists was Ronald A. Fisher. Notably, biologists Lancelot Hogben (long an opponent), J. B. S. Haldane, and Julian Huxley moved away from the category in the 1930s. See, for instance, chapter 9, 'False Science', in Kevles 1995, 129–47.

imported the term from Germany, where its connotation was shaped in the wake of defeat by Walter Rathenau, the German industrial and political leader. Rathenau sought to make a case for companies collaborating to promote efficiency within industrial sectors. He expressed this tactical need in convincing philosophical terms through the word 'Rationalisierung', professing trust in the potential of science and engineering as the source of new employment and the broader belief in radical innovation.[48]

The vision obtained a high profile in 1927 following a World Economic Conference convened by the League of Nations. A meeting of interested organisations led to the commissioning of a book-length report by the British management thinker Lyndall Urwick, who had recently been working for the confectionary company of Rowntree, an organisation known for its pioneering use of such modern science as industrial psychology, and who was currently the director of a management institute in Geneva. Urwick's report included chapters dealing with industrial research and with the scientific method that could be employed to improve management in general, including the management of people.[49]

Rationalisation was widely expressed in terms of entrepreneurial inventors and large corporations. Its flagship was the chemical combine ICI, formed in 1926 from four hitherto separate chemical companies, and its prophet was Alfred Mond, who welded these into a world-scale national champion.[50] In his writings of the time, he suggested rationalisation was perhaps best understood to comprise three processes: scientific management of labour, the adjustment to equality of supply and demand, and the application of scientific research.[51] Mond made clear his vision of research married to industrial need: 'Whilst honouring science in all its branches, it is best to make it abundantly clear that we are dealing with the work of engineers, physicists, biologists and chemists who have contributed not so much to the theoretical development of Science as to its application to the needs of industry and humanity.'[52] Equally, the emphasis on scientific management was associated with the development of large and even hegemonic companies or cartels with

---

[48] 'An Age of Mechanisation', *The Times*, 14 June 1921. For reflections on this, see Hård 1998; Volkov 2012; Staley 2018.
[49] Urwick 1930. On Urwick, see Brech et al. 2010.
[50] The complex origins and early development of ICI are explored in a classic two-volume history by the business historian W. J. Reader; see Reader 1970 and 1975.
[51] Alfred Mond, 'Rationalisation and Industrial Relations', 'Industrial Relations' supplement, *The Manchester Guardian*, 30 November 1927, 7–8; Mond 1927.
[52] Mond 1927, 155.

enlightened 'scientific' management policies. As the president of the economic section of the British Association explained in his 1927 address, prices were administered at an industrial or national level, and planning was a watchword.[53] By applying scientific research and finding new uses for products, rationalisation could bring supply and demand into balance – as well as through its more brutal cutting of production and the creation of unemployment.

In a 1928 article on innovation, the Austrian economist Joseph Schumpeter reflected on the links between capitalism and rationalism.[54] This, of course, had been an argument of his older German contemporary Max Weber, and 'rationalism' was soon applied to science and religion by American sociologist Robert Merton.[55] Similarly, in a 1932 summary of the area, Robert Brady of the University of California explored the enlightenment values associated with rationalisation.[56] However, concluding his paper in the depth of the Depression, Brady reflected sadly on the contrast between universalistic aspiration and the realities of nationalist mercantilism.

Rationalisation was not exclusively an ally of capitalism. It was also widely associated with planning, experts, and state power. *The Economist* magazine reported on H. G. Wells' address to the 1937 British Association for the Advancement of Science meeting:

Geographers, chemists, physicians, and engineers have vied with economists, physiologists and psychologists to produce blue prints for a new revolution – to control our eating and shopping, our work, our amusements, and our dwelling – with an avoidance, however, of the problems of scientific warfare. Maps and statistics have been flourished by the very prototypes of Mr Wells's Samurai. All this is in the vogue. Out of war and depression were begotten rationalisation and planning.[57]

Aldous Huxley published his famous satire of rationalisation, *Brave New World*, in 1932. The antihero's name, Mustapha Mond, was a reference to Alfred Mond, and Huxley's novel followed the author's visit to ICI's great Billingham ammonia-fertiliser complex. As he reported in an article following his journey, Huxley was fascinated by the spectacle of the plant and lyrical about the intellectual and artistic accomplishment it represented. He did worry, though, about the ordinary workers treated as

---

[53] MacGregor 1927.
[54] Schumpeter 1928. Schumpeter's thinking on rationalism has been extensively studied. See Moura 2017.
[55] Merton 1936.   [56] Brady 1932.
[57] 'Things to Come' 1937. Quoted by permission of the Economist Group.

insignificant tools of the engineers' and scientists' artistry.[58] Replete with such anxieties, the subsequent novel parodied the alliance of modernity and applied science.[59] This relationship was an enduring issue for Aldous Huxley, who reflected in a 1932 broadcast published in the BBC's *Listener* that it was not science itself that should be questioned but those who were applying it.[60] His was an early expression of an argument that would become well known. Pure science is morally neutral. Only through application did it become 'bad'.

Military Applications and Applied Science

Repeatedly, the related stories of the violent past and uncertain future coloured the interwar interpretation of applied science.[61] Even before the Depression, the years after World War One saw grave concerns about disillusion with the excesses of success. 'Science' had shaped the recent conflict and would, prophets warned, permit more terrible weapons in the next. The still very raw recent experience of poison gas lay at the heart of the alternative interpretation of science.[62] This fundamentally new weapon became the symbol of the scientist's, notably the chemist's, contribution to modern warfare. The American author Will Irwin compared the importance of the date of the introduction of chemical warfare to the Western Front, 22 April 1915, to that of Independence Day, 4 July 1776.[63] He cited commentators who had suggested that, with a favourable wind, a few large bombs containing America's latest invention, Lewisite, could have killed Berlin's entire population.[64] In 1923, renowned British novelist John Galsworthy, author of the recently published *Forsythe Saga*, released a pamphlet so damning that *Nature* devoted a whole column to its reply. Galsworthy regretted that science was more 'hopeful of perfecting poison gas than of abating coal smoke or curing cancer' and emphasised, '*Destructive science has gone ahead out of all proportion.*'[65] Science's failure to address the challenge of gas became a 'representative anecdote' for its general tendency to outrun the ethics that might control its use. It established an essential background to the

---

[58] Huxley 1931, 52.
[59] Huxley 1932a. On the book's engagement with rationalisation, see Sexton 1996, and on its relation to 'modernity', see Baker 2001.
[60] Huxley 1932b. This was broadcast as 'Science and Civilisation – II', https://genome.ch .bbc.co.uk/149df7a378bf4dc38cd8f3b8e8177fae (consulted January 2016).
[61] Richard Overy entitled his history of interwar Britain *The Morbid Age*; see Overy 2009.
[62] See, for instance, Girard 2008.   [63] Irwin 1921, 25.   [64] Irwin 1921, 38.
[65] Galsworthy 1923, 1–2 and 3; emphasis in original. For the reply, see 'Science in Civilisation' 1923.

more generalised anxieties that came to be associated with 'science' as a whole.

Research in the 1930s period of 'rearmament' was, of course, secret. Nonetheless, readers would have noticed that newspapers were carrying increasing numbers of advertisements for men with training or expertise in applied science to join government service. A parliamentary debate on air defence in 1938 led to widespread anxiety about a revolution in applied science and its implications.[66] This climate drew on already long-standing worries as well as sad memories of the recent war. Ethnographic reports of the British, carried out by the newly established 'Mass Observation' organisation, confirmed spread of a sense of anxiety about the consequences of science, and deep ambivalence about its benefits and risks.[67] Even before World War One, the physicist Frederick Soddy had raised the frightening possibility of atomic bombs, and H. G. Wells publicised it in a well-known novel.[68] In retrospect, we can see the inter-war talk about such a development and its use following the template laid down by gas.

Under the headline 'Destructive Science', the popular newspaper *The Daily Mirror*, covering the 1922 meeting of the British Association for the Advancement of Science, asked whether applied science would destroy the world. The emphasis was clear: 'It is *applied* science that matters to the average man.... "All the fault of the world's politicians", answers the scientist. Perhaps but the fact remains that science is today horribly the slave of the destructive mania in men's minds.'[69] More editorial assaults were launched by the *Mirror*'s columnist Richard Jennings, a well-known intellectual signing 'WM', evoking the memory of the English 'Arts and Crafts' designer William Morris.[70] The pages of *The Daily Mirror* were not the only forum for such reflections. An internationally performed play, *Wings over Europe*, took the development and control of an atomic

---

[66] See, for instance, 'Premier Refuses Air Defence Inquiry', *The Yorkshire Post*, 26 May 1938, British Newspaper Archive/British Library Board. For the debate itself, see *Hansard* HC 25 May 1938.

[67] See also Madge and Harrison 1939; also the wartime report Mass Observation, 'Report on Everyday Feelings about Science' (October 1941), no. 951.

[68] Frederick Soddy, 'Some Recent Advances in Radioactivity: An Account of the Researches of Professor Rutherford and His Co-Workers at McGill University', *Contemporary Review* 83 (May 1903): 708–20. Wells drew on Soddy's prophecies in his famous 1914 novel, *The World Set Free* (Wells 1914). See Willis 1995.

[69] 'Destructive Science', *The Daily Mirror*, 9 September 1922, British Newspaper Archive/British Library Board. Emphasis in original.

[70] WM [Richard Jennings], 'One Good Bang', *The Daily Mirror*, 26 October 1935, www.Britishnewspaperarchive.co.uk/The British Library Board; also see 'World May Blow Up', *The Daily Mirror*, 25 October 1935, British Newspaper Archive/British Library Board .

bomb as its subject. The danger of such weapons was also explored in novels by such well-known writers as J. B. Priestley (*The Doomsday Men*) and Harold Nicholson (*Public Faces*). A modern viewer of Alexander Korda's 1936 feature film *Things to Come*, scripted by H. G. Wells, would be struck by the convergence of discourse about the devastating conse-quences of gas warfare and the fears of nuclear warfare, then and later.[71] Even C. P. Snow wrote in a 1936 article in the influential *Spectator* magazine that the priority of applied science should be 'the abandonment of scientific efforts toward destruction'.[72] Both peace and the standing of science itself were at stake.

The Church

The most sustained questioning of the benefits of a modern society rationalised on the back of applied science came from the Church and religious thinkers. Between 1911 and 1931, membership of the Church of England fell, if only slightly, while Roman Catholicism grew substan-tially.[73] Against Wells' positive interpretations of 'modern civilisation', the Catholic novelist and conservative thinker Hilaire Belloc responded with a book reflecting on the loss of morality and faith since the Reformation.[74] Belloc's riposte sold fewer copies than Wells' *Outline of History*, but religion's appeal was real. Religious leaders led further questioning of 'the application of science'.[75] In 1924, the independent Anglican newspaper *The Church Times* appointed its first lay editor, Sidney Dark. He exposed the anxious debates about the place of science and reflected on the race between new devices and opportunities, and the ability of society and the Church to respond. There was a rich discourse on which to draw. Some, such as the moderator of the Church of Scotland, warned of a world civilisation based on applied science but with no place for God.[76] In a 1937 lecture reported by *The Scotsman*, the professor of divinity at Edinburgh complained that secularism, driven by liberal education and applied science, united modern intelligentsia.[77]

---

[71] Willis 1995; Thompson 2022. See also the classic study of nuclear fear from an American perspective, Weart 1988, 17–35. *Things to Come* is available on the web at www.youtube.com/watch?v=atwfWEKz00U (accessed January 2023).
[72] Snow 1936. Quoted by permission of *The Spectator*.
[73] Wolffe 1994, 69. For an extended treatment of English Christianity between the wars, see Hastings 2001.
[74] Belloc 1926.   [75] See Bowler 2010.
[76] 'World Conference: Stewardship and Church Finance', *The Scotsman*, 23 June 1931, British Newspaper Archive/British Library Board .
[77] 'Lee Memorial Lecture. Opposition to Christianity. Prof. Baillie's Analysis', *The Scotsman*, 24 May 1937, British Newspaper Archive/British Library Board. Baillie

He seemed to be referring to communism and secularism in general, while the other risk he saw was in the demonic abuses of Nazi Germany. The message, but not the underlying assumptions, had been different in the famous 'science holiday' call from Edward Burroughs, the Bishop of Ripon. In a sermon during the British Association meeting in 1927, he suggested that a decade's holiday from scientific research would allow time for humanity to catch up and focus on working together.[78] The next day, Burroughs explained to readers of London's *Evening Standard* that he should not be taken as an enemy of science, nor was he, for instance, attacking work on evolution. Indeed, rather than suggesting the cessation of research, he was calling for the remaking of humanity as a higher priority.[79]

Many religious thinkers cited applied science as not 'bad' but a challenge that needed to be met. Among Anglicans, William Inge, a professor at Cambridge and then Dean of St Paul's, was a distinctive figure. Harking back, his thought bore many parallels to Coleridge and the distinction he had made between a more elevated pure science and a necessarily less enduring applied science.[80] An enthusiast for eugenics, an academic, and perhaps the best-known theologian of the time, Inge was an influential agency for promoting thought about science within the church. Dealing with 'the age of science' in his 1930 volume *Christian Ethics and Modern Problems*, Inge suggested that applied science was the inevitable result of the modern confluence of science and industrialism. Although neither of these was dangerous, the combination was so transformative that it confronted society with completely new challenges.[81] Inge's pervasive influence can be felt in the Christian student movement's similar diagnosis, published and widely reported in 1932, as unemployment had once again risen steeply.[82] Even when not their principal concern, churchmen could often seem worried about applied science.

During the interwar years, usage of the twin categories of 'pure' and 'applied' took separate trajectories in the press. In the wake of the horrors of war, widely associated with the use of science and exemplified by poison gas, there were attractions to a re-emphasised distinction between the twins. If the linkages across science seemed essential to some scientists, disconnection would be crucial from the standpoint of public relations. Nor was this just a matter for public intellectuals. Worrying about

addressed the anti-Christian thrust of contemporary political ideologies as well as applied science.

[78] 'Truce from Science', *The Scotsman*, 5 September 1927, British Newspaper Archive/British Library Board. On this sermon, see Pursell 1974 and Mayer 2000.

[79] 'Bishop of Ripon on the Science "Holiday"', *The Evening Standard*, 6 September 1927.

[80] For a comparison of the thought of Coleridge and Inge, see Swiatecka 1980, 117–30.

[81] Inge 1930, 208.   [82] Compare Demant 1932 with Inge 1930, 204.

the status of science and blaming a British public poorly educated about science, a 1938 letter to *The Times* also complained. Noting, "'Better", many say, "to abolish 'science' altogether than suffer from poison gas'", the correspondent diagnosed the 'root of the trouble' as the widespread failure to distinguish between pure and applied science.[83]

The distinction between pure and applied science and their positions in a moral hierarchy were promoted by Oliver Lodge, whom we encountered earlier as a 'Maxwellian'. An influential and widely loved scientist, in 1930, he was chosen by the readers of *The Spectator* magazine as one of the country's five 'best brains' (he came second to George Bernard Shaw and well ahead of Winston Churchill in fourth place).[84] Lodge, of working-class origins himself, and developer of one of the first radio tuners (a coherer), had a strong social conscience and respect for the greatest intellectual achievers. For him, a perhaps unlikely follower of the aesthete and socialist John Ruskin, science was a calling. Repeatedly, his theme was a hero of the past, such as Isaac Newton, a 'heaven-born' and 'epoch-making' man.[85] He distinguished between such a figure and the smaller characters who pursued science to their personal ends as if it were like any other trade. Having lost a beloved son in World War One, he condemned the military uses of science, decrying the tendency by which nations competed to find the most efficient means of killing. Indeed, Lodge wrote to the Bishop of Ripon after the sermon to express his support for the underlying message.[86] A decade later, the *Mass Observation* surveys emphasised many informants' worries that science was getting 'ahead' of people's ability to cope.[87] So while it is easy to read the Bishop's sombre message as the provincial conservatism of a cleric left behind, his distinction between a civilised pure science and a culturally threatening applied science was quite conventional.[88]

Exhibitions and Discussion: Separating Pure and Applied Science

Conceptually, there was an attraction for the scientist to disavow responsibility for the application of his work. The vast 1924 British Empire Exhibition, constructed on a World's Fair scale, embodied this

---

[83] M. D. Hill, 'The Word "Science"', *The Times*, 9 June 1938, © M. D. Hill News UK and Ireland Ltd 9 June 1938.

[84] 'The Five Best Brains' 1930.   [85] Lodge 1893, 7.

[86] Lodge 1926, 33. For the letter to the Bishop of Ripon, see Lodge to Ripon, 9 September 1927, OJL 1/67/1, Cadbury Research Library.

[87] See, for example, Madge and Harrison 1939, 8–19. See also Hinton 2013.

[88] See, for instance 'Destructive Science'.

distinction. The Palaces of Industry and Engineering displayed applied science. The chemical manufacturer Nobel Industries showed a decorative arrangement of explosives in the former, while an awe-inspiring sixteen-inch naval gun dominated the latter.[89] Pure chemistry sat alongside its industrial applications in the Palace of Industry to emphasise the dependence of chemistry's commercial and technical developments on pure science.

Entirely separately, a display commissioned by the Department of Scientific and Industrial Research treated 'pure science' in the Government Pavilion. Organised by the Royal Society, this was generally about discovery rather than invention.[90] At its heart lay the work of Cambridge's Cavendish Laboratory and the search for new particles.[91] Nonetheless, visitors did need reassurance that scientists were not wasting taxpayers' money. The inclusion of electronic valves among the 'pure science' displays emphasised the message, expressed in the *Handbook*, that inventions grew out of humanity's curiosity-driven search for natural knowledge. Even if it was proving difficult to specify, there was wide acceptance of the need to distinguish between harmless pure science and dangerous applied.[92] The Depression of the early 1930s and associated worsening unemployment heightened the criticism of the latter.

The 1932 presidential address at the British Association for the Advancement of Science, normally an occasion for celebration, proved instead to be an occasion for sharing anxiety. The speaker was the leading engineer Alfred Ewing, who reflected on the future of the individual artisan.[93] His talk contrasted pure science's wonderful achievements and engineering's frightening consequences. While Ewing phrased his disillusion in terms of engineering, a review of his subsequent collection of essays by the chemist and prominent science journalist Alexander Smith Russell raised analogous doubts explicitly in terms of the goodness of applied science.[94]

---

[89] See photographs of the Vickers display in 'Equal to Six Trafalgar Squares: The Palace of Engineering', *The Illustrated London News*, 24 May 1924, British Newspaper Archive/ British Library Board, and a drawing in British Empire Exhibition, SR Jones Collection, SC/GL/SJ/002/ k1280266, London Metropolitan Archives. For the explosives display, see Nobel Industries Limited 1924.

[90] British Empire Exhibition 1924, 63; Royal Society 1925. See Boyle 2020, 101–4.

[91] Royal Society 1924, 143. On the atomic structure displays, see Hughes 2016.

[92] Rutherford famously dismissed the possibility of the uses of nuclear energy, perhaps more to protect his research from suspicion than out of complacency; see Jenkin 2011.

[93] Ewing 1932.

[94] Russell 1933. Even in the letter pages of *The Times* the theme would recur; see M. D. Hill, 'The Word "Science"', *The Times*, 9 June 1938.

Meanwhile, pure science could be portrayed as much more safe for civilisation. Having established his journal *Isis* before the war, the Belgian George Sarton fled to America from the experience of World War One under German occupation. For him, the discipline of the history of science was deeply interconnected with humanism and had little to do with practice.[95] Sarton's enthusiasm to promote better values through a proper interpretation of the past was paralleled in Britain. The visitor to Oxford can still find an initiative inspired by a wish to integrate science and culture through the history of science. The Ashmolean Museum, built in the seventeenth century, is the world's oldest dedicated museum building. For approaching a century, it has housed the Museum of the History of Science, fought for and founded by the passionate museophile Robert Gunther through the 1920s. The three key players were the collector Lewis Evans, Gunther himself, and his patron, Oxford's distinguished Regius Professor of Physic, the Canadian William Osler, familiar to Americans as one of the founders of the Johns Hopkins Medical School. Osler was horrified by the war and its barbarity.[96] In October 1915, a few months into the gas war, he spoke at Leeds School of Medicine and described to the assembled students his recent nightmare, which drew on the fear of both poison gas of the present and the use of atomic weapons in the future.[97]

Osler proposed to counter the association of science with barbarism by strengthening the link between civilisation and the humanities. He was particularly interested in its older history, from the time of the Greeks up to the Middle Ages. Thus, addressing the Classical Association in Oxford in 1919, he condemned barbarism and praised the importance of cultivation through study of the humanities.[98] Linked to the occasion, Gunther had mounted an inspiring exhibition of ancient, mediaeval, and early modern instruments. These were from the collection of Lewis Evans, the brother and travelling companion of Arthur Evans, excavator of Knossos, whom Cathy Gere has studied with such profit.[99] She has shown that Arthur had intentionally interpreted Minoan civilisation as a haven of peaceful and feminine culture, contrasting with the recently released barbarity in Europe. In his catalogue of Lewis Evans' collection, Gunther was sure to make an equivalent point. 'Scientific instruments are the chief means employed by civilized man to improve, and to broadcast his civilization,' he wrote on the first page.[100] Gunther would argue that the scientific instruments of the Renaissance were the

[95] See Sarton 1931. On Sarton, see Merton and Thackray 1972; Pyenson 2007.
[96] On Gunther, see Simcock 1985; on Arthur Evans, see Gere 2009.
[97] Osler 1915, 798.   [98] Osler 1919.   [99] Gere 2009.   [100] Gunther 1925.

northern European counterparts of Italian painting and a worthy model for today.[101] This material culture approach was quite different from the intellectual connoisseurship of George Sarton. However, the two men shared a commitment to transmitting the beauty of the past and avoiding the more recent enterprise of 'applied science'. Here was a definition, by exclusion, of a distinctive civilisation.

To counter criticism in the United States, a string of books about chemistry accentuated the positive.[102] In Britain, popular exhibitions showed the benefits of applied science.[103] In 1931 a huge exhibition at the Albert Hall celebrated the centenary of Michael Faraday's discovery of electromagnetic induction, the basis of the electric motor. Mounted by the electrical industry, the exhibit showed Faraday and historic relics associated with his work, surrounded by concentric circles of wonderful contemporary machines to which it had led. Based on late nineteenth-century biographies, the exhibition demonstrated both the totemic significance of Faraday's pure science and its blessing of the lineage of the electrical industry's applied science.[104] At the opening, broadcast on the BBC and widely publicised in the press, General Smuts described Faraday as the 'patron saint' of electrical science, indirectly responsible for modern developments in 'pure and applied' electricity.[105]

The Faraday exhibition lasted only a couple of weeks. However, a few hundred yards south was the Science Museum. Within a short time of the new building's formal opening in 1928, it had become the most attended museum in the Empire and the first with over a million visitors a year. Such attendance meant that in the decade leading up to World War Two, it would attract a significant portion of the country's population of less than 50 million.[106] The Science Museum contrasted vividly with its Oxford counterpart. Though it did mount a few history-oriented temporary exhibitions, such as of the Royal Institution's historical collections, many others celebrated the work of the Research Associations, modern research on ultra-low temperatures, the future of plastics, and

---

[101] 'Collection of Historic Scientific Instruments. Public Opening by the Earl of Crawford & Balcarres … May 25 1925', f. 52 Old Ashmolean Letter Book, vol. 1, Archive of the Museum of the History of Science, Oxford.

[102] See Ede 2002.   [103] See Boyle 2020, 98–104.

[104] This interpretation is based on Frank James' careful study. See James 2008. For the treatment of Faraday in late nineteenth-century biographies, see Cantor 1996.

[105] 'Faraday the "Patron Saint" of Electricians', *Western Daily Mail*, 24 September 1931. The opening speeches were reprinted in full in Paterson et al. 1931.

[106] A 1927 silent film shows four leading science museums of the 1920s, including the Science Museum and its counterparts in Munich, Paris, and Vienna. This has recently been set to music and is available on the web. See www.jeanphilippecalvin.com/museums-of-the-new-age (accessed May 2020).

the challenge of smoke abatement.[107] In November 1938, a few months before war began, it defied the nay-sayers with an exhibit on science in the army.[108] After a 1932 visit, C. L. R. James, the Trinidadian anti-colonialist, journalist, and historian, felt awe and admiration on seeing the Schneider trophy–winning Supermarine seaplane, persuaded by the beautiful shape and well-written label.[109] The government's investment in the celebratory 'cathedral of applied science' had been successful.[110]

Applied Science and New Media

The ideology of applied science as the base of a problematic western civilization was widely held and most influentially articulated through discussions amongst intellectual friends, expressed in print and through the new medium of the wireless. Following correspondence between Julian Huxley and the political scientist George Catlin (husband of Vera Brittain), *The Realist: A Journal of Scientific Humanism* launched in 1929.[111] The literary editor, Gerald Heard, and the editor, A. G. Church (Secretary of the Association of Scientific Workers), were both friends of the Huxley brothers. At the heart of affairs was the magazine's publisher, Harold Macmillan, a maverick Conservative MP committed to social reform. He had been a contemporary at Eton of J. B. S. Haldane, keeping up a friendship, and had a well-established interest in science.[112] The founders hoped to reach out to a self-educating community thirsty for knowledge, with an intended audience of 'the bank clerk … the trades union official, the small professional man'.[113] Indeed, the daily press quickly praised the magazine's distinguished and provocative articles, though the 'stodgy and old-fashioned appearance' was criticised.[114]

---

[107] For a complete list of temporary exhibitions at the Science Museum, see Morris 2010, 317–24.
[108] Quoted from the poster for the exhibit.
[109] James 2003, 4. At the beginning of the twenty-first century, the Supermarine, which would be developed into the 'Spitfire' fighter, is still on display in the Science Museum's Aircraft Gallery.
[110] The description of 'cathedral of applied science' was used in a report of the formal opening. 'A New Palace of Machinery: The King Opens Science Museum', *The Manchester Guardian*, 21 March 1928.
[111] For discussion between Catlin and Huxley, 25 December 1927, Box 9 folder 5, Papers of Julian Huxley, Rice University Special Collections and Archives, Woodson Research Center. For the entitling of *The Realist*, see Catlin 1972, 119. For a more detailed account of *The Realist*, see Bud 2017.
[112] Church to Catlin, 27 July 1928, Box 142, George Edward Gordon Catlin fonds, the William Ready Collection of Archives and Research Collections, McMaster University.
[113] Catlin 1972, 102. Quoted courtesy of Colin Smythe Ltd.
[114] 'The Realist', *The Yorkshire Post*, 7 November 1929, British Newspaper Archive/British Library Board.

Through essays on such topics as humanism and science, editors of *The Realist* emphasised an intellectual commitment to understanding the modern world. J. B. S. Haldane began a piece in the second volume (November 1929) with a reflection on the dependence of western civilisation on applied science.[115] *The Realist* offered potential advertisers an association with cutting-edge thinking, science and industry's contributions to progress and happiness, and famous writers.[116] Nonetheless, in the wake of the Wall Street Crash, subsidies withered, and, after just two volumes, the magazine failed.

A replacement for print, the recently available resource of the BBC radio broadcast now offered an opportunity to the *Realist*'s devoted circle. The friends were fortunate that just when they needed help, a newly appointed producer provided support and leadership. Mary Adams had been a professional botanist and held left-leaning political views. She operated skilfully within an organisation whose central place in British society meant that the cultural evangelism of dedicated producers and Lord Reith, its leader, had to be tempered by an awareness of public scrutiny (see Figure 6.2).[117]

Adams introduced a 1933 volume affirming, 'The practical applications of science order our civilization.'[118] During the period 1930–34, she commissioned series after series that addressed science's challenges to modern society, drawing on the same community as had animated *The Realist*. Gerald Heard, formerly the magazine's literary editor, hosted fortnightly scientific programmes, whose last series was entitled 'Science in the Making'.[119] Heard was a close friend of Aldous Huxley, and would later develop a special interest in consciousness and psychic research in post-war America. At the BBC, his very personal broadcasts took a quizzical, questioning approach to the remarkable phenomenon of scientific advance.[120]

---

[115] Haldane 1929, 49.

[116] 'The Future of the *Realist* (from the Advertising Manager's Point of View)', enclosed with A. G. Church to G. Catlin, 9 December 1929, Box 142, George Edward Gordon Catlin fonds.

[117] For the cultural evangelism of the BBC's first Director-General, see Briggs 1995, 12. For a more detailed treatment of the science programmes, see Bud 2017.

[118] Adams 1933, 11. See Sally Adams, 'Adams [née Campin], Mary Grace Agnes (1898–1984), Television Producer and Programme Director', in *Oxford Dictionary of National Biography*. On Adams' career and contribution, see Jones 2012; Murphy 2016, 181–88; and Jones 2020.

[119] On Heard, see Falby 2008 and www.geraldheard.com/. Heard published his 1934 talks as *Science in the Making* (Heard 1935). This series was a later development of a 1931 BBC series discussed by Jones 2020.

[120] See, for instance, Heard 1931.

Figure 6.2 Mary Adams, pioneer producer of science programmes on the BBC. © BBC Archive.

Under Adams' leadership, nearly every week, there was a programme about science on the BBC's single national channel, at peak time, typically at 7:30 pm on Wednesdays. Many constituted substantial edited series on such topics as science and religion, science and civilisation, and science as a social activity. Contributors came from across a broad spectrum of opinions. For example, the six-part 'Science and Civilisation' broadcast early in 1932 included both J. B. S. Haldane and Hilaire Belloc, the Catholic controversialist known for sparring with H. G. Wells.

The next year, the BBC decided on national surveys of industry, religion, and science. Julian Huxley was recruited to visit laboratories and reflect on his observations with such friends as the physicist Patrick Blackett. Huxley, struck by the amount of research conducted in industry, sought to tease out the relationships between pure and applied science. Though challenged on his assumption of a 'linear model' and willing to accept the occasional two-way relationship between the two, he came down to a sequence going from pure research to development. Thus, towards the end of his talks, Huxley admitted he had re-evaluated his earlier assumptions that 'I used to imagine that important new discoveries always started as pure science, and gradually filtered down into practice, via applied science.'[121] Though he still believed 'that was the usual way', he was willing to accept there were exceptions. Huxley offered an entirely conventional view of the time and expressed the language and model that pervaded contemporary DSIR reports. Nonetheless, through his broadcast and the transcripts published in both magazine articles and the volume *Scientific Research and Social Needs*, he added the reach of radio, the authority of a Huxley on the BBC, and the permanence of a book.

With its breadth of perspectives, the radio confirmed the reality of applied science. Collectively, these widely popular broadcasts placed it at the heart of issues and arguments central to the day's concerns. However, Adams' tenure in radio was also destined to be short. Responding to press claims of left-wing bias, the BBC reorganised its talks department, and in 1936, Adams moved to the new television service. There, after another war, she would nurture the talents of a young David Attenborough.

The printed works of the biologist Lancelot Hogben served as a sequel to the radio programmes promoted by his close and long-standing friend Julian Huxley. Two were especially popular, *Mathematics for the Million*, published in 1935, and its sequel, a history of science, entitled *Science for the Citizen* (1938). Beyond the intended breadth of its appeal to those committed to society making the best use of science, *Science for the Citizen* was remarkable as a successful account of applied science, offering a

---

[121] Huxley 1934, 204. Quoted by permission of the Rationalist Association. For this series, see Friday 1974.

mixture of history and equations.[122] Despite their demanding style, with challenging quizzes at the end of chapters, Hogben's books won huge, unexpected popularity. By 1940, *Mathematics for the Million* had sold 150,000 copies, and the recently published *Science for the Citizen* had managed 50,000 in less than two years.[123]

While these books addressed interests similar to the radio programmes Huxley had promoted, the more left-wing Hogben argued in *Science for the Citizen* that pure science was a derivative of applied science. This inversion of the more typically cited sequence reflected the direct impact of the Soviet historians who had descended upon the 1931 history of science congress in London to present a materialist view of the relations between action and thought.[124] It was also an expression of a broader trend. Remarkably, during the period 1900–40, though neither before nor after, the relative frequencies of 'applied science' and 'history of science' fluctuated together in 'British English' publications searched by Google and represented in 'Google n-grams', annually rising and falling in step with similar frequency of use. Even bearing in mind the caution with which such a corpus must be used, this pattern emphasises that the past of applied science was already a well-appreciated concept and retold story.[125]

Through the 1930s, science's utility served as an essential quality for scientists concerned with social relations. These included a wide range of researchers, academics, businesspeople, teachers, and journalists. The zoologist Solly Zuckerman ran a dining club called the 'Tots and Quots', including J. B. S. Haldane, Lancelot Hogben, and J. D. Bernal. A note in the Zuckerman archive suggests the growing emphasis on applied science was the topic of an early discussion topic (dating possibly from 1932).[126] In 1938 the British Association formed a new division of social and international relations of science, incorporating the British Science Guild. The establishment editor of *Nature*, R. A. Gregory, was its ruler, but Bernal was its prophet. His *Social Function of Science*, published the

[122] Hogben 1938, 9.
[123] For sales figures of Hogben's books, see Bowler 2009, 110 and 112.
[124] On this well-known event, see Werskey 1978; Chilvers 2007.
[125] On history of science between the wars, see Scheinfeldt 2003. Although there was no British society devoted to the history of science per se until after World War Two, the discipline was the subject of many lessons in schools. See Mayer 1997. On the opportunities and limitations of 'n-grams' as a source, see O'Sullivan et al. 2019; Pechenick et al. 2015.
[126] Tots and Quots records, SZ/TQ/1/1/18, Papers of Solly Zuckerman, University of East Anglia, Archives Department. On the Tots and Quots, see Ralph J. Desmarais, 'Tots and Quots (act. 1931–1946)', in *Oxford Dictionary of National Biography*.

next year, would be the 'bible' of the new movement.[127] True to his socialist beliefs, Bernal hailed the potential of applied science and criticized the government for insufficient support. The term, scattered throughout the book, appears on more than thirty occasions. The culture that infused policy also put its reality, power, and fearsomeness beyond question.

Conclusion

By the late 1920s, there was a broadly accepted narrative about science as part of contemporary culture. Applied science symbolised and described a new world of opportunity and modernity. It seemed to be the mechanism by which pure science research converted into a multitude of new gadgets, some wonderful but some frighteningly dangerous. This conversion was seemingly out of popular control. To some, such as Alfred Mond, R. B. Haldane, and Arthur Balfour, it was too slow due to the historical inadequacy of British industry. Such sentiments sustained the arguments, and the budgets, of the DSIR.

To others, such as articulate churchmen, *The Daily Mirror* columnist Richard Jennings, and the writer Aldous Huxley, badly applied science provided visions of a modernity that was changing dangerously and too fast. Characterising modern civilisation, the recent experience of World War One and the preparation of weapons for the next conflict were threatening its standing. Such rapid progress was challenging the ability of citizens, politicians, and scientists to adapt. The tension between these interpretations powered cultural storms whose waves would crash, time and time again, through the century. For scientific institutions, these would translate into urgent issues of communications and policy. Yet, at the same time, popular alarm reinforced a conviction in the potency of applied science held by the public, scientists, and policymakers alike.

---

[127] A brief but telling account of the image of Bernal in the 1930s was given by the historian and philosopher of science Stephen Toulmin in 'Progressive Man', *New York Review of Books* 6 (3 March 1966): 20. The standard biography is Brown 2006.

Stage 3

After World War Two

7   Co-existence through Growth

Introduction

World War Two is renowned for the scientific expertise deployed by the combatants. As soon as it could, Britain celebrated the achievements of such projects as 'Tube Alloys' (atomic bomb) and the automated decryption of German codes, the jet engine, radar, and penicillin development.[1] Under ministerial pressure, in 1967, the Post Office launched a series of stamps depicting the latter three successes together with television as part of an intended international rebranding of Britain.[2] Even if, to the historian, each of these developments is a complex multinational story, accounts of their distinctively British parentage circulated widely. We can see the British view of the time in the still frequently used film *Penicillin*, funded by Imperial Chemical Industries.[3] Made in 1944 to counter American claims, the film captured a sense of national achievement.

Despite pride in national achievements, pre-war investment in civil and military research was frequently described as woefully inadequate.[4] Previously, it seemed to campaigners of the 1940s that austerity had stunted the growth of a new society, and successive governments had foolishly restricted funding such public goods as science. During the early post-war years, the meaning of applied science did not change,

---

[1] Some of the important, if subsequently less well-known, triumphs of wartime industrial research were discussed at a meeting of the British Association for the Advancement of Science, published as British Association for the Advancement of Science 1945.

[2] Jones 2004, 163–76.

[3] A slightly later edit of the film *Penicillin* and its transcript can be seen on www.youtube .com/watch?v=-7HmDfgaakg (accessed December 2022). About the film, see Bud 2007, 70–71.

[4] See, for example, the outcome of the 1944 symposium at Nuffield College, 'Problems of Scientific and Industrial Research', published as Nuffield College 1944. In recent decades, the achievements and strategies of the 1930s have been re-evaluated. See, for instance, Finney 2000. See also the review article by English and Kenny 1999 and Edgerton 1997.

and, indeed, hopes for its benefits grew, though its adequacy as a tool used in isolation was increasingly questioned. Governments massively increased investment in scientific research of all kinds, inspired by hopes of leap-frogging the industrial, military, and medical advances of others.[5]

Across many of the world's countries, the era saw historically high growth rates and increasing prosperity. In Britain, scientists interpreted and exploited the experience using traditional rhetorical tools. Weighty issues were at stake: national survival in the face of nuclear threat, economic reorganisation and the transformation of manufacturing industry, the generation of electricity to meet rapidly increasing demand, the sustenance of the 'science-base', and the substantial enlargement of the professional engineering community. When talk of decline was a significant political reality, more applied science was widely proposed to solve the national malaise.[6]

Expansion was indicated by the changing size of industrial research. In 1945/46, British companies reported spending on research and development four times more than just seven years earlier in 1938 and a dozen times more than in 1930.[7] National expenditure on research and development, which doubled during the years of rearmament between 1930 and 1938, doubled again between 1938 and 1950 and increased by 40 per cent in just the three years between 1955/56 and 1958/59.[8]

This chapter addresses the questions of how post-war policymakers reshaped the models of applied science and technology, which coexisted in education and research policy and in the management of innovation. More than elsewhere in the book, it deals with the import of usage from overseas. In the years after World War Two, the United States held a hegemonic scientific and cultural place, which meant that its language and issues came to be influential across the Atlantic. They were, of course, not just passively absorbed, and we can see how British authorities, steeped in domestic experience and tradition, reconstituted as they ingested. The case study of the UK Atomic Energy Authority (UKAEA) shows how sharply rising investment in research sustained a variety of models of the knowledge required for industrial development and temporarily eased their coexistence.

For many, the deployment of atomic weapons instilled further terror in the power of science. Meanwhile, the nation's leaders and scientists

---

[5] On the technocratic hopes of post-war British governments, see Edgerton 2006.
[6] Tomlinson 2000. See also Hennessy 2018, 9.
[7] Federation of British Industries 1947, 4.
[8] Melville 1962, 2–3. Between 1933/34 and 1938 expenditure on defence doubled. See Thomas 1983.

frequently cited the peaceful uses of nuclear energy to counter the identification of atomic research with mass destruction.[9] Such arguments offered modern parallels to the efforts for a positive cultural inspiration that had followed World War One. There was also a human side to the linkage between the two eras. Men and women shaped by pre-war experiences came to the fore as science interpreters. Mary Adams became the 'Head of Talks' for BBC television. Julian Huxley won the founding-directorship of the international cultural body UNESCO, and his friend, the 'Tots and Quots' member Joseph Needham, took charge of its scientific division. At the November 1945 founding meeting in London, the British Minister of Education suggested that science needed to be part of this international cultural organisation because the world required its accountability.[10] Mathematician Jacob Bronowski, filmmaker Humphry Jennings, and architect Misha Black, in their youth, had all been involved in establishing the Cambridge science-promoting cultural magazine *Experiment*, founded in 1928. Professionally established in the post-war years, each became important in the spirit-raising 1951 Festival of Britain and in expressing a new post-war culture.[11]

In a 1962 paper for the left-wing Fabian Society, biologist John Maynard Smith reflected on the post-war ascendancy of a generation of scientists. Dismissing the condescension of a vocal minority as irrelevant, he suggested that one should look instead at the positive views of the majority.[12] A published variant of Smith's paper reflected that now science and scientists were getting newspaper coverage somewhere between 'holiday travel and fashion', and made news in unprecedented ways: 'Now, the cult of personality builds up for us pictures in the newspapers of powerful, shadowy figures to whose tune the politicians dance.... These men make decisions that affect us all. They speak in the name of science, and their words become gospel: only an expert can disagree with an expert.'[13] Elite scientists had become leading lights of a culturally coherent social movement whose discourse was well versed and whose members were influential promoters of science and expertise.[14]

---

[9] Laucht 2012.
[10] On science at UNESCO and the roles of Huxley and Needham, see Petitjean et al. 2006. For the speech of Ellen Wilkinson at the founding conference, see Archibald 2006, 39.
[11] On *Experiment*, see Donaldson 2014.
[12] Untitled paper by Maynard Smith, May 1962, the Fabian Society Science Group papers, Papers of Patrick Blackett, PB/5/3/1, GB117, Royal Society, Library and Archives. This was apparently published as Russell 1962. On this article, see Vig 1968, 87.
[13] Russell 1962, 10. Quoted by kind permission of *New Left Review*.
[14] On these people before World War Two, see Werskey 1978.

New Institutions

None of the numerous policy papers about science in the closing stages of World War Two and its aftermath proved definitive or pathbreaking. There was no match for the institutional and cultural revolutions of the previous war, such as the Department of Scientific and Industrial Research (DSIR); for Britain's 1942 Beveridge Report for the fundamental reconstruction of social security; or for America's 1945 science policy statement, *Science the Endless Frontier*. Nevertheless, high-level discussions about British science and industry's future expressed extensive political engagement. A private meeting at Nuffield College Oxford in 1944 drew on the contributions of ninety-two eminent scientists and economists. Its outcome was a brochure that arrived on the Chancellor of the Exchequer's desk, in time for the budget, under the title: 'Problems of Scientific and Industrial Research'. At its heart lay the distinction between fundamental and applied research and the dependence of the latter upon the former.[15] The recommendations expressed the consensus growing since the previous war: a call for greater state participation in scientific and industrial research and a network of research associations.

No dramatic action followed directly on the Nuffield College plea. However, reflecting its sentiments, the government did make a limited number of immediate post-war organisational innovations, going beyond the litany of 'more'.[16] I wish to point to three that would shape subsequent debates over the established 'applied science' category. One was a sequence of high-level advisory councils. In wartime, the government had established a Scientific Advisory Committee. A new peacetime administration replaced this with two committees lasting until the 1960s. The Advisory Council on Scientific Policy and the Defence Research Policy Committee were responsible for civilian and defence research advice, respectively.[17] Whereas the military committee was naturally secretive, the civil science committee's annual reports provided information to the public and important indicators of elite scientific thinking to the historian. Indeed, its considerations defined the official scientists' consensus. The next chapter deals with the late 1960s conflicts of its successor, the Council for Scientific Policy, over concerns at the heart of government about the standing of applied science.

[15] Nuffield College 1944; see Bud 2018.
[16] A further major innovation that was framed in terms of 'fundamental research' was the Colonial Research Fund, established in 1940, which was, for some years in the 1940s, the country's second largest sponsor of civil research. Clarke 2018.
[17] On the Defence Research Policy Committee, see Agar and Balmer 1998; see also Agar 2016 on its successor. On the Advisory Council on Scientific Policy, see Gummett 1980.

An entirely different legacy of this period expressing concern with development and invention was the 'National Research Development Corporation' (NRDC). Created by the post-war Labour government in 1948, its formation had responded to national resentment. Whereas British scientists had first developed penicillin, American companies had reaped the profit. Its title reflected Labour's determination to maximise the benefits from research's outputs. Economic problems and the administration's subsequent fall initially stunted the early hopes for the NRDC. Still, these would revive under the later Labour government of the 1960s, before a long demise and final abolition by Margaret Thatcher in 1981.[18] A third post-war initiative was the institutionalisation of research into atomic power, which would become the new reference point for the potential of applied science. At the operational level, this included the Harwell laboratory established in rural Oxfordshire in 1945, complemented by the semi-autonomous administering body, the UKAEA, of 1954. These powerful and high-profile new organisations would mark a new era in the country's cultural history.

Politicians and Applied Science

Politicians took these initiatives in the confident belief that, in supporting science, they had the public ear and were expressing widespread impatience. It was now they who recited the allegorical stories of scientific heroes. Two were favourites. According to one, American revolutionary hero and scientist Benjamin Franklin supposedly responded to a question about the significance of his studies of electricity with the retort, 'What is the good of a newborn baby?' The other was the story of Michael Faraday, showing his discovery of electrical induction to Prime Minister Gladstone and reportedly commenting, 'Soon you will be able to tax it.'[19] Typically, Food Minister Lord Woolton invoked both when addressing the British Association for the Advancement of Science in 1945. He not unusually amalgamated the two accounts, attributing the reply of Benjamin Franklin to Michael Faraday, as he emphasised the importance of applied science.[20]

---

[18] Keith 1981.
[19] Cohen 1946. It is striking that it was during the immediate post-war period that Harvard's distinguished historian I. B. Cohen explored the history of the two frequently cited anecdotes. In a later article, Cohen pointed out that they were frequently confused. See Cohen 1987.
[20] Woolton, 'The Future: What Science Might Accomplish', in British Association for the Advancement of Science 1945, 142–46, see 145. The proceedings of this meeting dealing with the post-war world invoked the name and therefore the story of Faraday eight times.

The decline of imperial power cast the need for a reinterpreted past into sharp relief. In January 1957, the liberal-conservative politician and publisher Harold Macmillan was hurtled unexpectedly into the post of British Prime Minister. His first challenge was the legacy of the failed occupation of the Suez Canal. Despite a swift military victory, the British and French had immediately withdrawn under American pressure. The government leadership changed suddenly, and Harold Macmillan came to power. The debacle occurred precisely two centuries after the beginning of the Seven Years' War, which had launched Britain's global leadership. Seeing 'Suez' as the end of the age of the British empire, many felt they needed a fresh national narrative. The new leader, who had published *The Realist* magazine in his younger days, gave a speech broadcast to the nation on his first night in office and printed the next day in the press. Dismissing those who wrote off the nation, he reassured his people that their accomplishments in science and invention, such as the steam engine or the recent development of the world's first nuclear power station at Calder Hall, showed they would overcome their current difficulties.[21] For his audiences and readers, the role of scientists in 'the war effort' was lived experience, captured, for instance, in a commissioned account co-authored by J. G. Crowther, the doyen of science journalists and close associate of many wartime scientists.[22] Radar, operational research, the jet engine, and penicillin were then matters of special pride. In the post-war years, such tales of scientific accomplishment promised a bridge from a glorious past and austere present to the post-war future.[23]

The rhetoric adopted by the Prime Minister was also becoming a familiar publicity tool. In 1954, civil servants at the 'Central Office of Information' had discussed a memorandum addressing the anxiety that Britain should be written off, by means of a high-profile short film about nuclear power. To promote awareness of Britain's early importance in the 'basic research in nuclear physics' and its position 'in the van of the application to industrial power', the head of the Films Division proposed the title 'Atomic Achievement'.[24] Capitalising on the Queen's opening of

---

[21] This speech was widely covered and often reprinted in regional and national newspapers. See, for instance, 'Britain Is Not on Way Out – Mr Macmillan', *Belfast Telegraph*, 18 January 1957, British Newspaper Archive/British Library Board. 'Prime Minister Calls for End to Defeatist Talk', *The Times*, 18 January 1957. For the writing of that speech, see Catterall 2011, 615–16.

[22] Published as Crowther and Whiddington 1947.

[23] See, for example, Bud 1998; Forgan 2003. The film is discussed by Hughes 2021.

[24] Appendix to Langston to Mr Hadfield and Mr Watson, 13 January 1954, INF 12/668, TNA.

the world's first full-scale nuclear power station in October 1956, the Central Office of Information indeed commissioned the film, cited here in the introduction.[25] It is striking how the civil servant and the new premier were drawing on a similar discourse.

Beyond the national narrative, there was intense public discussion over two separate institutional issues long at the core of the concerns addressed by applied science: the production of trained people, on the one hand, and of useful knowledge, on the other. One related to the best context for training, the other over the role and organisation of vast new government-run laboratories intended principally for atomic and military research. In immediate post-war Britain, the shortage of skilled labour dominated many urgent debates. Later, cultural and managerial issues over the management of knowledge came to take centre stage. Management of local traditions and institutional issues would entwine with reflections on research imported from the United States, where the problems were similar, though far from identical.

To some, the discussions of training and of knowledge production related to quite different categories. Sir Harold Himsworth, Secretary to the Medical Research Council, emphasised what seemed to him an important distinction. He reflected to a government enquiry into science's organisation that 'there could be no greater error than to confuse demarcations devised for one purpose (e.g. teaching) with those required for another (e.g. research)'.[26] There might have seemed 'no greater error', but terms flowed effortlessly between contexts.

Therefore, this chapter needs to deal with shifting meanings in three dimensions: the specifically British problems with education, transatlantic policy discussions over research policy, and the debates local to Britain over the management of research funding.

Education. Education. Education

The management of people and the breeding of skills dominated science policy concerns in government and the press during the 1940s and

---

[25] Whether Calder Hall was technically first in the world to produce nuclear power for the electric grid was contested. The Soviet Union had connected a reactor at Obninsk to its grid in 1954, but the power was only 5 MW. The reactor launched at Calder Hall in 1956 produced 60 MW, and this was therefore claimed to be the first full-sized plant to be connected to the grid. Of course, such claims and counterclaims were part of commercial and Cold War propaganda.

[26] Committee of Enquiry into the Organisation of Civil Science, 'Minutes of a Meeting Held at the Treasury, S.W.1 on Thursday, 12th July, 1962 at 10.30 a.m.', p. 1, T218/485, TNA.

1950s. Existing bodies proved remarkably flexible and able to grow through the support of the University Grants Committee (UGC). The so-called dual support of funding (through both UGC and external funders such as the research councils) ensured that the education policy of the UGC would carry through into research support. Moreover, its ambitions discussed in terms of multiplying certain kinds of skill also entailed decisions about knowledge, particularly the lines between 'pure science', 'applied science', and 'technology'. This focus led to criticism that while, on the one hand, the military priorities of the Cold War were absorbing too much of the nation's precious resources, on the other, the UGC was a victim of academic traditions and insufficiently focused on industrial skills.[27]

Continuing a century-old pattern, such debates about categories connected closely with institutional choice in training and education. Recall the early twentieth-century settlement: the universities and associated 'university colleges' taught applied science and prepared students for professional leadership in careers ranging from acoustical engineering to textiles.[28] Meanwhile, approximately 150 local authority-run technical colleges offered training in several hundred trades, from armature winding to upholstery. About thirty also gave 'higher technological education' and conducted research.[29] Between these ends of a professional spectrum were the technological institutes affiliated with universities since the beginning of the twentieth century. Manchester (the Municipal College of Technology), Glasgow (the Royal Technical College), and Imperial College London were always cited.[30] In the background, their importance was emphasised by the oft-cited success of the American MIT model.

As early as February 1944, in the midst of war, the Minister of Education formed a committee to look at 'higher technological education'. The choice of chairman, Lord Eustace Percy, highlighted the enquiry's roots in long-standing debates. Percy had been an education Minister in the 1920s, and while, during the interwar years, his wishes had been overwhelmed by economic crises, now he saw a new opportunity. Despite his aristocratic lineage (his distant ancestor 'Hotspur'

---

[27] See, for instance, Rose and Rose 1969.
[28] Lists of the careers that might follow from education in university and technical colleges were prepared by H. Lowery, principal of the South-West Essex Technical College and School and member of the Percy Committee as part of the Percy Committee deliberations. See Lowery to Bragg, 25 August 1944, Papers of William Lawrence Bragg, 46A/36, the Royal Institution.
[29] See Ministry of Education 1945, 7.
[30] See, for instance, Association of Scientific Workers 1944, 25.

featured in Shakespeare's Henry IV), Percy had long fought to obtain parity of esteem between graduates of technical colleges and universities.[31] Currently, he was vice-chancellor of Durham University, where J. F. W. Johnston had once taught.

Percy's 1945 committee report began with the observation that 'The annual intake into the industries of the country of men trained by Universities and Technical Colleges has been, and still is, insufficient both in quantity and quality.'[32] The committee would have liked both to promote great new technological universities and to upgrade locally run technical colleges to be research-oriented schools of technology.[33] In parallel, a small committee on 'Future Scientific Policy' (the Barlow Committee) produced a 1946 report on scientific 'man-power', commenting that British scientists had achieved estimable advances in fundamental understanding but that 'we were not always so successful in those applications of science which lie in the field of engineering and technology'.[34] The echo of Arthur Balfour, almost to the word, two decades earlier, was now an urgent call to action for a new mechanism and skills to translate scientific expertise into industrial development. The committee estimated that the country would not meet its need for 70,000 scientists by 1950: only 55,000 would be available at current production. To meet new needs, the civic universities would need to double their output, and the Barlow Committee emphasised the challenge.[35] Together, the two reports recommended three developments: first, the expansion of the universities; second, the expansion and development of existing technical colleges; and, third, the development of super technological bodies. This balanced compromise described, roughly, what did transpire. Such a description obscures, though, profound differences in educational philosophy associated with these institutions and the intensity of disagreement about needs and identity.

Fervent pre-war debates about the training of engineers resumed. Focus upon the roles of civic universities and the enhanced technical colleges expressed the social reality of the categories of applied science and technology and their relationship. No wonder the UGC wrote in its

---

[31] For the history of the Percy family and Lord Eustace Percy's place in the lineage, see https://en.wikipedia.org/wiki/House_of_Percy (consulted May 2020).
[32] Ministry of Education 1945, 5. In 1930 Percy had expressed his principles in a widely cited book, Percy 1930. For the general debates on technology education, see Davis 1990.
[33] Ministry of Education 1945, 5.
[34] Lord President of the Council 1946, para 30, p. 10. See Gummett and Price 1977 and McAllister 1987, 129–39.
[35] Lord President of the Council 1946, 22.

1947–52 report, 'No subject has been so much discussed as that of technology, not only by industry, professional associations and other bodies and the Press, but by ourselves and our Technology Sub-Committee.'[36] The report described the distinction between genres of organisation in the interlinked terms of discipline and technology.

The Percy report had seen technology as 'both a science and an art'. This two-fold character had underpinned the justification for a variety of institutions training its practitioners:

> In its aspect as a science it is concerned with general principles which are valid for every application; in its aspect as an art it is concerned with the special application of general principles to particular problems of production and utilisation. Universities and Technical Colleges must deal with both aspects; but Universities have regarded it as their duty to select and emphasise the science aspect, and Technical Colleges the art aspect.[37]

To some, this balance between both functions and institutions was self-evident. A similar distinction appears in a document published by the scientists' trades union, the Association of Scientific Workers, in March 1944, at the same time as the launch of the Percy Committee.[38] However, pressures for systemic change were coming from the sides of both technology and applied science. Churchill's wartime advisor, Lord Lindemann, argued loudly for new British equivalents to *technische Hochschulen*. He had studied in Germany and, during the 1930s, had brought a group of his now-distinguished fellow pupils to Oxford. There, they not only established a world-leading school of low-temperature physics but were also articulate in promoting science education in a technological context.[39] Ultimately, rather than achieving its original objective, the campaign for a post-graduate technological institute would lead to the new college at the University of Cambridge, Churchill College. While that initiative was redirected, in 1956, the government upgraded ten existing technical institutions to university-level 'Colleges of Advanced Technology'.[40]

The universities had retaliated against the promotion of competing dedicated technology schools. Civil servant Edward Playfair, Sir Lyon's great-nephew, argued within the UGC that the German model was inappropriate for Britain. After all, he suggested, the British universities

---

[36] University Grants Committee 1953, 58.
[37] Ministry of Education 1945, para. 24, p. 10.
[38] Association of Scientific Workers 1944.
[39] See, for instance, the newspaper articles collected by Simon 1960.
[40] On the campaign that ultimately resulted in Churchill College, see Walsh 1998; on the Colleges of Advanced Technology, see Burgess 1970.

did teach technology that their counterparts in Germany and Switzerland excluded.[41] By 1951, Lord Percy himself had changed his mind about the possibility of satisfying the need for the distinctive needs of technologists through existing institutions. Now he felt that the universities could and should evolve to meet the needs of industry.[42]

Even in training for the chemical industry, leaders perceived a need for a technological, as opposed to narrowly scientific, emphasis. Much of a chemical plant is a complex manufacturing environment in which engineering, and increasingly chemical engineering, was important. Strikingly, Imperial Chemical Industries sponsored a multi-volume *History of Technology* in the 1950s. Wallace Akers, the company's head of research and leader of Britain's wartime atomic bomb project, was a firm supporter. Before he died in 1954, Akers had read every word of the *History*'s first volume.[43] So technology was being tightly defined and differentiated through these national discussions about education conducted widely in the press.

In the movies, wartime Brits look very pleased with themselves. Their avatars, David Niven and Trevor Howard, were great actors. At the same time, drama frequently represented military hierarchy as blinkered conservatives, and industrialists were invisible. Whether or not the image was accurate, young policymakers, politicians, and scientists were far from self-satisfied. A complaint in a 1946 memo about the Barlow committee's composition caught the need for change: 'youth and adventure are conspicuously missing'. Even this plea conceded that the recently appointed committee would 'not be unduly tender to claims of seniority' because it included Blackett and Zuckerman and, indeed, *The Economist* editor Geoffrey Crowther.[44] Soon these acerbic intellectuals would become the leaders of a new generation addressing the old problems.

Research, Secrecy, and Organised Science

The education debate was rooted in the long-standing challenges of adapting British traditions to international patterns. In contrast, the talk about research, which emerged slightly later, was part of a new

---

[41] E. W. Playfair to Mr Petch, 'G. F. (56)/1 – Technological Education', T/03, UGC 7/ 878, TNA.

[42] Clarke 1951.

[43] Singer 1960. On chemical engineering in 1950s Britain, see Divall and Johnston 2000, 153–217.

[44] M. T. Flett to Mr Nicholson, Mr Pimlott, 13 February 1946, CAB 125/546, TNA. McAllister explores the tensions between radicals and conservatives on this committee in McAllister 1987.

conversation circulating internationally. The United States, with its overwhelming industrial research, government expenditure, and academic system, had set a benchmark for developing machines and capacities, weapons, and knowledge. Moreover, through Marshall Aid and NATO, its support of European countries framed Western Europe's reconstruction on American lines.

The conduct and communication of research in the Cold War context gave special weight to the terminology used to reflect the various secrecy levels. For a century, communication of applied science had been considered more open than technology, which had been consistently regarded as private. Thus, in 1952, the veteran industrial research leader Kenneth Mees urged a distinction be recognised between corporate research in science and technology. Custom and practice suggested that the former be published, but the latter kept secret.[45] Of course, some applied research in industry had always been hidden. However, after World War Two, governments required a veil of institutionalised secrecy over stretches of science greater than ever before. In recent years, this has been a matter of extensive study.[46]

Britain's laboratory responsible for biological warfare provides an interesting case study. While the Porton Down Microbiological Research Establishment ensured that 80 per cent of its work was published, secrecy permeated the institution's culture.[47] If this was not always made obvious, the historian Brian Balmer has observed that it was sensed even by those excluded. He has reflected on the Open Day held in 1968: 'By calling attention to secrets, a particular type of absence is created – something akin to seeing a fleeting glimpse of an object out of the corner of your eye.'[48] Even more in the United States, the regulation of security became a major use, and indeed determinant, of the classification of science.[49]

When drawing up British secrecy rules in 1946, Michael Perrin, the polyethylene discoverer and wartime mandarin, explained new legislation to the recently appointed head of Britain's nuclear research, John Cockcroft. Perrin wrote, '"Basic scientific information" may be freely disseminated; so may information in connection with plants or processes, so long as they are on the laboratory scale.'[50] The instruction was explicit that this treatment was required to mitigate anxieties in the United States

---

[45] Mees 1952, 972.   [46] See Daniels and Krige 2018.   [47] Bud 1993, 113.
[48] Balmer 2015, 43. Material from Brian Balmer, 'An Open Day for Secrets: Biological Warfare, Steganography, and Hiding Things in Plain Sight', published 2015, Palgrave Macmillan, reproduced by permission of SNCSC.
[49] See Daniels and Krige 2018.
[50] Michael Perrin to John Cockcroft, 13 March 1946, AB 16/4876, TNA.

about any bill on atomic energy that did not have a 'clause dealing with the disclosure of information'. Although this related specifically to research on nuclear weapons and power, it reflected discussions over a broader challenge.

As the Manhattan Project became a realistic programme for building an atomic bomb in 1944, the American government felt it should offer a public account and in August 1945 published a report on the bomb's development. Author Henry DeWolf Smyth protected national security by emphasising physical principles already known to the Soviet Union, at the expense of the huge and diverse industrial challenges that had to be overcome to produce working bombs.[51] Thus, with the larger technological issues hidden, the public saw the bomb as pre-eminently a scientific achievement.

At the same time, during hot and cold conflicts with Nazism and, subsequently, Communism, the 'freedom of science' was adopted as a defining quality of liberal democracy. There was thus an attempt to protect the communication of scientific research as 'public'.[52] This approach was foreshadowed in 1943 when James Bryan Conant, the chemist-president of Harvard and a leader of America's scientific war effort, lectured on 'science and society in the post-war world'.[53] He harkened back to a theme, the symbiosis of science and industry, explored years earlier when he had been rather impressed by the work of the English Marxist J. G. Crowther. Now, though, Conant added the evolution of free institutions. Collectively, he identified them as pillars of civilisation, and beyond them, free communication was critical.

At the very end of the war, Conant wrote to the *New York Times* explaining the vision of science he and his close colleagues had adopted. Conant was anxious to demarcate the boundary between pure and applied science, emphasising that while there was 'no difference in methodology or techniques, as to goals there is as much difference as between red and blue'.[54] Later in that letter, Conant translated this distinction into the 'unorganised and undirected' character of pure research, and the 'organised' nature of applied research. The emphasis on such a distinction, it has been argued, served in the United States to protect the autonomy of basic research and its practitioners' freedom to

---

[51] Kaiser 2005, 33. See also Wellerstein 2021.
[52] See Wolfe 2018. Hollinger 1995 deals with the political use of the scientific method.
[53] Conant 1943. The earlier article citing Crowther was Conant 1939.
[54] James Bryant Conant, 'National Research Argued: Dr Conant Favors Federal Subsidies but Wants Freedom for Science', *New York Times*, 13 August 1945. Quotation by kind permission of Ellen Conant.

communicate their work.[55] However, Conant was also pitching into a debate about the soul of science, already rumbling in Britain for some years. Should scientists be free to follow their curiosity, or should they address the challenges that society prescribed? In 1940 the Hungarian immigrant Michael Polanyi, now a professor in Manchester, and his allies organised the 'Society for Freedom in Science', challenging left-wing scientists, such as Bernal, for whom the mission orientation of applied science might serve as the template for the entire scientific enterprise. The Society for Freedom in Science further developed into a study group within the CIA-funded international 'Committee for Cultural Freedom', which met in 1950 in the still-devastated German city of Hamburg.[56]

Long-established distinctions came to the fore at the Hamburg meeting. Following a similar path to Conant, Polanyi sought to distinguish between pure and applied research. Pure science, he argued, was an aspect of communication and activity within a self-managing community.[57] On the other hand, applied science was carried out within a structured hierarchy. Such arguments became a familiar theme of post-war science. Although often treated as secret in the Cold War era, applied science could also be made public. Accounts of advances made a good cover story, obscuring the non-disclosure of more specifically useful information.[58] Accordingly, pure science could be widely published; 'technology' was generally kept secret, or patented, with applied science hovering at the boundary.

Meeting the Transatlantic Challenge

The United States provided cultural capital, as well as markets and financial loans. Now the world's newly dominant power, it attracted both resentful admiration and inevitable scorn.[59] In Britain, local newspapers reported to regional audiences speeches about the American challenge by various leading industrial figures, such as the director of the Hosiery and Allied Trades Research Association and the chairman of ICI. The

---

[55] Daniels and Krige 2018, 227.
[56] See McGucken 1978 on the Society for Freedom in Science; and Aronova 2012 for the Hamburg meeting.
[57] Polanyi 1956, 242; Nye 2011.
[58] Most recently, see Daniels and Krige 2018. On secrecy and science more broadly, see Galison 2004 and Galison 2010. On the exchange of information in biological engineering, see Bud 2013a.
[59] See the description by Edward Shils, 'The Intellectuals – (1) Great Britain', *Encounter* 4, no. 4 (April 1954): 5–16, on 9.

message tended to be the same: the British excelled at pure science, but America exploited 'our' contributions through its primacy in applied science.[60] Perhaps regrettably, America's research achievements appeared worthy of emulation.

The American policy proposal *Science – The Endless Frontier* by distinguished engineer Vannevar Bush and frequently referenced as the Bush Report was published in Washington at the end of July 1945.[61] Within a few days, its sequel, the explosion of the atomic bombs over Hiroshima and Nagasaki, seemed to confirm its global significance. Although the report did not lead immediately to the civilian National Science Foundation, it did serve to promote concepts of distinct pure and applied research within the United States and internationally. Immediate anxieties revealed themselves in the discussions within the subcommittees that had led to *Science the Endless Frontier*. One, under the chairmanship of Johns Hopkins University president Isaiah Bowman, addressed the question, 'What can the Government do now and in the future to aid research activities by public and private organizations?'[62] Like Conant, members of the group were insistent on the distinction between pure and applied.[63] Printed in the Bowman subcommittee report, this became a 'perverse law' that 'applied research invariably drives out pure'.[64] Faithfully following the analysis of the Bowman subcommittee, the Bush Report was adamant that national welfare would follow from successful applied research in industry, which, in turn, would be nutrified by basic research, mainly conducted in universities with federal funds.

Bush's report is famous for establishing basic research as a primary subject of American science policy. Within months of its publication, the British were discussing it too.[65] Alexander King, the scientific attaché in Washington, circulated a six-page memorandum dominated by news of

---

[60] See, for example, 'Local Jottings', *Nottingham Evening Post*, 4 December 1950, British Newspaper Archive/British Library Board; 'U.S. Now "King of Industry"': Britain Needs More Applied Science', *The Northern Whig*, 18 January 1951, British Newspaper Archive/British Library Board.

[61] From a huge literature, see Kevles 1977; Reingold 1987; Backhouse and Maas 2017. The 2020 edition contains the reports of the subcommittees that went into the final document. See Bush 2020.

[62] See Committee on Science and the Public Welfare 2020.

[63] This is discussed at length by Backhouse and Maas 2017. See also 'Detailed Minutes of Meeting of Steering Committee', 3 January 1945, Bowman folder 9, Box 25, Collection 14–17–2354, Papers of Henry Guerlac, Rare and Manuscript Collections, Carl A. Kroch Library, Cornell University.

[64] Committee on Science and the Public Welfare 2020, 91.

[65] Dr Bush's report was commended in the parliamentary debate 'Scientific Manpower and Resources' on 30 November 1945 by the Conservative politician William Wakefield, *Hansard* HC Deb. 30 November 1945.

it. This was one of the first documents submitted to the newly constituted British 'Committee on Future Scientific Policy' in January 1946.[66]
Quick as the translation across the Atlantic might have been, in London,
the lesson drawn was distinctively different from Washington's reading.
In 1947 the parliamentary scientific committee commented, 'Whereas,
therefore, Dr. Bush recommended that in the United States more attention should be paid to fundamental sciences, in this country more attention should be paid to applied sciences.'[67] The British used the report to
reflect on their military and industry's difficulties in developing new
products.

Over the first post-war decade, whereas British research funding grew
radically, at times at 10 per cent per annum, its underlying principles
were not a matter of public discussion. Instead, the tendency was to build
on the inherited institutional ecosystem. Implemented through great
government laboratories, although many of these were new, particularly
those of the UKAEA, this growth shaped the discussion of applied
research. Beyond bureaucracy, meanwhile, science was changing.
Before the war, DSIR had largely funded applied science in universities
and in its own laboratories, while pure science was generally of such low
cost that the universities could afford to fund it themselves. In the 1950s,
the picture changed fundamentally, as increasingly expensive pure
research also demanded support from DSIR. In a 1968 book, Norman
Vig pointed to a ten-fold growth in academic research grants across the
six years 1958/59 to 1964/65.[68] Meanwhile, the contribution of DSIR to
industrial research had become less and less significant compared with
industry's own input. The diversity of demands clouded the agency's
identity and mission, and the conflicts it engendered underpinned calls
for bureaucratic clarification and change.[69]

It was not until the late 1950s and early 1960s that the debates of the
mid-1940s were rejoined. The new discussions led to the abolition of
DSIR and its replacement by research councils, generally responsible for
science in universities, both pure and applied, and another agency
responsible for the government's support of industrial research. These
conclusions were recommended by an enquiry known (after its chairman)
as 'The Trend Committee'. Underlying its approach was the previous

---

[66] Alexander King, 'Organisation of Scientific Research in the United States Government',
circulated by M. T. Flett to the Future Scientific Policy Committee, 15 January 1946,
CAB 132/52, TNA.
[67] Parliamentary and Scientific Committee 1947, 20. Quoted by kind permission of the
Parliamentary and Scientific Committee.
[68] Vig 1968, 61.
[69] Committee of Enquiry into the Organisation of Civil Science 1963, 30–43.

work of a Treasury Committee chaired by Solly Zuckerman from 1959 to 1961 (hereafter Zuckerman Committee).[70] An unpublished epilogue to the report of this body reflected that at the end of World War Two, the chief problem had been to increase the number of scientists and engineers. Naturally, then, questions of organisation had taken a back seat. By the end of the 1950s, however, its problems were proving to be acute.[71]

Zuckerman as a skilled bureaucrat, committee chair, and biologist was able both to portray himself as a prophet and to draw on his own lifelong experiences and policy framework.[72] His report dealt at length with applied research. We can compare its thrust with the discussions around the DSIR four decades earlier. Then the distinction between different aspects of science was regretfully accepted as an imposition by business. Now it was analysed as a real phenomenon to be examined and measured. The committee reported that there were five categories: 'These are pure basic research, objective basic research, applied (project) research, applied (operational) research, and development.'[73]

Three ancestries are immediately apparent.[74] One is the four-stage distinction articulated by Zuckerman's friend and fellow biologist Julian Huxley in the early 1930s, a formalisation of a widespread British consensus. This model was enhanced by the category of operational research, developed by another colleague on a range of committees, Patrick Blackett.[75] Finally, the Bush Report's distinction between basic and applied research had enriched and validated Zuckerman's approach. Early on, his committee's secretary had circulated an abstract of a recent paper from the United States Congress on the organisation of science and technology in America. The first witness to those trans-Atlantic hearings had been Alan Waterman, director of the National Science Foundation (NSF), who had explained the Bush distinction between basic and applied research at length. The abstract of this testimony, transmitted to the Zuckerman Committee, emphasised the NSF's standardisation of terminology.[76]

---

[70] The two reports referred to here were (first) Office of the Minister for Science 1961, chaired by Solly Zuckerman, and (second) Committee of Enquiry into the Organisation of Civil Science 1963, chaired by Burke Trend.

[71] [Unpublished], 'Report of the Committee on the Management of Research, Epilogue: The Problems of Organisation', 3 July 1961, T225/1977, TNA, para. 4, p. 2.

[72] See Zuckerman's autobiographical volumes, Zuckerman 1978; Zuckerman 1988.

[73] Office of the Minister of Science 1961, 7.

[74] The approach taken here benefits from Godin 2006.

[75] On Blackett and Operational Research, see, for instance, Ormerod 2003; Nye 2004.

[76] G. W. Robertson, 'Extracts from the 32nd Report by the U.S. Committee on Government Operations – August 1958', M.R. (59) 29 n.d. (first sent to Zuckerman for approval 27 April 1959), SZ/MR/2/2 enc 28, Papers of Solly Zuckerman, University

The Zuckerman Committee's final report drew upon these conventions and the evidence of change. It also took up Bernal's quantitative approach to offer measures of the 'pure research', 'applied research', and development supported by the British government. From these it raised questions about the current institutional arrangements hidden in the unpublished 'epilogue'.[77] In particular, it questioned the linkages between parent government departments and the laboratories that supplied the research they needed. In a foretaste of the later 'Rothschild Committee', the epilogue questioned the enduring relevance of the 'Haldane Principle' that separated research councils' work from the day-to-day priorities of counterpart government departments. It reflected on the possibility of transferring laboratories from research councils to parent departments.[78] The distinction between scientist-driven basic research and user-driven applied research lay at the heart of this analysis. These issues would come back to haunt official discussions during the decade to come.

The subsequent 'Trend Committee' inherited the five categories of R&D identified by Zuckerman and his colleagues.[79] In particular, the Treasury witnesses to the new hearings supported this classification of research. Building on it, they suggested, 'One possible approach would be to distinguish between "pure basic" research on the one hand and, on the other, "applied" research directed to specific objectives of Government policy.'[80] At the same time, the Committee itself worried about the unstable foundations of such an approach. In discussing the Treasury's testimony, the point was made that the demarcation between basic and applied research was so uncertain that it could not sustain an acceptable organisational upheaval.[81] Nonetheless, the Committee recommended dismantling DSIR and replacing it with several specialised scientific research councils, including one responsible for applied research.

British discussions, in turn, had wider ramifications, as they were influential internationally. Moreover, in the background lay American

of East Anglia, Special Collections. For the hearings and testimony of Waterman, see United States House of Representatives, Subcommittee of the Committee on Government Operations 1958, 4–8.

[77] [Unpublished], 'Report of the Committee on the Management of Research, Epilogue: The Problems of Organisation', 3 July 1961.

[78] [Unpublished], 'Report of the Committee on the Management of Research, Epilogue: The Problems of Organisation', 3 July 1961, T225/1977, TNA. This is also to be found in 'CSA's Private File', SZ/MR/4, Papers of Solly Zuckerman.

[79] Committee of Enquiry into the Organisation of Civil Science 1963, 23.

[80] Committee of Enquiry into the Organisation of Civil Science, 'Minutes of a Meeting ... 14 June, 1962', T218/485 TNA.

[81] Committee of Enquiry into the Organisation of Civil Science, 'Minutes of a Meeting ... 14 June, 1962'.

support for Western European reconstruction through Marshall Aid, coordinated by the newly formed Office of European Economic Cooperation (OEEC), which became the global OECD in 1961. From the early 1960s, this body's thinking developed with British science policy and drew on the Zuckerman translation of American thought. The head of science and education was the same Alexander King, who in 1946 had briefed the British government about American policy. Under his guidance, what started as a working group on 'scientific information' interpreted science broadly and took an ambit covering the entire process incorporating the earliest research and product development.[82] This committee then sponsored the creation of a framework giving enduring bureaucratic life to post-war conversations.[83] The detailed formulation fell to another, younger man and British cartographer of science. Christopher Freeman, inspired by Bernal and his *Social Function of Science*, had been working on British industrial research statistics.[84] Through the OECD statistical reports it engendered, known as the 'Frascati manuals', his work led to the internationally accepted formalisation of the categorisation of pure research, applied research, and development.[85] Clearly, this model was a close relative of the recent Zuckerman Committee report. Therefore, talk about 'applied science' and 'applied research' in Britain during the post-war years related to both local traditions and a cosmopolitan language.

The Great Laboratories

Vannevar Bush's *Science – The Endless Frontier*, worried about the balance between basic and applied research. In Britain, during the early post-war years, it was the relative lack of overall scale that was a perennial worry. A study of British industrial research conducted by DSIR in 1957/58 dismissed questions of categories. It urged respondents to aggregate whatever internal classification they might use to enable a big picture of 'R&D' in British industry. The survey highlighted the long-established concentration in a few companies and industries.[86] Above all, it pointed to the weakness of industrial research in Britain compared with the United States. American industry spent five times as much as its

---

[82] King 2001, 339.
[83] On the measurement underpinning science policy, see Godin 2003.
[84] For Freeman's debt to Bernal, see his VEGA lecture, 'Bernal and the Social Function of Science – Science Video', http://vega.org.uk/video/programme/86 (consulted January 2021). See also the website freemanchris.org (consulted March 2023).
[85] Godin 2003; Godin 2007; Freeman and Soete 2009.
[86] DSIR, Economics Committee 1958.

British counterpart. Even taking account of the larger size of the transatlantic ally, American firms were three times as research-intensive. There were, of course, great British industrial research centres. In 1951 the Glaxo company opened new laboratories initially to accommodate over 200 people.[87] Combines such as the electrical giants GEC and AEI, and the chemical corporations, headed by ICI, already boasted substantial laboratories building on their pre-war experience. However, the actual working of research departments and their functions remains much less clear than the rhetoric. With the sole exception of AEI (formerly Metrovick) in Manchester, there is a paucity of studies of British private sector industrial research laboratories equivalent to the book-length histories of American and Dutch institutions.[88]

Military establishments were common to all countries but particularly important in Britain. Her defence effort was an order of magnitude smaller than that of the United States, but in the 1950s, several times larger than that of other European countries. Numbers can be mind-numbing, but a few indications of their scale are important to understanding their challenges. A 1958 review showed that expenditure on research in government defence departments, at about £100 million, approached the entire civil R&D of private industry in the country.[89] A different estimate for 1960/61 found that, collectively, the defence departments were spending £241 million per year on 'R&D', of which 19 per cent was applied research and 80 per cent development.[90] The numbers in the two estimates were different, but the message was not. The nation was short of skilled scientists, and many, possibly too many, were in nationally dominant intramural laboratories. The Zuckerman Committee report suggested that in 1959 the forty-four non-nuclear military establishments employed 1,900 professional scientists.[91]

---

[87] See Jephcott 1965.
[88] Three unpublished doctoral dissertations at University of Manchester deal with the Metrovick laboratories: Niblett 1980; Cooper 2003; and Whitfield 2013. Sally Horrocks devoted part of her doctoral thesis on British food research to the major industrial laboratories, Horrocks 1993. British public sector laboratories have also been addressed. Three PhD theses have been conducted on research for the public-sector Post Office: Ward 2018; Boon 2020; and Haigh 2021. On post-war non-nuclear military laboratories, see Bud and Gummett 1999.
[89] Russell 1962, 13. Quoted by kind permission of *New Left Review*. The figure given for civil R&D by private industry was £124.3 million. The estimates were from the Association of Scientific Workers.
[90] Office of the Minister for Science 1961, 20–21.
[91] Office of the Minister for Science 1961, 20.

As in the past, the great in-house government laboratories gave meaning to the category of applied science in the public sphere.[92] Civil departments were devoting 45 per cent of their effort to it, 50 per cent to development, and, like the military, very little to fundamental research.[93] Above all, Harwell, established as an atomic research laboratory by the government in 1945, was well known. By its height in 1962, Harwell was employing 6,000 staff, having grown tenfold in less than a decade.[94] The only other laboratory with a similar public profile was the even larger aeronautical research laboratory at Farnborough.[95] In 1962, it had a total staff of 8,500. Their names became part of the nation's culture and a source of pride, but soon also a worry.

By the 1960s, there was concern that these laboratories were too important, and the entire network was widely criticised. Under the pseudonym 'Peter Russell', the left-wing biologist Maynard Smith dismissed 'a massive expenditure on defence whose exact significance cannot be ascertained, an inadequate and poorly organised private industrial research effort, and an autocratic, uncontrolled and unco-ordinated government expenditure on civil research'.[96] Similarly, a few years later, Blackett pointed out that the Ministry of Technology employed a considerable proportion of the nation's stock of qualified scientists and engineers working in research and development.[97]

Looking back in 1980, the government's former chief science advisor Sir Alan Cottrell described the two decades from 1945 as the 'romantic era' of science policy: 'In everything to do with pure science, or in non-commercial aspects of applied science, it all went brilliantly. But in one area it failed almost completely. This was in the use of science to strengthen the national economy by invigorating our industries.'[98] Such expressions highlight the problems of research establishments poorly integrated with industry. The size of these laboratories had grown massively during the early Cold War. However, it had become questionable whether they and the science they embodied would offer the benefits once assumed.

---

[92] Bud and Gummett 1999, 1–28. This book contains essays on the development of the important non-nuclear defence laboratories after World War Two. Although household names in Britain, internationally, however, they were not well known amongst the public. In the *New York Times*, for instance, the dominant single association of the Harwell laboratory was with spy Fuchs and the defector Pontecorvo.
[93] Office of the Minister for Science 1961, 20. The figures relate to the year 1959–60.
[94] For figures from the mid-1950s, see Cockcroft 1956; for 1962 figures, see Cottrell 1998. On the history of Harwell, see Hance 2006.
[95] Nahum 1999.   [96] Russell 1962, 16.
[97] Blackett 1968, 1108. Blackett suggested a total of 16 per cent.
[98] Cottrell 1980, 182. Quoted by kind permission of the Royal Institution.

Atomic Energy

The post-war era's stand-out research project was nuclear power development, dwarfing all other civil research endeavours. It was also deeply interconnected with research on atomic weapons and the production of fissile plutonium to arm them. Nuclear power research can serve as a proxy for the other large early Cold War research projects. Its scale and public prominence during the 1950s make this a useful case study to examine the balance between contemporary representation of 'applied science' and technology.[99]

Later commentaries on atomic energy in the 1950s often cite the naive aspiration for 'electricity too cheap to meter'. Nevertheless, this quotation from the US Admiral Lewis Strauss was a product of a specific trans-Atlantic context. In Britain, the more mundane aspiration was to keep the lights on.[100] Naturally, as Macmillan's speech illustrated, governments could turn nuclear achievement to public relations advantage. The programme was, however, begun and conducted as a response to an impending crisis. In 1951 the post-war Labour Government was voted out of office, and the Conservatives returned on a programme of ending austerity, at a time when wartime rationing was still in place for numerous products. Providing enough electricity was going to be a challenge. In every decade since the 1880s, demand had doubled.[101] Now its provision would depend on traditional coal-fired power stations and the new oil-fired generators. But both oil and coal were in short supply. Government portrayed nuclear power as the necessary alternative. While it was also developing atomic weapons, they, and the underpinning research, were kept secret. In contrast, the civil programme was in the public eye, shaping attitudes to science.[102]

Until the early 1970s, Britain produced more nuclear energy than any other European country, generating almost as much as the United States and more than the rest of the European Economic Community combined.[103] In 1963/64, the R&D budget for atomic power was

[99] For detailed studies of the individual military laboratories in this period, see Bud and Gummett 1999.
[100] Minister of Fuel and Power 1955.
[101] This argument was laid out in the White Paper of 1955, which paved the way to the new nuclear regime. Minister of Fuel and Power 1955.
[102] Although the two programmes were publicly presented as separate, the first two nuclear power stations, Calder Hall and Chapel Cross, were intended also to provide plutonium for the weapon project.
[103] In 1973 the British were producing 5,959 MW, compared with a total European Economic Community (including the UK) total of 11,719 MW. See Pocock 1977, 214. The nuclear installed capacity of the Soviet Union climbed rapidly in the early

£45 million. This commitment compares with the entire Research Council budget of £40.3 million or the Ministry of Aviation's R&D expenditure of £31.2 million.[104] But, more important than the value of the exact figure is the esteem it indicated. At the Conservative party conference in the autumn of 1955, the chancellor, R. A. Butler, announced that he would prioritise successes and gave as his first example atomic power.[105]

A few years after Macmillan's inaugural speech, C. P. Snow (then a junior Minister of Technology) reflected on the key resources inherited by the new Labour administration. One was the NRDC; the other was the UKAEA and, specifically, Harwell.[106] To the citizen of the 1960s, Harwell expressed the glamour of applied science. From its foundation to 1959, it was headed by the Nobel Prize–winning Cambridge physicist John Cockcroft, who would serve as the face of nuclear science in the post-war era.[107] To the modern historian, the relationship with the development centre in North-West England provides insight into both the specific experience of atomic power and the general relationship between applied science and technology. We see here two models of innovation sometimes in fretful competition, and often in fruitful tension.

The British atomic bomb required plutonium and, therefore, a factory to produce it through the radioactive transmutation of natural uranium.[108] Since this process also released heat, the plutonium factory could be further developed into a power station. Even before this further step had been thought through, the initial factory team started work. It was controlled by the 'industrial group', headed by Christopher Hinton. Used to managing complex concerns, he had formerly been the engineer in control of Britain's vast munitions manufacturing enterprise during World War Two and, beforehand, a senior engineer at ICI in the North-West of England. For his headquarters, he sought a redundant plant close to his former chemical industry colleagues, which took him to Risley, near Warrington in Lancashire. By 1957 he was employing 14,000 staff in a network of sites across the region.[109] A few years earlier, in 1954, the government's research body, production-development

1970s but reached 4,700 MW in 1975. See Zheludev and Konstantinov 1980, 35. Pocock 1977, 162, points out that in 1963 the United States' installed capacity was still less than that of the British, but approaching it.
[104] Gummett 1980, 39.
[105] 'Call to Expand Success and Curb Excess', *The Times*, 7 October 1955.
[106] 'Notes of an Address given by Lord Snow, CBE, on Friday, 3rd December 1965', p. 3, Box E64, Papers of Patrick Blackett.
[107] Hartcup and Allibone 1984.
[108] On the design of the first British atomic bomb, 'Blue Danube', see Aylen 2015.
[109] On Hinton, see Crowther 1959; Gowing 1990.

facilities, and the Aldermaston site dedicated to nuclear weapons development had, collectively, been put into a new organisation, the UKAEA. By the early 1960s, this was employing 41,000 people under the control of Cockcroft in Harwell, Hinton in Risley, and William Penney, who directed Aldermaston.[110]

At one level, the new organisation proved a great success. A novel method for breeding plutonium from uranium, considered much safer than what was done at the earlier American plutonium factory at Hanford, produced fuel for the new atomic bomb. Graphite slowed the neutrons to propagate the fission (the moderator) and gas, rather than water, cooled the reactor. After Harwell further refined the system to sustain a power station, in 1956, UKAEA launched the pioneering nuclear power station at Calder Hall on the coastal strip bordering Cumbria's Lake District.[111] Over the next fifteen years, this was followed by eleven more similar 'Magnox' power stations in Britain, two overseas, and the development of a new generation of advanced gas-cooled reactors.[112] Even though, in 1957, a plutonium factory close to Calder Hall experienced the world's worst nuclear accident before Chernobyl, and only a safety device attached to a chimney limited the disaster, there was little national concern about the location of reactors at the time.[113] Instead, anxiety focused upon Britain's atomic weapons, which aroused fierce animosity and led to the formation of a powerful opposition under the flag of the Campaign for Nuclear Disarmament (CND).

Before exploring the problems of a massive new institution confronting unprecedented technical challenges and the clash of values, it is worth emphasising the personal opportunities UKAEA provided. The leading metallurgists working in the British nuclear power programme were typical of the 'New Men' identified by C. P. Snow in his novels.[114] Leonard Rotherham, the son of a miner, was a physicist who worked in industrial research in the steel industry before World War Two, became the first head of metallurgy at Farnborough, and then was recruited by Christopher Hinton as his deputy at Risley. At Harwell, the head of the

---

[110] Williams 1983, 37.
[111] For more on the significance of Calder Hall, see Pocock 1977, 35–36.
[112] In addition, a number of other experimental reactors of various designs were built. For an overview of the programme, see Butler and Bud 2018.
[113] Although there was no nuclear explosion, there was widespread radioactive contamination of the soil in a wide area, with consequences for farmers. See Arnold 1992; Wynne 1992.
[114] Snow 1954. Snow was explicitly referring to atomic scientists in his characterisation. The link to Snow's 'New Men' was being made by others at the time. Thus, half the 1966 programmatic volume about Bath University was devoted to the section 'New Men and New Tasks'. See Walters 1966.

department of metallurgy was the Glaswegian Monty Finniston, son of an immigrant drapery salesman.[115] The relationship between Risley and Harwell might have seemed a classic linear model. Hinton explained in 1953:

The Research department at Harwell examines alternative designs and processes, selecting, with such discussion as is necessary, the best of these for development to the stage where it can be turned by the Production organisation into an industrial project. When the Production Division has pioneered this industrial development and obtained satisfactory operating experience, the further development will be turned over to industry.[116]

Perhaps sounding clear, this formulation represented the result of numerous negotiations. At Harwell's heart was 'applied science', though 20 per cent was considered 'fundamental science'.[117] At Risley, the focus was on development, but about 10 per cent of its 'R&D' was not directly related to production and design commitments.[118] An elaborate treaty specified the respective responsibilities between the northern and southern departments. Thus, at the first liaison meeting between Risley and Harwell in 1951, the Hinton team raised the relationship between the R&D branch of the development site and the research centre. Rotherham explained:

A completely clear division of responsibility between R.&D. B[ranch] and Harwell will not be possible and in all fields collaboration will be necessary. This is especially true where special facilities are necessary such as those in Reactor Physics. Careful thought is necessary to decide the best method of arranging for reactor development studies. Harwell will clearly be responsible for pointing the main lines of worthwhile study but where a reasonable clear time can be established it seems to me that Risley should pick up at an earlier stage than previously.[119]

Conflicting allegiances expressed through professional styles, personal trust, and interests distinguished the two groups in practice. Each

---

[115] On Rotherham, see Quayle and Greenwood 2003; and for Finniston, Petch 1992.
[116] Hinton 1953, 368. Quoted by kind permission of *The Bulletin of the Atomic Scientists*.
[117] For the emphasis of Harwell's research, see Cockcroft 1954, 191; Gowing and Arnold 1974, 208. See also 'Technical Steering Committee: Minutes of the Meeting Held on Monday 7th December, 1953', AB 12/170, TNA. For a participant's-eye view of applied science at Harwell, see two essays by Cottrell, who came there in 1955 as deputy head of the metallurgy department: Cottrell 1992 and Cottrell 1993; and for Risley research, see L. Rotherham, 'Item 1. Co-ordination of Research and Development Work', appended to 'Notes of Harwell-Risley Technical Policy Meeting', 18 December 1951, AB 9/156, TNA.
[118] On the development of research at Risley and the strained relations with Harwell, see Gowing and Arnold 1974, chapter 18, 'Research: Harwell's Role'.
[119] L. Rotherham, 'Item 1. Coordination of Research and Development Work', 2.

believed that it was the crucial creative hub, with the other in a merely supporting role. In a draft of an uncompleted volume of official history, John Hendry has argued that 'Harwell believed it was the design authority and Risley merely the construction contractor. Risley believed it was the design authority and Risley merely the construction contractor.'[120] As the two centres planned new generations of reactors, a fundamental disagreement appeared over the role of pilot plants. Echoing debates in the American Bowman Committee a few years earlier, the 'pilot plant' was firmly planted on the 'applied science' side of the line dividing it from technology and production.[121] Hinton argued that such expensive distractions were unnecessary, promoting misconceptions and inducing laziness amongst the engineers responsible for them. By contrast to Hinton's scorn for pilot plants, Harwell built fourteen small reactors. Therefore, the two sites embodied both the different visions of two forceful leaders and contrasting approaches to innovation. In the first meeting of the Harwell/Risley coordinating committee in 1951, Hinton raised a concern clearly of profound significance to him. The minutes recorded: 'Sir Christopher Hinton said that he was not at all happy about the performance of the Division in improving processes. It seemed to him that the Organisation was unexpectedly good at pioneering but had not met with success in carrying out improvement.'[122] So behind the celebratory stories of British success, there were deep conflicts in research policy.

As public figures, Cockcroft and Hinton received equivalent numbers of mentions, but Harwell was far more prominent than any of the northern factories. Thus, from their very first treatment, the makers of the film *Atomic Achievement* emphasised the central place of Harwell in Britain's atomic energy research.[123] It credited 'British science and technology' but gave too much weight to technology for the sponsoring official. In his response to a subsequent draft, the production's manager at the government's Central Office of Information argued to superiors that Harwell 'is the brains of the group. A. E. A. Risley and the Industrial Group is the brawn.'[124] The UKAEA marketing followed through that

---

[120] Hendry 1991, 25. I am grateful to John Hendry for a copy of his paper.
[121] Hinton 1953, 367. The argument over pilot plants has been discussed by Hendry 1991, 30. See also Rotherham and Mcintosh 1956.
[122] 'Minutes of the First Harwell-Risley Technical Policy Meeting, Risley 18.12.1951', p. 4, AB9/156, TNA.
[123] Rayant Pictures Ltd, '"Atomic Achievement": Draft Treatment', 27 February 1956, INF 12/668, TNA. The film itself accordingly emphasised Harwell's place in the research trajectory.
[124] 'Britain's Atomic Achievement, "Nuclear Nation": A Paper by Central Office of Information Films Division', 4 April 1956. INF 12/668. This paper reproduces the

vision, and the 'Britishnewspaperarchive' contains over 10,000 articles citing the laboratory in the period 1946–59. By comparison, fewer than half the number of uses of 'Risley' were to be found in the same corpus over the period.[125] Crucially, coverage was illustrated as well as factual. Soon after its creation, in 1948, Harwell was the subject of a full-page picture article in the *Illustrated London News*.[126] Three years later, the huge Festival of Britain displayed fascinating and attractive drawings by the cartoonist Lawrence Scarfe interpreting work in its laboratories (see Figure 7.1).[127] Work in those novel spaces was also the subject of programmes shown by the newly established television service as early as 1951 and 1952. Later in the decade, the well-known industrial photographer Walter Nurnberg would be commissioned to make compelling images of work and machinery there (see Figure 7.2).[128] No wonder Harwell became the icon of British ambitions in applied science. The prominence given to the laboratory was even the topic of a 1958 reflective essay in *The Observer*.[129] Television, Scarfe cartoons and Nurnberg photographs, and the more general public relations efforts of UKAEA had ensured that Harwell was easy to visualise, if not to understand.

In part, the explanation for Harwell's greater visibility resembled David Kaiser's account for the interest in the science, rather than the engineering, associated with the atomic bomb. Science, already known internationally, could be publicised more readily than the details of manufacture.[130] Certainly, Harwell had to become more security-sensitive after it transpired in 1950 that one of its research leaders, Klaus Fuchs, had been a Soviet spy.[131] Nonetheless, the official historians suggest that secrecy concerns were responsible for the fact, much resented by Risley, that Harwell always got the credit for British nuclear

---

text of Raylton Fleming to Mr Mayne, 28 March 1956, INF 12/668. D. B. Mayne was the chief production officer of the Films Division, while Fleming was the production controlling officer for this film. Gowing and Arnold 1974, 237.

[125] 'Britishnewspaperarchive' searches conducted January 2022 indicate 4,718 articles containing 'Risley' compared with 10,867 articles containing 'Harwell' for the period 1946–59.

[126] See 'Atomic Research in Britain. First Pictures Harwell Atomic Research Establishment', *The Illustrated London News*, 31 July 1948, British Newspaper Archive/ British Library Board.

[127] These are held now by the Science Museum, London.

[128] For the Nurnberg photographs, held by the National Science and Media Museum, see, for instance, the 1958 image of work on the Harwell pilot reactor, BEPO, Figure 7.2. Programmes were broadcast on 14 February 1951 and on 4 July 1952.

[129] John Davy, 'Science and Prestige', *The Observer*, 14 September 1958.

[130] Kaiser 2005.

[131] The complications have been explored by Simone Turchetti in his study of the Italian physicist Bruno Pontecorvo; see Turchetti 2012.

Figure 7.1 Temple of applied science: 'Isotope Production Lab', by Lawrence Scarfe, as shown at the 1951 Festival of Britain. © Science Museum Group.

success.[132] Thus, security, bureaucracy, and the classification of knowledge had come together in the evaluations of the press.

Of course, such tensions were not unique to the one industry whose records are, by chance, accessible. Contemporary American experience also demonstrated widespread and deep conflicts over the applied science model's relevance to recent high technology.[133] Moreover, UKAEA was not just another British company. The scale and visibility of its operations made it quite exceptionally well known. Thus, Hinton was perhaps the most famous engineer at work in Britain, bestriding 'the British engineering profession of his day like a colossus', in the words of his Royal Society obituary.[134]

The cultural divide was emphasised by Hinton's subsequent career. When in the 1960s, both he and Rotherham had left UKAEA, they

---

[132] Gowing and Arnold 1974, 237.
[133] For the tensions within the American Department of Defense between the 'R&D'-oriented Clifford Furnas and the engineer Frank Newbury, see Converse 2012, 407–14.
[134] Gowing 1990, 235. Quoted by kind permission of the Royal Society.

Figure 7.2 At Harwell, scientists taking readings from the loading site of the 6 MW BEPO pilot gas-cooled reactor, commissioned in 1948. Photograph by Walter Nurnberg. © Walter Nurnberg/Science Museum Group.

continued to collaborate and promote 'technology'. Rotherham became the vice-chancellor (effectively CEO), and Hinton took the chairing role of chancellor of the new technological University of Bath. This was the descendant of the Bristol Trade School launched a century earlier by Canon Moseley when seeking an alternative to applied science.[135] Cockcroft, meanwhile, had become the first master of the science-oriented Churchill College, Cambridge.

---

[135] Buchanan 1966.

There were numerous moves to elevate the public status of 'Engineering', in line with the German esteem for *Technik*. Thus, engineers' subordinate place was under new criticism in the venerable and prestigious Royal Society, devoted to science. For a decade from the mid-1960s, the Society was riven with conflict, resolved only by the foundation in 1975 of an entirely independent body, the 'Fellowship of Engineering', whose first president was Christopher Hinton.[136] Shortly after, Monty Finniston would chair an enquiry into the professional standing and organisation of engineers. His report reflected on an ancestry stretching back to the nineteenth century. The practice of teaching scientific principles and then applications built 'into engineering formation a dichotomy between "theory" and "practice" which did "not arise in courses built upon a philosophy of "Technik"'.[137] Once considered the solution, applied science was now being denounced as the problem.

Conclusion

The terms 'applied science' and 'technology' migrated within the country through government reports, reflective essays for general audiences, and newspaper articles, and internationally across borders. Consequently, the observer can see the interplay of old and new connotations, assumptions, and resentments. Influential people, whose language and interpretations had been shaped by inter-war experiences, were deploying such terms and concepts as 'applied science', 'applied research', 'technology', 'technologists', and 'pure science'. Not surprisingly, earlier aspirations influenced many of the post-war arrangements.

The institutional solutions to which politicians turned were familiar: large central industrial and military research laboratories, and more training in universities or in specialised technological schools. Yet the debates' urgency and centrality and the outcomes' vigour were unprecedented. At a time of rapid growth, both applied science and technology served as bases for desperately needed engineering training. Significantly increased in scale, research came to be bureaucratically administered as never before.

New committees and influential reports formalised old language to establish a rhetoric formally identifying categories of basic and applied research and development. The demands for new levels of scale and secrecy in the Cold War intensified the search for meaning in the new

---

[136] Collins 2010; Collins 2016, 82–107.
[137] Committee of Enquiry into the Engineering Profession 1980, 90, para. 4.35. © Crown Copyright.

organisational ecosystem. Government engagement entailed levels of bureaucracy that reified such concepts. At OECD, the Frascati manuals identified separate and measurable categories. These served the purposes of security and planning.

Such terms would sustain statistical collections, but how useful were they in the everyday world? The experience of the development of atomic power shows the radical disjunction in philosophies between the laboratory at Harwell and the industrial centre in the country's North-West. This was paralleled by the call for better recognition of 'engineering', leading to the Fellowship of Engineering (and later the Royal Academy of Engineering). Common to the two stories was the larger-than-life figure of Christopher Hinton. Yet, while funding was exploding, competing models could coexist.

By the 1960s, increasingly hard decisions about the use of enormous sums of money had to be made. As Philip Gummett has pointed out, expenditure on civil science in 1964 was an order of magnitude larger than in 1947.[138] Accordingly, following the report of the Trend Committee, the new Labour government replaced its existing science advisory committee with two complementary successors: the Council for Scientific Policy (CSP) within the Department of Education and Science, with responsibility for the science budget, and a separate Advisory Council for Technology within the Ministry of Technology. Therefore, applied science under the CSP was set against technology promoted by a different department. These were bureaucratically distinct silos and fundamentally different systems of classification. The next chapter follows the conflicts that ensued.

---

[138] Gummett 1980, 171.

8 From Applied Science to Technological Innovation

Introduction

During the late 1960s, the terminology, as it had developed in the previous half-century, became deeply problematic. For a decade, the use of 'technology' had risen in public discourse and as a term of art. While funding for research was growing vertiginously, the twin models of technology and applied science had coexisted in public policy and popular debate. When public money became tighter, tension became acute. Many politicians ceased to believe that by drawing on government-supported applied science, industry would convert pure science into the basis for revolutionary improvement. The remarkable alliance of public and bureaucratic interest in supporting the brand broke down. As a result, in the 1990s, parliamentary references to 'applied science' were just a seventh of their occurrence in the 1960s.[1] The use of language in the public sphere had changed radically.

This chapter asks what processes erased applied science from public view. As elsewhere in the book, we shall examine the nexus between public opinion, politicians, bureaucrats, and experts. The process involved intellectuals with ideas about a new industrial revolution, widespread anxiety about the implications, economists promoting the concept of 'innovation', political direction, and government-sponsored committees. No one factor would have been potent in isolation.

When first the prospect, and then the reality, of much slower growth in government funding had to be confronted from the 1960s, open bureaucratic conflict broke out. Thereafter, during the last third of the twentieth

---

[1] This count is based on a search on https://hansard.parliament.uk/ conducted in February 2021. Parliamentary usage fell from 142 occasions across the 1960s, to 19 across the 1990s (and 25 in period 2000–2009). Usage of the term 'applied research' also fell slightly in the same period, from 90 to 88, but then to 61 in the first decade of the twenty-first century. For a more extended treatment of genres of science as brands, see Bud 2016. The Google 'ngrams' graph shows a similar story. Between a peak in 1962 and 1999 there was a three-fold reduction in Google's measure of 'British English' use of 'applied science'.

century, it was, increasingly, private businesses, rather than public organisations, which were responsible for extracting practical value from science. In 1984, the science-policy analyst and academic Jarlath Ronayne reported of the distant past, 'At one time, it was customary to categorize scientific research as either pure or applied.'[2] As the state's industrial strategy underwent several transformations, the language of science policy was changing decisively.

New Industrial Revolution

At the heart of this chapter is a political conflict – between a sequence of emergent industrial policies, to which science would be just an input, and the traditional approach, centred on applied science. However apparently esoteric, this dispute did not exist in cultural isolation. On the contrary, it drew upon wider public concern about the nature and import of a 'second industrial revolution'. Across the public sphere during the 1950s and 1960s, 'the second industrial revolution' referred to the transformation that would be wrought imminently by automation, nuclear power, and new plastic materials.[3] In discussing this, the public and political discourse came to be principally about 'technology' rather than 'science'.

A few maverick intellectuals had shaped interpretations ready for this moment, linking technology as a culture to industrial revolutions. During the 1920s, the British sage Patrick Geddes and his American protégé Lewis Mumford had distinguished the imminent 'neotechnic age' from its coal and steel-based predecessor, 'the paleotechnic age', much as that had superseded wood-based cultures. In the interwar years, other writers had picked up the language of a 'second industrial revolution', and aspirations for the better life it could bring, widening the discussion.[4] Scholars on the European continent explored it: in

---

[2] Ronayne 1984, 34. Quoted by kind permission of Jarlath Ronayne. For a review of changing thinking about science policy in the late twentieth and early twenty-first century, see Martin 2012.

[3] The 1950s public usage ran contrary to historians' usage of the term 'second industrial revolution', which refers roughly to the period 1870–1914. London School of Economics historian D. C. Coleman dismissed the terminological impertinence of historically uninformed commentators; see Coleman 1956, 1. For the only overview of the modern second industrial revolution discourse of which the author is familiar, see Homburg 1986. I am grateful to Ernst Homburg for bringing this to my attention. For the Dutch discourse in particular about the second industrial revolution, see De Wit 1994. The only historical treatment that touches on the automation discourse in Britain, in particular, is Hayes 2015.

[4] The entire May 1930 issue of *the Annals of the American Academy of Political and Social Science* was devoted to the new industrial revolution.

German, for instance, Edgar Salin (Talcott Parsons' Heidelberg disser-
tation supervisor), and, in French, George Friedmann, the Marxist
sociologist of work.[5] They provided a template for expectations after
World War Two.

In the 1950s, the anticipation of a 'second industrial revolution'
escaped from the typewriters of a few thinkers and entered the mass
media and public discussion. Most significantly, the founder of cybernetics, Norbert Wiener, entitled a chapter of his 1950 *Human Use of
Human Beings*, 'The First and the Second Industrial Revolution'.[6]
Politicians, trade unionists, writers, and journalists picked up his usage.
In 1952, Kurt Vonnegut launched a distinguished career with his widely
read novel *Player Piano*, about a dystopic society that followed the
second industrial revolution.[7] The term was linked to the new and
terrifying concept of 'automation' by the influential management
writer John Diebold.[8] Fuelled by news of computers and machinery
for material handling, discussions went on quickly to 'catch fire' in
American industry.[9]

In October 1955, the United States Congress held hearings on
'Automation and Technological Change'.[10] Testifying to the committee,
the trade union leader Walter Reuther warned that the talk of the second
industrial revolution was so common, it was in danger of being
dismissed.[11] The issue moved to the political stage in July 1956. In
Germany, the SPD party debated the 'Zweite Industrielle Revolution'
at the Munich meeting, which decisively redirected its programme away
from traditional socialism.[12] Such usages were internationally infectious.
The SPD document on the 'Second Industrial Revolution', by the
engineer Leo Brandt, who would drive German atomic research, cited

---

[5] Friedmann 1936; Salin 1956. On Friedmann, see Vatin 2004; and on Salin,
Schefold 2004.

[6] Wiener 1950; Wiener 1965, 27–28. See Kline 2015.   [7] Vonnegut 1953.

[8] Diebold 1952, 2. See Hounshell 2000; Resnikoff 2022. Harder at Ford has also been
credited with the coinage. It is of course possible that Harder and Diebold made the
journey to automation independently.

[9] Sterling F. Green, 'Automation: Is a Reckoning Coming?', *Washington Post and Times
Herald*, 15 May 1955. See Pollock 1957. Appropriately, this book's translators from his
German original (W. O. Henderson and W. H. Challoner) were themselves well-known
British historians of the first industrial revolution.

[10] See United States Congress, Subcommittee on Economic Stabilization of the Joint
Committee on the Economic Report 1955. Also, United States Congress, Joint
Economic Committee 1956.

[11] See 'Statement of Walter P. Reuther, President, Congress of Industrial Organizations' to
United States Congress, Subcommittee on Economic Stabilization of the Joint
Committee on the Economic Report 1955, 97–114.

[12] Orlow 2000, 108.

many contemporary British developments.[13] Appropriately, there was an English translation of this German manifesto, for it was in Britain that the metaphor had most purchase.

Amongst the British, the experience of the first industrial revolution was a significant theme in national identity. It was the topic of a six-part series on the BBC's principal radio station, the 'Home Service', at 7:30 on Wednesday evenings. The presenter, professional historian Asa Briggs, wrote a brochure about the broadcasts linking talk in the public sphere of an imminent second industrial revolution with his historical account of the first.[14] In its series introduction, the BBC's listings magazine *The Radio Times* also emphasised the connection.[15] As academic historians took a close interest in the early nineteenth-century experience, in parallel with the newfound press usage, their interpretation of its meaning changed. Whereas once they had emphasised its rapidity and terror, now interpretations highlighted the significance for growth.[16]

As early as January 1951, *The Observer* newspaper ran a three-column profile of the ideas and personality of Norbert Wiener.[17] At the end of April that year BBC radio also broadcast a talk by him on the theme of the current industrial revolution.[18] Day to day, the movement found its prophet in Walter Puckey, chairman of the Institution of Production Engineers. His newspaper articles captured the combination of hope and fear evoked by the very word 'automation' and its revolutionary implication.[19] In June 1955, Puckey's institution organised a major conference, titled 'The Automatic Factory'. Invited to speak there, the Swedish observer Daniel Viklund found the British commitment to the new technologies expressing more than technical concerns. His hosts

---

[13] On the aspirations for redemption through technology, international parallels, and developments among social democratic parties, see Costa 2020. Also see Trischler and Bud 2018. I am grateful to Helmuth Trischler for advice on the German discourse.
[14] Briggs 1957, 3.
[15] 'Where We Came In', 9 January 1957, 7:30 pm BBC Home Service, https://genome.ch .bbc.co.uk/b41b0723d7544394ba00ff3c5025972e (accessed November 2021).
[16] Cannadine 1984.   [17] 'Profile: Dr Norbert Wiener', *The Observer*, 28 January 1951.
[18] 'The New Industrial Revolution', 30 April 1951, Third Programme 2:05, repeated 2 May 1951 at 22:55, https://genome.ch.bbc.co.uk/4ceadca7167e44579d924b93dfa50bef (accessed November 2021).
[19] See, for instance, Walter Puckey, 'Automation – What Does It Mean?', *Truth*, 17 June 1955, British Newspaper Archive/British Library Board. Puckey's name appears thirty-five times in the periodicals of just the two years 1955–1956 as captured by the Britishnewspaperarchive corpus (consulted February 2021). His speeches have been kept at the Institution of Engineering and Technology, NAEST 146/3. See also 'Obituary: Sir Walter Puckey', *The Times*, 10 October 1983.

seemed to believe it could assure their country of a leading position in the new age.[20]

Viklund was perceptive. Deployment in Parliament, whose mandate is to think about the future rather than the past, is an illuminating indicator of cultural pertinence. Its speeches used the term 'industrial revolution' just five times across the entire nineteenth century. Usage rose to about fifty in the decade of the 1920s. During the 1950s, the number of occasions approached 400; in the two subsequent decades, about 500 uses were recorded.[21] The peak of public discussion was in the middle of 1955, during the debates over automation. Atomic power also had a new high profile because of the publication of British government plans to build numerous reactors, and the August 'Atoms for Peace' conference in Geneva. In July, the mass-circulation conservative *Daily Mail* newspaper published a lecture on the forthcoming industrial revolution by the president of the Board of Trade.[22] According to the *Manchester Guardian*, even Conservative Prime Minister Anthony Eden was using talk of the first and the new industrial revolutions to ensure current experience contrasted with the past. The first time had been associated with well-known narratives of suffering as well as of achievement. This time, the British would do it right.[23]

Such an evocation of the nation's experience by the Conservative leaders was associated with a new willingness to intervene in industry and use the term 'technology'. In 1961 the Minister for Science, Lord Hailsham, announced how the Department of Scientific and Industrial Research (DSIR) was advancing 'technology in the industrial sphere'.[24] Shortly after the 1957 launch of the Macmillan government, the newly appointed Minister responsible for supplying the armed forces, Aubrey Jones, had sent a memorandum to Macmillan suggesting that the government reorganise its civil and military development activities as a new 'Ministry of Technology'.[25] Though the administration did not follow up on Jones' suggestion, its formal submission highlights the potential of a

---

[20] Viklund 1955, 818.

[21] Results from https://hansard.parliament.uk/ (accessed May 2019).

[22] 'Industrial Revolution Coming – Thorneycroft', *The Daily Mail*, 6 July 1955. See also the earlier Ronald Hurman and Harold Pendlebury, 'Factories That Think', *Daily Mail*, 11 November 1954, which reflects on automation and the second industrial revolution.

[23] See the reports 'Fresh Start for the Commons. Premier's Theme: Atomic Age in Industry. Painless Revolution?', *The Manchester Guardian*, 23 January 1956; and 'This Nuclear Age', *The Manchester Guardian*, 24 January 1956.

[24] See contribution of Lord Hailsham to the House of Lords Debate, 'Technical and Scientific Manpower and Research', *Hansard* HL Deb.15 November 1961, cols. 714–27, on col 722.

[25] Jones, 1985, 85. The memorandum is reprinted on pp. 145–46.

new linkage between language, vision, and institutions. In power from 1964, the Labour government was much more committed to both state intervention and technology. It would ditch the long-standing complex of applied science rhetoric.

Mintech

The Labour party had a strong tradition of industrial intervention and close links with science. Thus, in July 1955, the leader Hugh Gaitskell and his deputy, the economist Harold Wilson, attended a private workshop at Nuffield College, Oxford, dealing with the economic challenges of the next decade. There, *The Manchester Guardian*'s financial editor, R. H. Fry, who had a close relationship with Wilson, celebrated the changes already happening in the British economy. However, his speech warned that the second industrial revolution would bring about social and economic turbulence, which unions, businesses, and the government would each experience.[26]

Shortly after the Nuffield College symposium, in the summer of 1955, Wilson himself picked up the theme. He was speaking to the Fabian Society at the Labour party's annual meeting, which also happened to fall just a few days before the American congressional hearings on automation. Wilson warned his audience not to fixate on the still-troubled legacy of the past. Contemporary Britain was in the throes of a new industrial revolution even faster than its predecessor. His party had the responsibility to take it seriously; otherwise, 'while they [the Conservative party] are talking of nuclear power, we shall be thinking of coal and cotton'.[27] Labour had to ensure that capitalists enabled automation to proceed quickly enough; that restrictive practices did not impede the revolution; but also that the changes benefitted workers.

Reference to science marked the language around the new revolution, but the moral drawn was not the need for more breakthroughs. Instead, 'technology' was the coming term, describing the requirement for government-driven development. This term, too, migrated from its roots in education through elaborate discussion within the Labour party. Many

---

[26] Fry 1956. This is the published transcript of Fry's talk at Nuffield College. See R. H. Fry, 'The Next Ten Years in British Economic Policy', IV Industrial Organisation, for discussion, 3 July 1955, Nuffield College, Oxford, Nuffield College Archives 12/1/314. The Archive also lists participants in the meeting. I am grateful for the help of Nuffield College in obtaining copies of this material. For Fry's background and relationship to Wilson, see his obituary, William Clarke, 'Richard Fry', *The Guardian*, 30 January 2002.

[27] 'Labour preparing "Shape of Things to Come": Approach to New Industrial Revolution', *The Manchester Guardian*, 11 October 1955. Copyright Guardian News & Media Ltd 2022. I am grateful for permission to quote from Robin Wilson.

of those involved had been discussing the role of expertise since the 1930s. Gaitskell, leading the Labour party in the late 1950s, had been a member of the 'Tots and Quots' discussion group and supported forming an advisory study group to shape the party's future science policy. Its distinguished membership included P. M. S. Blackett, Solly Zuckerman, J. D. Bernal, Ritchie Calder, and C. P. Snow. Though under Gaitskell, managing the interface between politics and 'experts' proved conflict-ridden, during the subsequent leadership of Harold Wilson, such advisors would have real influence.[28]

Two issues were to be wedded: a new authority for science within industry, and the economy's management through planning. The question of a skilled workforce, and thus education, was a continuing challenge. Getting companies to do more, and the right, research, and using its products throughout organisations was another. Therefore, a working paper for a March 1963 document explained:

There are two main types of Government policy on industrial research, development, and the application of developed ideas:

(a) the 'passive' type, which assumes that 'industry knows best', and confines itself to giving assistance at special points of difficulty, whose existence industry makes known to Government:

(b) the 'active' type, which assumes the possibility of using the injection of scientific and technical resources as a means of changing industry (even against its will) ... we do not believe industry knows best and therefore favour policies of type (b).[29]

In the years of opposition (1951–64), the interaction between separate ministries responsible for planning, research, development, labour, and education was a matter of repeated discussion. An idea for a super-ministry of science, including training, would have continued the pre-war belief in applied science. That was not, though, the approach of Blackett, the most active member of the Labour party group.[30] Armed with the status of a Nobel Prize for physics, he began by calling for much more industrial research and the formation of a 'Civil Industrial Research Development Authority' in 1959. In thinking further about how industry, research, and innovation would interact, Blackett tended to deploy

[28] See Horner 1993; Favretto 2000; Ortolano 2009.
[29] Labour party: Paper for Group on Civil Research and Development meeting on 28 March 1963, Box 2, BVB/1/26, Papers of Vivian Bowden, Special Collections Division, University of Manchester Library. Quoted courtesy of the Labour party/ University of Manchester Library. I am grateful for the support of the Labour History Archive & Study Centre, People's History Museum, Manchester, in obtaining permission to quote.
[30] Kirby 1999.

the term 'technology' and, in January 1964, proposed a 'Ministry of Industry and Technology'. By September, he was suggesting speed and a simpler, smaller 'Ministry of Technology'.[31] The following month's general election brought the party to power.

The new Prime Minister, Harold Wilson, grafted Blackett's talk of 'technology' to his long-standing 'industrial revolution' concerns. Speaking before his election to the 1963 Labour party conference, he began with a passionate evocation of the revolution's urgent reality.[32] As in his speech to the Fabian group eight years earlier, Wilson warned of 'Luddism' and restrictive practices. He reminded his listeners of the revolution in work currently underway in the United States where they could produce an entire car, 'and I mean an American motor car, with all the gimmicks on it – without the application of human skill or effort'. The consequence could be mass unemployment, and the solution was to take 'technology' seriously. He emphasised then the importance of providing much-expanded scientific education. His use of an unusual hybrid term, 'scientific industrial revolution', highlights his characteristic balancing act. At the meeting's end, the party adopted a confused document. Its title offered no deference to technology, firmly reading 'Labour and the Scientific Revolution'.[33] Yet the theme was managing an industrial revolution, to avoid unemployment and to bring up to modern standards traditionally strong but now failing industries such as shipbuilding, textile machinery, and machine tools. In this discussion, we see the dominance of the newly reformulated concept of technology, allied to the British memory of the industrial revolution. There was a strong echo of the anxieties of the early 1930s, expressed, for instance, by Alfred Ewing and Alexander Smith Russell. Nonetheless, the past concerns to differentiate between public and private spheres, which had sustained applied science, were being swept away.[34]

---

[31] Patrick Blackett, 'Government Participation in Industrial Research and Development', 16 July 1959, revised 20 September 1959, PB5/4/2, Papers of Patrick Blackett, the Royal Society; P. M. S. Blackett, 'The Case for a Ministry of Industry and Technology', 17 January 1964, PB5/4/2, Papers of Patrick Blackett; Patrick Blackett, 'The Case for a Ministry of Technology', September 1964, PB5/4/2. These papers are discussed in the obituary by Lovell 1975, especially 77–79.

[32] Wilson 1963, 134. The quotations from that speech are reprinted by kind permission of the Labour party and of Robin Wilson. See also Horner 1993; Coopey 1991a; Coopey 1991b; Edgerton 1996. It should be emphasised that none of these treatments apparently take account either of the precedent of Wilson's 1955 Fabian speech or of the preceding Nuffield College meeting.

[33] Labour Party 1963.

[34] Godin argued that the appeal of 'technological innovation' as a category in the 1960s was its appearance as a holistic, all-encompassing category, though the individual measures it covered were frequently uncoordinated. See Godin 2019, 152.

Wilson's actions after the election showed a decisive turn away from science-centrism. Following earlier suggestions of Blackett, the new government established a Ministry of Technology ('Mintech') incorporating those two national weapons, the National Research Development Corporation (NRDC) and the United Kingdom Atomic Energy Authority (UKAEA), together with research centres from DSIR. The reorganisation split these decisively from the management of science. The new Ministry also took responsibility for four technologies associated with the new industrial revolution: computers, electronics, telecommunications, and machine tools.[35] Mintech replaced the 'industrial research and development' research council recommended by the Trend Committee. Other recommendations of the committee informed the new government, which corralled education and academic research in a new Department of Education and Science (DES). The ideologies of technology and pure/applied science now had separate sponsoring departments. Of the two, Mintech would become progressively more powerful, speaking for technology and industry. Therefore, it was left to the science councils to develop the applied-science agenda, demonstrating their extrinsic benefits to society.[36]

A speech by C. P. Snow, given shortly after he had become one of the first ministers in Mintech, explained why the government had alighted upon 'technology'. It exemplified the close interdependence between the new department, Labour's analysis, and carefully chosen language: 'because we thought, in the special conditions of this country, it was important to throw emphasis on technology at a governmental level from the start in order to redress the balance between technology and the applied sciences'.[37]

This rhetoric was particularly striking as it contrasted with the language Snow himself had used just a few years earlier in his 'two cultures' address. Then, this former chemist had scarcely referred to 'technology'. Now, 'talking the talk' did not suffice to save his job, and Snow proved to be a short-lived minister, as did his boss, the trades-unionist Frank Cousins. Instead, the charismatic Tony Wedgwood Benn would lead

---

[35] Coopey 1991a. The continuing resonance of the automation rhetoric of the previous decade was indicated by the 1964 Reith Lectures, broadcast on the BBC and published as *The Age of Automation* (Bagrit 1965). The producer of this series was Kenneth Hudson, who had recently coined the term 'Industrial Archaeology', largely referring to the relics of the first industrial revolution.

[36] Council for Scientific Policy 1966a, 9–10.

[37] 'Notes of an Address Given by Lord Snow, CBE, on Friday, 3rd December 1965 at the Caxton Hall, London, on Technology and the Nation's Needs', Box E64, Papers of Patrick Blackett. Quoted by kind permission of Curtis Brown and the Royal Society.

the expanding Mintech in the mission to rescue the British economy.[38] In a 1968 talk, he recognised the frequent criticism of the separation of academic science from technology but argued that the first was the target of as much spending as the country could afford, whereas the latter represented investment.[39] To an outsider, this distinction might seem invidious because it specifically discounted the claims of applied science. However, Benn's conscious omission, as much as the words committed by Snow, illustrated the competing demands of 'applied science' and 'technology'. Sir Richard (known as 'Otto') Clarke, the senior civil servant leading the Ministry for much of its life, documented the development of Mintech. He argued that it went through three phases, and at each stage, it might have been, but was not, renamed as its responsibilities widened from technology to engineering and, finally, to industry as a whole.[40] Progressively, the lines between public and private spheres were blurring.

Labour complemented the institutional entrenchment of technology with its education policy. It kept separate the former technical colleges, art schools, and teacher training colleges from the universities; amalgamating ninety-four smaller institutions into thirty new-style, vocationally oriented 'polytechnics' under the control of local authorities. Again, through speeches and policy, the British government's ambitions for a distinctive technological education had been made plain.[41]

The Economists and Innovation

By the late 1960s, the role of applied science as the motor of industrial policy was being loudly questioned in both Britain and the United States. After all, despite the unprecedented expenditure, commentators observed that growth rates had fallen behind that of less generous competitors.[42] The defence was often a reaffirmation of faith. Frederick Jevons, a biochemist and one of the first science studies professors, dismissed this strategy. Prefacing the 1972 book *Wealth from Knowledge*, he warned, 'Too often stirring professions of faith in basic

---

[38] Quoted in Coopey 1991b, 128.   [39] Benn 1969, 158.
[40] Clarke 1973a; Clarke 1973b. See also Coopey 1993 and Edgerton 1996. Edgerton also reflects on the radical widening of the Ministry's scope.
[41] Vig 1968, 140–56; see also Pratt 1997. The vision behind this aggregation and the continuation of the so-called binary system was expressed by the Minister of Education in his programmatic Woolwich speech of April 1965, at www.hepi.ac.uk/wp-content/uploads/2016/08/Scan-158.pdf (accessed January 2022).
[42] See, for instance, comments of Solly Zuckerman, 'Central Advisory Council for Science and Technology: The Task of the Council', 20 January 1967, AC (67) 2, Papers of Solly Zuckerman, SZ/CACST/1.

science are coupled with dire prophecies that technology will die on the vine if starved of the rising sap of new ideas from undirected research.'[43] The book by his team, a study of award-winning innovations, suggested that many factors were critical to industrial success, the most important of which was a product champion within an organisation.

The analytical shift was brought about by the intervention of economists who hitherto had shown little interest in technical change. An influential minority was becoming interested in the category of 'innovation', incorporating not just invention but also its commercialisation. So widely was this associated with the great Austrian Joseph Schumpeter that his reputation obscured the contributions of others, such as Rupert Maclaurin of MIT.[44] Nevertheless, innovation seemed to serve as a point of common interest for various academic disciplines. As John Langrish, a progenitor and co-author of *Wealth from Knowledge*, pointed out in 1988, this new term brought together those seeking to justify scientific research and others whose principal concern lay in stimulating the economy.[45]

Early in the twenty-first century, the late Benoît Godin explored the arguments within the economics of innovation in considerable detail.[46] His work made three points critical for the understanding of applied science in the second half of the previous century. First, economists formalised a model of the technological 'innovation' process, understood as the commercialisation of 'invention'. Their approach built on the so-called linear model of the successive stages of first pure and then applied research followed by development, long implicit and often explicit in the work of writers, politicians, and policymakers.[47]

Second, Godin showed how attention shifted from the origin of an invention to the process of commercialisation. Against the preceding emphasis on discovery driving innovation, widely characterised as 'science-push', there emerged a refreshed focus on 'need-pull' (or later demand-pull) – according to which scientific research was only a small,

---

[43] Jevons 1972, xii. Material from J. Langrish, M. Gibbons, W. G. Evans, and F. R. Jevons, *Wealth from Knowledge. Studies of Innovation in Industry*, published 1972 Macmillan Press Ltd, reproduced with permission of SNCSC.
[44] On the importance of Maclaurin and his eclipse by Schumpeter, see Godin 2008 and Backhouse and Maas 2016. On Schumpeter, see McCraw 2009.
[45] Langrish 1988, 115.   [46] Godin 2017. See also Godin 2006.
[47] The linear model has attracted considerable attention. Edgerton 2004 suggested it was an artefact of retrospective criticism. Others have strongly disagreed. The literature responding to Edgerton up to 2010 (such as Godin 2006) is surveyed by Balconi et al. 2010. In contrast to Edgerton's scepticism about the historic appeal of this analytical framework, Godin traces a detailed genealogy with a North American emphasis. For more recent work, see Kaldewey and Schauz 2018, 253–56.

if often necessary, component of the innovation process.[48] For example, in the late 1950s, two distinguished academic economists aligned with the Labour party, Bruce Williams and Charles Carter, produced three books about science in industry. They emphasised the importance of inculcating an interest in science and its uses across an organisation and not just in the research laboratory.[49] Thus, the stage downstream of a science-based discovery became a new priority for understanding innovation. A few years later, the team from Jevons' department at the University of Manchester emphasised complexity and variety. Generally, the evidence from their detailed empirical study of recent prize-winning British developments made the team wary of general claims for the push of more discovery and more frequently accorded primacy to the 'pull' of need.[50]

Third, Godin pointed out that the 'need-pull' model's dominance, in turn, had a transitory existence as it was superseded in the 1980s.[51] From the 1990s, a more complex interpretation of integrated institutional systems, centred on the firm but including government bodies and universities interrelating to sustain innovation, became fashionable. Thus, a report on British science and innovation policy, entitled *The Race to the Top*, from the Minister, Lord Sainsbury, in 2007, suggested an entirely new perspective in terms of a system that affected a country's innovation rate:

industrial research; publicly funded basic research; user-driven research; knowledge transfer; institutions governing intellectual property and standards; supply of venture capital; education and training of scientists and engineers; innovation policies of government departments; science and innovation policies of RDAs [Regional Development Agencies]; and international scientific and technological collaboration.[52]

This approach to industrial policy had moved a long way from 'applied science'.

Science Challenged

While technology was coming to the fore of government attention, there was anxiety about the escalation of science funding. Lord Hailsham could boast in 1961 that in the four years since 1957, funding for new

---

[48] On the history of these approaches, see Godin and Lane 2013.
[49] Carter and Williams 1957; Carter and Williams 1958; and Carter and Williams 1959.
[50] Langrish et al. 1972. For the origins of this work, see Jevons 1972. See also Langrish 2023, 209–15. See also Georghiou et al. 1986; and the review Langrish 1988.
[51] See also Rothwell 1992.   [52] Sainsbury 2005, 4, © Crown Copyright.

post-graduate training grants from DSIR had mounted by over 70 per cent.[53] Such an increase seemed even to leaders of science hard to sustain at a time of increased government ambition, reduced resources, and recurrent financial crises. As early as 1965, Vivian (Lord) Bowden, engineer and Labour's Minister for Education and Science, warned that the age of rapidly increasing growth in research funding was coming to an end.[54] From its initial report, the following year, the Council for Scientific Policy (CSP) recognised the challenge was the 'levelling off of the growth rate'.[55] In the late 1960s, the DES science budget's annual growth rate fell from 12.5 per cent to less than 6 per cent.[56]

As change felt disconcertingly slow and uncertain to economists, it felt threatening and fast for many others. In reviewing a 1956 collection entitled *What Is Science?*, I. B. Cohen complained that the lead article by Bertrand Russell dealt almost exclusively with the worries that applied science threw up and the limited future of the human race.[57] In the past, Russell had dealt with significant cultural challenges engendered by science and transcending the individual country.[58] Now elderly, he provided continuity with the early twentieth century and expressed anxieties inspiring the young. In the coming years, particularly from 1962, the foci of protest lay in the abuses for which science was blamed and the untrustworthiness of the institutions with which it was associated. Urgent issues were following each other rapidly: radioactive strontium-90 released into the atmosphere by hydrogen bomb tests and absorbed by children's bones; of the insecticide DDT, made by chemical companies and endangering bird populations; foetus-harming thalidomide, made by pharmaceutical companies; environmental damage from waste produced by the chemical industry; and revelations of factory-farming of animals.

A study of British newspaper content has documented the change in the treatment of science and technology during the 1960s. The emphasis

---

[53] Lord Hailsham to the 1961 House of Lords Debate, 'Technical and Scientific Manpower and Research', col. 720. Council for Scientific Policy 1966a.

[54] Bowden 1965. For completeness it should be mentioned that Bowden's full name was Bertram Vivian Bowden, though he was rarely referred to as such. See also Edge 1995, 6.

[55] Council for Scientific Policy 1966a, 3.

[56] For the falling growth rate in the late 1960s, see statement to the House of Commons, 'Civil Science', *Hansard* HC Deb. 21 July 1969, col 1364; Ince 1986, 30.

[57] For the review, see I. B. Cohen, 'A Shared Excitement', *New York Times*, 8 January 1956.

[58] The historian of the concept of technology in the United States, Eric Schatzberg, has pointed to Ellul's, Marcuse's, and Mumford's critiques as shaping American opinion in the 1960s and in bringing the idea of technology out of control to public attention. Schatzberg 2018.

had moved towards warning of risks rather than celebrating achievements.[59] Such a shift was visible even in science-supporting media, including the magazine *New Scientist* and *The Manchester Guardian* newspaper. Their campaigning journalists challenged, for example, the use of antibiotics to promote the growth of animals.[60] In 1963, American trade journal *Chemical & Engineering News* described the transformation of science coverage from the 'Gee Whiz Age', before World War Two, into the 'Reportorial Age' after the war, to an era the author called the 'Interpretive Age', which was far more critical.[61] A few years later, the OECD published an eminent committee's report calling for new responsiveness in science policy to public unease, expressed across its industrialised member states.[62]

Such generalised anxiety became a direct challenge for science academics on account of student behaviour. Although British universities doubled in size between 1960 and 1970, the numerous recruits to the student body were not turning to science.[63] Their absolute numbers were still rising, but science enrolments were not growing as quickly as in other subjects. The proportion of all students studying 'science and technology' had already fallen from 45.9 per cent to 40.6 per cent in just five years (1962–67), and analysts expected the decline to continue, perhaps down to a quarter within a few years. As early as 1965, the scientists' anxieties were sufficient for chemist Frederick Dainton to launch an enquiry on behalf of the CSP. In part, the blame fell on the premature requirement for specialisation, but there was another newer problem in the background to this long-established pattern. 'The young' seemed to find science and technology 'too materialistic', and there was possibly a 'real repugnance' to the moral issues they created.[64] The CSP interpreted what was widely called 'the swing against science' as converting public attitudes into a practical problem for science's leadership. The economist Harry Johnson described his perception of the challenge to a meeting of the CSP in January 1970. As science became less popular, and the side effects of applied science more questioned, expenditure on research, he argued, would be increasingly challenged.[65] Public interest

---

[59] Bauer et al. 1995; Bauer et al. 2006.
[60] See the comments by Bernard Dixon in Reynolds and Tansey 2008, 76–78. See also Bud 2007.
[61] Lessing 1963.   [62] Brooks 1971.
[63] Figures on the decline of science from Council for Scientific Policy 1966b. During the 1960s the number of full-time students (in universities and polytechnics) increased from 200,000 to 430,000. Perkin 1972.
[64] Council for Scientific Policy 1968, 80.
[65] Council for Scientific Policy, Minutes of Meeting held 30 January 1970 [CSP (70) 2nd meeting, p. 4], ED 215/51, TNA.

in the applied science brand was waning.[66] We need now to examine the bureaucratic battles that further marginalised it as a category within government.

Bureaucratic Defeat and Innovation

Funding joined student attitudes and economic change at the boundary between public disgruntlement and scientific ambition. In February 1971, the MP and recent Secretary for Education and Science, Shirley Williams, expressed the new mood as she linked disappointment with the economic growth rewards of research, public concern over the social costs of science, and the end of rapid funding growth.[67] Science's leaders responded by emphasising their contribution to the economy and the potential for more. The claims for the ways pure science could power innovation through applied science became more ambitious. But, equally vociferously, these were denounced.

In the United States, too, the discourse questioned the linkage between innovation and investment in science. The Department of Defense, the leading supporter of research, produced Project Hindsight, whose first interim report appeared in 1966. It distinguished between three stages – 'undirected science', 'applied or directed science', and 'technology' – in developing weapons systems. The findings were disconcerting for fundamental science. It classed 92 per cent of the events leading to 'development' as 'technology', and almost all of the significant 'science' was 'applied science'.[68] In the summer of 1968, the National Science Foundation (NSF), established to support fundamental research, was allowed for the first time to support applied research. It set up the 'Interdisciplinary Research on Problems of Society' program, which led to the RANN (Research Applied to National Needs) programme.[69] NSF also retaliated against 'Hindsight' by commissioning its own 'Project Traces', which showed the importance of fundamental science to technological breakthroughs.[70] In 1965 the National Academy of Science organised a report titled *Basic Research and National Goals* for Congress, following that up, two years later, with *Applied Science and Technological Progress*. This found that the most significant innovation in the 'pursuit of modern (as opposed to older) applied science is the big

---

[66] Stewart 1989; Mandler 2015.
[67] Shirley Williams, 'The Responsibility of Science', *The Times Saturday Review*, 27 February 1971.
[68] Sherwin and Isenson 1967.   [69] See Belanger 1998.
[70] Illinois Institute of Technology 1968.

mission-oriented industrial or Government laboratory'.[71] Despite the acclamation, the association between the laboratories and the category proved now to be a weakness. This would become particularly apparent in Britain, where the great industrial and government laboratories had even greater local significance, and the country's economy failed to thrive.

While scientists fought to demonstrate their continuing relevance, engineers competed for increased standing. In 1964, the same year as Britain established its Mintech, in the United States, the Economic Report to the President suggested that federal support could connect better to universities and private business to 'accelerate the technological progress of our civilian industries'.[72] The same year again, the National Academy of Science established the National Academy of Engineering. The new body engaged in discussions within the federal administration, supporting the Assistant Secretary for Science and Technology in the Commerce Department, metallurgist Herbert Hollomon. A strong prot-agonist of a need-pull rather than a science-push model of innovation, he commissioned an investigation, 'Technological Innovation: Its Environment and Management', from a panel chaired by Robert Charpie, director of technology at the Union Carbide Corporation. The Charpie Committee's final report put research and invention as only 10 per cent of the cost of innovation. This widely cited conclusion accorded 'R&D' a far more modest role than hitherto.[73]

British developments mirrored and, in many cases, responded to their American models. The Charpie report raised the profile of a new committee established to link the often competing interpretations of the research councils' advisory committees (in the Department of Education and Science) and Mintech. The principal concerns of the new body were innovation and the place of science within it. Chaired by the stalwart of science policy, Solly Zuckerman, the Central Advisory Council on Science and Technology (CACST) was established in 1967. With the participation of economist Bruce Williams, who was interested in linking research conducted in the government's name to customer needs, this was somewhat more open to contemporary economic thinking than the scientists' earlier Advisory Council on Scientific Policy.[74] Following the Charpie Report, in 1968, the CACST published its British version,

---

[71] National Academy of Sciences 1967, 3.   [72] President of the United States 1964, 14.
[73] United States Department of Commerce 1967. This was cited, for instance, by Blackett in a paper for a Select Committee of Parliament. See Blackett 1968, 1108. For a critique but also discussion of the report's significance, see Stead 1976.
[74] This was pointed out in Gummett 1971, 120 and 229.

Technological Innovation in Britain, under the supervision of a subcommittee, including Mintech champion Patrick Blackett.[75] His presentation to a Parliamentary Select Committee expressed the tensions under which a scientific leader was then operating. Although Blackett described a conventional linear model of technological innovation beginning with pure science and then applied science, he emphasised the critical importance of production and marketing.[76] Scientific research was not the deserving determinant of success in the innovative process.

Inevitably, the changing emphasis in talk of innovation threatened the research councils. As early as 1960, the Advisory Council on Scientific Policy had asserted with all the authority it could command that 'technological progress depends ultimately upon the work in science in the universities and technological institutions'.[77] The 1967 Second Report of its successor, the CSP, was also a defiant statement of fundamental science's value. 'Basic research', it asserted, 'provides most of the original discoveries and hypotheses from which all other progress flows.'[78] But would the government continue to believe? In 1968, the government rejected the Science Research Council's request for extra support, including a substantial contribution to a new European accelerator and a promise to sustain the council's other activities at a growth rate of 8 or 9 per cent from 1972 onwards.[79] The issue of commitment had become particularly acute.

Over the next two years, there was a remarkable battle of wills between the CSP, responsible for the research councils, and Zuckerman's CACST. An overlapping membership complicated the conflict. Each was familiar with the other's next move. Thus, it became known to the scientists that Zuckerman favoured the transfer of the hitherto independent Agricultural and Medical Research Councils into their respective government departments and out of CSP's and DES's sponsorship. The proposed loss of independence was not just a matter of bureaucratic tidiness but a changing model of innovation, from science-push to need-pull.

Preparing for his 1969 third report, the CSP chairman circulated the council with a copy of the half-century-old Haldane report, clearly emphasising the claims for independence of the Agricultural Research

---

[75] Central Advisory Council for Science and Technology 1968. For discussion of the Charpie report by the CACST, see AC (67) 5th meeting, 8 June 1967, SZ/CACST/1, Papers of Solly Zuckerman.
[76] Blackett 1968, 1108.   [77] Advisory Council on Scientific Policy 1960, 5.
[78] Council for Scientific Policy 1967, 12; Gummett and Williams 1972.
[79] For the 300 GeV accelerator as the context for the Byatt and Cohen report, see Jevons 1976, 731. For the accelerator arguments, see Clarke 1968 and Gibbons 1970.

Council (ARC) and Medical Research Council (MRC). Quickly learning of this defensive action, Zuckerman was outraged and protested to the Cabinet Secretary.[80] When news of the conflict reached the ears of parliamentarians, the Prime Minister had to reassure the House of Commons that he had no plans to make such changes in the management of science as Zuckerman was contemplating.

At the heart of the CSP approach to its anticipated 1969 report was a defence of the research councils' role by showing how science led remorselessly to innovation. The committee's Civil Service secretaries, Harry Cohen and Ian Byatt, drafted a paper first presented to the council in December 1968. Their ambition was to find a quantitative link between the number of major scientific discoveries that science funding might make possible each year, and the probability that these would lead to substantial industrial change.[81] The ambitious plan was evaluated at the University of Manchester, where research on the process of innovation had already begun. However, Jevons and his colleagues treated Byatt and Cohen's enterprise with scepticism, arguing that the linkage between curiosity-oriented investigations and their economic benefit was complex if nonetheless important.[82] Although brought to a published conclusion as a stand-alone paper, the CSP abandoned the work, admitting instead that the relationships it had wanted to flaunt were much more complicated than it had first assumed.[83]

In any case, by the time the council eventually published its 'Third Report' in 1972, chronic strategic considerations had given way to acute concern and tactical defeat. Science-push was giving way to need-pull. The first to drive this administratively was Tony Wedgwood Benn, the Labour Minister of Technology who was anxious about the vast research

---

[80] For the information about CSP plans, see D. L. Jones to Solly Zuckerman, 29 October 1968, CAB168/218, TNA. Zuckerman objected immediately to the Cabinet Secretary; Zuckerman to Burke Trend, 30 October 1968, CAB168/218, TNA. Prime Minister Wilson reassured Parliament that no decisions had been made, *Hansard* HC Deb. 21 November 1968, cols. 1561–62.

[81] 'An Attempt to Quantify the Benefit of Academic Research in Science and Technology – A Brief Summary', CSP (68)67 ADDENDUM 7th November 1968, TNA.

[82] For the report, see F. R. Jevons and A. W. Pearson, 'Feasibility Study of the Method Proposed by Byatt and Cohen for Quantifying the Economic Benefits of Scientific Research', September 1969, ED214/86, TNA. A published summary of the Manchester team's conclusion about the Byatt and Cohen approach is found in Langrish et al. 1972, 35–39. See also Frederick Jevons to Harry Johnson, 25 March 1969, ED214/86. Surprisingly, there is no evidence here, or indeed in informal conversations with John Langrish, that the Manchester professorship of Bruce Williams was anything more than incidentally significant. For subsequent reflections by the reviewers, see Gibbons and Johnston 1974 and Jevons 1976.

[83] Council for Scientific Policy 1972, Appendix D, 25–27. The analysis was published as Byatt and Cohen 1969.

establishment run by the UKAEA. The authority was making strenuous efforts to evolve into a more general research body, a British equivalent of the German Fraunhofer Society, which had also originated in the hope of drawing civilian benefit from institutes with military orientation. By 1969, a quarter of Harwell's budget was related to commercial activities.[84] Late in 1969, Benn determined to go a step further, proposing to convert the entire UKAEA and the NRDC into a single contract research organisation that only conducted research demanded by a paying customer.[85] Such a transformed body would have about 2,500 professional staff.[86] Zuckerman has written that he first heard the phrase 'customer-contractor principle' in the Minister's presentation of the plan to his committee.[87] Although Benn would shortly be out of office, the Conservative victors would build upon his approach when the Labour party lost the 1970 general election.

As the 1970s dawned, Zuckerman's campaign to transfer the ARC and MRC to the operating departments seemed to be within sight of success. The new Conservative administration commissioned a report on the management of government R&D from a fellow maverick, Victor Rothschild, formerly research coordinator of the Royal Dutch Shell group and currently head of the government's think-tank, the Central Policy Review Staff. At the time, he was developing his extensive collection of forty-five classifications of research and development, though he dismissed almost all. Pure and applied research were among the few survivors of his cull.[88] Rothschild's recommendation on the organisation of government scientific research was published by the government, bound with a competing interpretation by the new chairman of the CSP.[89] Ostensibly, the two documents were possible alternatives, but the government focused upon the Rothschild report. This made a formal distinction between pure research, whose funding stayed with the research councils, and applied research commissioned by the operating Ministries. Consequently, Rothschild suggested that rather than absorbing the two research councils, the relevant government

---

[84] Coopey 1991a, 120. For the general atmosphere and hopes of the time, see Vig 1968, 152. For the Fraunhofer institutes, see Trischler and Vom Bruch 1999.

[85] 'The Organisation of Government Industrial Research and Development Resources', Central Advisory Council for Science and Technology, Annex AC (69) 13th Meeting, Item 2, (21st October 1969): 2. 'Proposed Reorganisation of the Establishments of the Ministry of Technology and Atomic Energy Authority' (AC(69) 12th Meeting, Minute 2); Memorandum by the Minister of Technology, AC (69)43, SZ/CACST/4, Papers of Solly Zuckerman. See also Ministry of Technology 1970.

[86] AC (69) 8th meeting, Central Advisory Council for Science and Technology, Minutes, 17th June 1969, SZ/CACST/4, Papers of Solly Zuckerman.

[87] Zuckerman 1988, 424.   [88] Rothschild 1972.   [89] Lord Privy Seal 1971.

departments should appropriate those parts of council grants currently devoted to applied science. They should then use their funds to commission any needed research from whomever they wished.

Although administratively less aggressive than Zuckerman, Rothschild drew on similar assumptions. In place of the model of science driving innovation, applied science was a commodity produced to order, for use in the process of 'innovation'. The consequences were bureaucratically radical. The Ministry of Agriculture, Fisheries and Food took more than half the budget of the ARC.[90] In this new world, government industrial policy sought to shape applied science, which was, in any case, considered subsidiary to 'development'.[91] Though the immediate consequences at the laboratory bench were slight, the Rothschild proposals were deeply unpopular with scientists.[92] Because the research was now just one among several activities of operating departments, it lost its collective identity. Curiously, therefore, the very process of focusing on the distinctive instrumental qualities of applied science undermined the cultural and institutional coherence of the category across government. The 'natural' partnership of pure and applied science, or basic and applied research, was being swept aside by new industrial policies.[93]

As support for science funding fell, the research councils became more interested in engineering and technology. In the late 1960s, the Science Research Council sought to emphasise 'applied' rather than 'pure science'. However, the shift was not in itself enough to silence critics. In 1969 the council created an Engineering Board, and then in 1981 changed its name to 'Science and Engineering Research Council' (in 2023, its descendant is called the 'Engineering and Physical Sciences Research Council').[94] Meanwhile, responsibility for ensuring appropriate support for the now disaggregated category of 'applied research and development' was given to a committee with the acronym ACARD. The letter inviting council members reflected on the national importance of effective research and development, 'particularly the latter'. The new

---

[90] 'Select Committee on Science and Technology: Science Subcommittee, Memorandum by the Lord Privy Seal', CAB 164/1337, TNA.
[91] The respected scholar Niels Rolls-Hansen pointed out dangerous parallels between the internationally prevalent science policies of the 1970s and the Soviet, pre–World War Two policies that had spawned the rise of Lysenko; Rolls-Hansen 2006; Rolls-Hansen 2015.
[92] See Gummett 1980, 202–5. See also Williams 1973; Whitehead 1978; Parker 2016.
[93] See the paper by the former Universities and Science Minister David Willetts: Willetts 2019.
[94] See 'Science Research Council: Name Change to Science and Engineering Research Council', PC15/1374, TNA. In 1980, the Medical Research Council supported the foundation of the early biotechnology company Celltech. Dodgson 1990.

body was principally concerned with 'technology' rather than 'applied science'.[95] Indeed, in 1986, its report blamed Britain's failure to capital-ise on its scientific prowess on faith in 'the simple progressive model of scientific innovation'.[96]

Civil war amongst the science-policy fraternity was not over. As ACARD was disowning the 'linear model', a modern version was reasserted with a vengeance elsewhere in government. Jon Agar has shown how the Conservative government of Mrs Thatcher distanced itself from the Wilson governments' flirtation with 'technology', or even the earlier Haldane commitment to supporting private industry with publicly funded applied science.[97] At a time when she was recasting the government's role across all its functions, as a former chemist, Thatcher celebrated curiosity-driven research and did consider that the state should pay for this. She came to believe, though, that R&D close to the market should be the private sector's responsibility.[98] So, whereas the formal scientific advisory committees emphasised the importance of mediation between basic science and industrial practice, the Prime Minister's office disowned the public sector's responsibility for that process.

In the era of 'New Public Management' (NPM), the administration defined the 'private sphere' as the proper place for market-facing research, which some complained was being treated as a commodity.[99] Accordingly, the establishments that had so characterised the applied science estate since World War One transferred to the private sector. First, around 1990, went the civilian stations that DSIR had run. In 1995 the UKAEA was divided, and part privatised. A few years later, most non-nuclear military research and evaluation establishments were spun collectively out of the public sector as a private company called Qinetiq. In any case, military funding as a percentage of GDP would decline for many years because of the ending of the Cold War.[100] Physically and

[95] See, for instance, Frederick Peart to Sir Frederick Stewart, 21 September 1976, CAB 164/133e, TNA. The acronym ACARD was derived from the committee's full title 'Advisory Council for Applied Research and Development'.

[96] Chairmen of the Advisory Council for Applied Research and Development (ACARD) and the Advisory Board for the Research Councils (ABRC) 1976, 2, © Crown Copyright.

[97] Agar 2011 and Agar 2019.

[98] On the more general shifts in the role and nature of government, see Campbell and Wilson 1995.

[99] Boden et al. 2006, 125.

[100] In 1984 the UK defence expenditure was 5.5 per cent of GDP. This fell to around 2 per cent in the second decade of the twenty-first century. See https://data.worldbank.org/indicator/MS.MIL.XPND.GD.ZS?locations=GB sourced from the Stockholm International Peace Research Institute (accessed January 2023).

journalistically, the old bodies that had exemplified applied science were losing visibility.[101] The state's hopes for an industrial policy based on state-provided applied science that had begun at the beginning of the twentieth century seemed to be fading at its end.

So, although Margaret Thatcher's premiership ended in 1990, and Conservative government philosophy evolved after that, the administration had entrenched many earlier initiatives.[102] Worldwide, research conducted by business had come to dwarf its public sector counterpart. In Britain, in 2015, the value of 'R&D' undertaken by private firms was £22 billion, while the government's activities were worth but a tenth of that.[103] In the United States, the standing of applied research within the National Science Foundation fell. Its budget share dropped from 9 per cent in 1977, at the end of the RANN era, to 5 per cent in 1984.[104] An NSF conference, 'The Categories of Scientific Research', in 1979 signalled American loss of confidence in the reality of old forms of classification.[105] In Britain's Parliament, too, the use of language changed at the end of the twentieth century. Whereas references to applied science collapsed between the 1960s and the 1990s, the use of 'technology' was in the ascendant. Usage in Parliament rose from a smaller number than 'science' in the 1960s, 5,819 occasions (compared with 7,914 for science), to 70 per cent more than science (11,295 occasions compared with 6,674) in the 1990s.[106]

In 1994 Michael Gibbons, one of the Manchester team responsible for the Byatt and Cohen scheme's dispatch, and other science-policy analysts, published *New Production of Knowledge: The Dynamics of Science and Research in Contemporary Societies*.[107] The authors argued that the traditional discipline-bound organisation of science (Mode 1) was giving way to another model of the search for knowledge in the context of application (Mode 2). The old demarcation lines between pure and applied, scientist and audience, were becoming irrelevant. While many historians

---

[101] A report published as this book was being completed suggested the continuing declining place of the Public Sector Research Establishments (PSREs) in the national research and development landscape into the first two decades of the twenty-first century. See Nurse 2023. For the parallel process in France where too the old binary system of universities and government laboratories, in which government laboratories embodied applied science, was breaking down, see Larédo and Mustar 2004.
[102] Boden et al. 1998; Boden et al. 2004.   [103] Hennessy 2018, 9.
[104] McNinch 1984, 8. I am grateful to Marc Rothenberg, formerly NSF historian, for sight of this document.
[105] United States National Science Foundation 1980. See too Hensley 1988.
[106] Results from https://hansard.parliament.uk/ (accessed August 2022).
[107] Gibbons et al. 1994. An argument related to the new production of knowledge was put at the same time by John Ziman. See Ziman 1994.

voiced suspicion about the novelty of the phenomena described by Gibbons and his colleagues, support came from a perhaps surprising direction. The distinguished physics-historian and Smithsonian curator Paul Forman published an analysis also suggesting an ideological change around 1980.[108] Until then, Forman argued, technology had been subsumed under science, but since then, technology had risen to dominance. He identified the changing power relationship with a shift from modernity to post-modernity. This was a transfer of emphasis from means and bureaucratic form to a culture more concerned with achieving goals. Few may have read Forman, but he was voicing an impression experienced by many. The term 'technoscience' became a popular symbol of the new dispensation.[109]

This book has argued that, in Britain, knowledge classification expressed distinctions between what has been considered 'private' and what 'public'. Science as public knowledge has been distinct from the privacy of technology. This clear distinction was rendered obsolete by the changing nature of what could be patented. In 1980, the US government changed its rules on exploiting federally funded research. Previous administrations had considered the discoveries it yielded as public property and circumscribed the rights of universities to commercialise them. Now, Washington encouraged private profit from the results of federally funded research. Such other countries as Britain followed with parallel processes. In 1981, Margaret Thatcher's government incorporated the NRDC within a new British Technology Group (BTG), encouraged academic scientists to pursue commercial exploitation with private industry, and then removed the BTG monopoly four years later.[110] Across the world, the narrative of science policy was changing by the early twenty-first century, and a new language of grand challenges defined by practical problems and commercialisation was emerging. In applications for the support of scientific research from the research councils, 'impact' and 'routes to impact' became critical funding criteria. The place of applied science would be renegotiated once more.[111]

---

[108] Forman 2007.   [109] On the history of the term, see Hottois 2018.
[110] Agar 2019, 69–73.
[111] See Flink and Kaldewey 2018. In the new British government's 2020 'Research and Development Roadmap', in which grand challenges have an important place, the term 'applied research' is used eleven times. The term 'innovation' occurs 200 times in the document. 'UK Research and Development Roadmap' published 1 July 2020, https://assets.publishing.service.gov.uk/government/uploads/system/uploads/attachment_data/file/896799/UK_Research_and_Development_Roadmap.pdf (accessed August 2022).

Epilogue

All science is not the same, nor is all industry.[112] The diversity of appropriate innovation models, including the importance, in some circumstances, of applied research, came to be widely accepted early in the twenty-first century. Within the expanding health field, 'translational research', a new category akin to applied science, prospered, and the role of the government as research sponsor has flourished. An advisor's 1987 memo to the Prime Minister, urging her to concentrate government support on 'fundamental science', had contained the caveat 'except where the state is the customer'.[113] Although historically that had meant, above all, the military sector, now healthcare was the area where the administration was itself a dominant and rapidly growing supplier of services and performer of research. Through the issues raised by research for the health sector, reflection on the classification of science was re-energised.

Institutional reorganisation came to express and promote a revived concern with the space between 'bench and bedside'. In the twentieth century, the Medical Research Council had dominated healthcare research in Britain. An in-depth study conducted by the former civil servant Maurice Kogan pointed out that despite rejecting the basic/applied dichotomy, the council did, in effect, emphasise curiosity-driven research. Its outstanding outputs were on the basic and biomedical side rather than clinical delivery.[114] Responding to such anxieties, in 1988, a committee of the House of Lords explored 'Priorities in Medical Research'. Apparently, industrialists were surprised by the small commitment of the National Health Service (NHS) itself to research and development.[115] The final report reflected that despite the council's success in winning Nobel Prizes, there were reservations 'where the MRC's responsibility for basic research shades into its responsibility for clinical research'.[116] Indeed, the existing system was defective 'in applied research and in the application of knowledge gained from research'.[117]

---

[112] Walsh 1984; Schmoch 2007.
[113] George Guise, 'Public Expenditure on Science', 24 July 1987, PREM 19/2477, TNA; and Guise to Thatcher, 25 May 1988, PREM 19/2479, TNA. I am most grateful to Jon Agar, who found this letter and identified the contribution of Guise for his help in understanding it. Agar discusses the significance of this letter in his book, Agar 2019. For Guise's own interpretation, see Guise 2014.
[114] Kogan et al. 2006, 86.
[115] Nick Black, quoted in Atkinson and Sheard 2018, 15. See also Davies 2017.
[116] House of Lords Select Committee on Science and Technology 1988, 28, para. 3.16, © Crown Copyright. On its importance, see Davies 2021.
[117] House of Lords Select Committee on Science and Technology 1988, 28, para. 3.17, © Crown Copyright.

The report reflected upon the relevance of the traditional categories of basic, strategic, and applied research, but found these rather lacking. To the latter two terms, it preferred two others, 'public health research' and 'operational research', often referred to within medical circles as 'health services research'. It also added clinical research to the spectrum, following basic research, concluding that the NHS was the key to improving the funding of 'applied research in medicine'.[118] The 1988 House of Lords report became a turning point.

Witnesses denounced the relatively tiny Department of Health expenditure on medical research, a fraction of 1 per cent of the total NHS budget.[119] So, within three years, the NHS head of R&D won a commitment to research of 1.5 per cent of NHS funding.[120] This enabled a remarkably increased salience of healthcare in the British research scene. Between the mid-1980s and 2016, while GDP had grown roughly four-fold, the value of medical-related research performed in the private and public sectors increased by an order of magnitude.[121] By 2009, health had already outstripped defence as the single largest beneficiary of British government-funded R&D.[122] During the following decade, in 2017, across all industry, 'pharmaceuticals' had become the largest sector for the performance of R&D in Britain (£4.3 billion), approaching 20 per cent of the total. This was so large that critics complained such an imbalance might not be sustainable.[123]

With these figures in mind, it was striking that about a billion pounds, significantly more than the MRC budget, and comparable to (if less than) the defence R&D budget, was spent by the new National Institute for Health Research (NIHR).[124] Founded in 2006, this organisation brought together the now-growing research budgets of the NHS. Its expenditure constituted something like 20 per cent of all government research spending, and along with the MRC, it accounted for a third of

---

[118] House of Lords Select Committee on Science and Technology 1988, 28, para. 3.18, © Crown Copyright.

[119] Davies 2021, 202. See also House of Lords Select Committee on Science and Technology 1988, 25, para. 2.72.

[120] Atkinson and Sheard 2018, 16.

[121] For 1985/86 figures for biomedical R&D, see House of Lords, Select Committee on Science and Technology 1988, 10, para. 1.27; for 2016 pharmaceutical industry R&D, see Office for National Statistics 2017; for public sector expenditure (including charities), see AMRC 2017; for a time-series of GDP, see Office for National Statistics 2023.

[122] Jones 2019, figure 12, p. 28. The figures presented are based on OECD data.

[123] Jones and Wilsdon 2018.   [124] On the origins of the NIHR, see Atkinson et al. 2019.

the total. It also gave a continuing unusual prominence to applied research supporting the NHS. The Chief Medical Officer, Dr (later Dame) Sally Davies, who served from 2010 to 2019, had driven the new institute's establishment, leading the campaign to win a rapidly growing and substantial budget. In an interview, she described how the new institute had justified its budget by drawing on modern economics, in arguing the need to remediate areas of 'market failure', and by respecting historical sensitivities.[125] Once the new body had become a reality, the government commissioned a report into its place in the institutional ecology, alongside the Department of Health and the MRC.

The review, led by the venture capitalist and public servant Sir David Cooksey, was interesting for its use of classical categories. It divided health research into 'basic research', 'applied research', and 'translational research' and dedicated an explanatory paragraph to each. The first two would have been familiar to Zuckerman's committee of half a century before. However, introducing 'translational research' was more innovative. Recently pioneered in the United States by the American Cancer Society and the National Institutes of Health (NIH), this term possessed a flexible meaning concerned with translating knowledge 'from bench to bedside'.[126] In the British context, Cooksey's report identified two translational problems of the current organisation: 'translating ideas from basic and clinical research into the development of new products and approaches to treatment of disease and illness; and implementing those new products and approaches into clinical practice'.[127] The first of these seemed close to the traditional challenges to which applied science had once seemed the answer. Recall that Chapter 4 of this book showed how the research interpretation was linked to the previous pedagogical concerns of the term's users by three considerations: the continuity of narrative, industrial policy, and links to pure science. Similar continuities characterise the early twenty-first-century discussion of translational research.

---

[125] Helen Compton, 'Better Together: Collaborating with NIHR Research Programmes', 9 November 2018, www.amrc.org.uk/blog/better-together-collaborating-with-nihr-research-programmes (accessed August 2022). Sally Davies explained the origin of the name of these grants in an interview at the Royal Society in 2016. 'A Life in Health: In Conversation with Dame Sally Davies', April 2016, https://royalsociety.org/science-events-and-lectures/2016/04/a-life-in-health/ (accessed August 2022).

[126] On the development of the concept of 'translational research', see, for instance, Woolf 2008; Lander and Atkinson-Grosjean 2011; Rushforth 2012. Todd Olewski has suggested that NIH introduced the term as part of an attempt to resolve the tension between its missions directed to basic and applied research. Olewski 2018.

[127] Cooksey 2006, 35, © Crown Copyright.

The Cooksey Report has been widely attributed with the circulation of the term 'translational' into a British context.[128] Cooksey himself was happy to illustrate its meaning with a story sacred to British science. He took the laboratory work of Florey and Chain and the industrial development of their American partners to convert penicillin from laboratory curiosity to a useful drug as a worked example of the category.[129] Indeed, both the MRC and the NIHR have been prominent advocates of the new class, each taking it as an essential organisational tool. Within the community of those addressing problems of linking rapid growth in basic science to operational change, the narrative had shown remarkable continuity with applied science of earlier years.[130]

Throughout the early twenty-first century, administrations of different parties have been enthusiastic about linking a public-sector life science budget to broader industrial prosperity.[131] A 2006 report to the Labour government, for instance, touted the research funding of the National Health Service as the centrepiece of its strategy to lift national R&D expenditure to 2.5 per cent of GDP by 2014.[132] A report to a subsequent Conservative government, and accepted by it, would come to describe the key to its science policy as the development of the Health Advanced Research Programme (HARP). The model was the American military-driven Defense Advanced Research Projects Agency (DARPA), particularly famous for its development of the internet.[133] Such talk about health research has been influential. In 2021 the British government announced plans to fund a scientist-led body to 'identify and fund transformational

---

[128] See, for instance, the Oral Evidence of John Bell and Alex Markham, to hearings on the 'Office for Strategic Coordination of Health Research (OSCHR)', House of Commons Innovation, Universities, Science and Skills Committee. 2008–9. On 8 June 2009, qq. 1–19.

[129] Cooksey 2006, 17; Rushforth 2012.

[130] The evolving meaning of 'translational research' has been mapped by Fort et al. 2017. The incorporation within social studies of sciences is studied by Crabu 2018. I am also grateful to Sanna Alas for sharing her Cambridge MPhil thesis dealing with the history of 'translational research'.

[131] The rhetorical interweaving of talk about 'health' and 'wealth' in the underpinning of the NIHR is a theme of Atkinson et al. 2019.

[132] Patricia Hewitt, 'Forward by the Secretary of State for Health', in Department of Health Research and Development Directorate 2006, 1. In 2017 official statistics reported that R&D expenditure represented 1.69 per cent of GDP. See Office for National Statistics 2019, 3. Changing methodologies make difficult comparison over the long term. Total R&D expenditure was recalculated in 2022 and the estimate increased. See Office for National Statistics 2022.

[133] Bell 2017, 14–18. DARPA has also variously been called ARPA across its lifetime; see Jacobson 2015.

science and technology at speed' to be known as the UK Advanced Research and Inventions Agency.[134]

Conclusion

Public talk in the 1950s of a new 'industrial revolution' broke the compelling momentum of building on an existing intellectual and institutional structure. Suddenly, the challenge was no longer to apply science as quickly as possible. Instead, the prospects of job-changing automation and cheap nuclear power prompted widespread discussion of a new era. They evoked challenging memories of the ambiguous past of Britain's society and economy. Science was no longer analysed in relative isolation. Whereas previously, applied science had been the favoured response to both the cultural and policy challenges of modernity, successive governments framed the new industrial revolution and the opportunity for a British renaissance in terms of the wider category of 'technology'. The formation of Mintech and the technology agenda were self-conscious responses to that sense. They provided an alternative narrative to the frustrations of disappointing benefits from rapidly growing science budgets.

Despite the protests of the CSP, the tide of argument increasingly threw doubt on the ways wealth flowed from the simple growth of scientific knowledge. The CACST committee, led by Solly Zuckerman, then Margaret Thatcher, and later the government's New Public Management, increasingly gave weight to the customers rather than producers of knowledge in the private and public sectors alike. General distaste expressed by the 'swing against science' and negative press further damaged scientific confidence, but science was not at the centre of discourse. Unlike the interwar period, technology garnered the greatest attention. As 'innovation' became the objective of science policy, the CACST's deliberations amounted to the intentional destruction of the survival strategies of the CSP within the machinery of government. More than an internal bureaucratic struggle, this represented a cultural denouement. No longer was industrial policy equated with science policy and the promotion of 'applied science'. As the Cold War waned and the political mood switched away from direct government engagement, many great laboratories were privatised and sometimes transformed. References in the press plummeted, even to such a survivor as the National Physical Laboratory. The practice of research shifted decisively

---

[134] Department for Business, Energy & Industrial Strategy 2021, © Crown Copyright. Also see aria.org.uk.

to private industry and moved out of the public eye. Having lost the support of government policy, institutional reality, and narrative authority alike, applied science was becoming less 'real'.

However, early in the twenty-first century, British governments once again emphasised connecting the interpretation of, and support for, science to industrial policy. This move represented a startling continuity with longer-term applied science traditions. The emerging category of translational research and the building of the NIHR highlighted the continuing hunger to manage the processes between laboratory and treatment. Above all, in the life sciences, a specific context justified the new term. Politicians resurrected old stories of triumphs past. For many, 'translational research' served as a localised development of what had long been called 'applied science'.

Conclusion

For over a hundred years, applied science seemed an attractive elixir, ideally suited to cure many of Britain's economic and cultural ills. Accordingly, it occupied an important place in national life. Citizens and politicians would daily encounter stories, institutions, and inventions that confirmed its benefits. From the 1960s, however, the category was superseded, and soon its promise seemed unconvincing and old-fashioned. This book has explored why it rose and why it fell.

Certainly, use of the term was international, and many of the cultural challenges it addressed were global, but its usage and associations were distinctive to individual contexts. In Britain, it served as an integral part of local cultural, academic, and political environments. Its users constructed and reconstructed the concept to build civilisation and prosperity in their own circumstances. Since popular and political meanings were shaped by widespread public discourse, this book has not been limited by the scientific projects and useful devices it has explored. The government policies, stories, and institutions it has also followed have been sites of private controversy and public discourse about the prospects for British industry and civilisation.

Politicians gave applied science a central place in British industrial policy, which amplified its importance and privileged its meaning. Governments were attracted by its enduring identity as public knowledge and a reputation as the source of revolutionary improvement. It promised modernity without threatening private knowledge or the free market. These qualities made it the legitimate and desirable object of government support, for both teaching and research. Thus, in the nineteenth century, governments were attracted by the promise of new cohorts of industrial managers, whose training included applied science, superseding 'rule of thumb' traditionalists, without infringing on the privately held secrets of engineering skill. In much of the twentieth century, applied science served as the basis of hopes for British-led disruptive innovation. Support was assured, during the early part of the century, by the influence of two leading philosopher-politicians associated with opposing parties, the Conservative Arthur Balfour (died 1930) and the Liberal R. B. Haldane (died 1928). After World War Two, policy discussions intended to inform the organisation of government science developed earlier reflections on the invention process into a classification of science – in which the categories of development, applied science, and pure science were carefully demarcated and often followed one another. However, this book emphasises that the concept of applied science had much richer associations too.

Beyond political economy, applied science proved to be a term useful for professionals and the public alike. Stories about it illuminated the puzzling interface between the esoteric activities of the very few and the changes experienced by the very many. Accounts of a distinctive 'applied' science, separate from 'pure' science, emphasised the links between abstract investigations and the lives and problems of citizens while protecting the reputation of pure science from the association with the agents of death, pollution, and destruction. At the same time, narratives of progress, such as the 1931 Albert Hall exhibition on the work and its consequences of the already mythologised hero Michael Faraday showed how applied science drew on pure science.[1] The efficacy of this model was seemingly proven by the radical technical change of the time and by the success of foreign laboratories, set through World War One experience and maintained by institutions created before and during the conflict.

During the mid-twentieth-century heyday of the concept, applied science provided a language and an interpretation for more than knowledge, gadgets, and education. Also describing the rapid changes in modern civilisation, management, and politics, applied science was central to enduring narratives about modernity and culture. For example, those looking forward to a planned, rationalised economy put it at the heart of what has come to be termed 'sociotechnical imaginaries'. Within these, vivid imagery and narratives of future development were made real by aircraft, telephones, radios, oil from coal, plastics, and numerous other inventions. As we have seen in the various projects to convert coal to oil, the stakes were high. In times of high unemployment, finding new industries was an urgent priority. The satire of the ideology of order and science by Aldous Huxley in *Brave New World* was nonetheless enduringly persuasive.

Of course, many people also feared that applied science threatened lives and culture through the devices and processes it made possible.

---

[1] See James 2008.

Particularly in the wake of World War One, there were deep anxieties about the consequences of applied science as exemplary of the darker side of modernity. Poison gas and then the atomic bomb were only the worst products. Nonetheless, even if society might consider applied science sometimes deeply unattractive, such anxieties emphasised its potency and reality.

At the same time, the place in the envelope of 'science' confirmed applied science's meaning and history within traditional concepts of civilisation. Association with the folklore of Faraday and Newton and the terminology of 'pure and applied science' assured continuity, truth, and utility. During the 1960s, Lyon Playfair served as a much-cited prophet whose message was urgently relevant, though he had died in 1899. Promoters accumulated cultural and political resources through promises that science could rescue the nation, reminding audiences of past achievements. Politicians, activists, and newspapers incorporated the romances of James Watt, Robert Stephenson, William Perkin, and Alexander Fleming within allegorical tales about the nation's progress.[2] Building on such epics, reports of recent triumphs like atomic power and the jet engine made a convincing case for applied science's continuing benefits.[3] Hence such stories had moral significance and a compelling message linking Britain's past to a successful future. Their historical truth is not here an issue, nor was it subjected to the critical scrutiny of numerous scholars at the time.[4]

Institutions built in the shadow of hopes and fears associated with applied science gave physical reality to the concept. The experience of encountering them gave the term public and political meaning. Initially, applied science helped structure massive volumes of the *Encyclopaedia Metropolitana*. Then, in the mid-nineteenth century, it described the valuable chemical services promising to rescue agriculture from the

[2] The highly popular young adults account *People in History* (1959) by R. J. Unstead was subtitled *From Caractacus to Alexander Fleming*. See Bud 1998.
[3] The Ministerial 'Foreword' to the 2020 Research and Development Roadmap continues this rhetorical tradition: 'The UK is internationally recognised for our leadership in research, the excellence of our scientific institutions, and the innovation in our economy. We can proudly claim to be the nation that gave the world the steam engine and the jet engine.' www.gov.uk/government/publications/uk-research-and-development-roadmap (consulted February 2021).
[4] One pioneering exception to the general support characteristic of the 1950s should be cited. John Jewkes, working with two assistants, studied sixty-one inventions; however, the methodology was criticised because the case studies were hand-picked and the histories were sometimes problematic. See Jewkes et al. 1958. Jewkes, it should be emphasised, identified himself strongly with opposition to planning and with free market economics and would later be president of the Mont Pelerin Society. See also Jewkes 1960.

repeal of the Corn Laws. Meanwhile, publishers and promoters popularised the term through advertisements and catalogues.

From the 1870s, new colleges, civic universities, specialised research laboratories, and research funding agencies of the early twentieth century provided imagery and a framework of hope. The neo-gothic edifice of Mason College towered over Birmingham's centre from 1880. Glaswegian engineers took a degree in applied science, having studied in the massive building of the Royal Technical College. At the beginning of the twentieth century, Yorkshire students built careers by training in the impressive Department of Applied Science at the University of Sheffield. The Department of Scientific and Industrial Research addressed urgent national problems ranging from smoke pollution to rotten Australian apples. The work of its scientist Watson Watt would underpin the successful deployment of radar in World War II.

In the mid-twentieth century, research in applied science was embodied in great government laboratories, for example, the National Physical Laboratory, the Royal Aircraft Establishment, and the Fuel Research Station, and the laboratories of such industrial behemoths as GEC and ICI. Such achievements as the BBC radio station and the Billingham factory demonstrated the model's success. After World War Two, the government built a widespread network of atomic energy and defence-related laboratories. Several were nationally famous, such as the Harwell atomic research campus and the Telecommunications Research Establishment at Malvern. These institutions intruded onto the pages of newspapers, into politicians' speeches, and into the lives and minds of citizens, providing training and frameworks for good careers. Moreover, those bodies developed the public accounts of synergy between pure and applied science leading to military security and industrial prosperity.

Today, the continuity between the discussions of pedagogy and research might seem surprising. Yet government policies, narratives, and institutions supported the lineage. During the nineteenth century, progressive reformers diagnosed the labour force's inadequate training as the underlying industrial problem and prescribed education in applied science. The prospect promised utility during professional training and work, complementing the deep roots in culture and scientific principles important to the new civic universities. Then, during much of the twentieth century, research promised analogous science-based solutions to industrial problems. Again, the inheritance of allegorical stories, heroes, and morals assured the parallels. There was also a cross-over in the talk about the relationship between applied and pure investigations. The progression from principles to applications, so characteristic of pedagogy, slipped easily into the understanding of research and fitted the experience drawn from heroic tales. Certainly, the relationship was generally dynamic and two-way rather than merely derivative, but this quality was often added as an exception. Moreover, the great civic universities whose origins lay in the panic over pedagogy now conducted applied research.

Never was applied science an unproblematic description of the necessary underpinning of prosperity. During the nineteenth century, disputes over technical education and the calls for 'technology' highlighted uncertainty about the adequacy of science as a description of the formal knowledge needed by industry. We have seen the debates amongst the press, the lay backers of new educational institutions, government, engineers, and scientists. In rhetoric about nineteenth-century education and twentieth-century research support, 'technology' was used as an alternative term to describe a broader knowledge system closer to the market and its complex needs than 'applied science'. Bureaucratic innovations and debates reinforced change in linguistic conventions. Competing professionals jostled for pre-eminence within such institutions as the Atomic Energy Authority. When new political-economic ambitions came to the fore in the last decades of the twentieth century, technology challenged and replaced applied science in general political and popular parlance. From the end of the 1960s, the government policies, narratives, and institutional relationships that gave 'applied science' meaning to policymakers and laypeople alike faded. Those who deployed this concept could find that the present was a foreign country.

The shift away began with stories. Well before World War One, particularly through 'City and Guilds' education, the competing term 'technology' had come into common use. Connoting a broader range of skills, rules, and knowledge than the strictly scientific, it was also less associated either with liberal values or with the borders between private and public goods. Meanwhile, Americans were coming to use the term in preference to applied science. Having already been used widely in education talk, after World War Two, 'technology' came to dominate the discussion of invention and the broader concept of innovation. A new generation of economists helped widen its use in Britain.

American influence on the broadening discourse was important, but so were domestic considerations. During the early 1950s, talk of a new industrial revolution that had begun in the United States crossed into Britain, where it had a special resonance. After all, this was the country where the first industrial revolution had begun, was still celebrated, and was remembered with pride. However, it was also a traumatic experience for many families and was costly to society and the environment. In the wake of imperial decline, engineers, the press, and politicians were putting special hopes in the prospects of nuclear power, automation, and plastics. The main political parties agreed that technology, if mismanaged, could lead to mass unemployment; properly deployed, it was the root of growth.

The Labour party, in particular, adopted talk of an imminent industrial revolution. Coming to power in 1964, it used a new vocabulary to describe the country's industrial and social problems and proposed a much more interventionist approach than had been traditional in peacetime. Bringing the term 'technology' into vogue, the Labour administration founded an increasingly powerful Ministry of Technology immediately after it was elected. It left 'applied science' in the bailiwick of a newly formed Department of Education and Science. Distinguished scientists such as Patrick Blackett and Solly Zuckerman, who occupied influential positions, suggested to their Labour-politician friends that research was just one input into innovation. Without the old inhibitions on the state's direct industrial intervention, arguments that investment in scientific research worked as the economy's driver lost potency. Even when governments changed, old policies did not revive. Following Lord Rothschild's recommendations in the 1970s, applied science was transferred from the research councils' remit to operating departments and sacrificed its coherence. Funding within departments declined as they found other priorities. When, early in the 1980s, new industries were heralded once again as a future hope, it was 'biotechnology' and 'information technology' that were at the leading edge of hopes for a further industrial revolution and foci of state support.[5]

Therefore, applied science ceased to be a key component of innovation policy. Yet the need to describe the topography between science and practice and the opportunity to bond diverse communities concerned with discussing science have remained. Indeed, these issues became urgent in biomedicine, which has been rapidly growing and newly dominant in the national research enterprise, both private and public. It has faced the challenge of linking 'bench to bedside' or, more generally, science to practice. The solution chosen by specialists has been to highlight the need for 'translational research'. The prominence of biomedicine within industrial policy has made this new term popular even for politicians.

This book has found that applied science achieved its specific meanings through service in Britain's national debates. This raises questions about

---

[5] On information technology around 1980, see Kline 2015. On biotechnology in Britain, see Bud 2010.

how the model applies to other English-speaking countries. 'Applied science' has been used widely in the United States, but the British and American discourses about science have been intellectually, industrially, and politically distinct.[6] For example, the great private industrial laboratories acquired a higher profile in the United States than in Britain, while government funding had a different role. Moreover, the leaders of the engineering profession, as described by Kline, took a nationally defined rhetorical tack. As a result, the interface between applied science and technology moved distinctively in the two countries, as expressed by the adage of two countries separated by a common language.

For much of the period of this book, Britain was the centre of a global empire. Science was an important part of imperial culture, and there is a rich literature on the colonial roles attributed to science and technology. In the 1860s, the Royal College of Science in Ireland served as a prototype for a pedagogical succession of training in pure science, then applied science and separate training in practice, subsequently used more widely in liberal science colleges across Britain. Later, even the 'Metropole' itself could migrate. For a time, Montreal's McGill University was perhaps the most important imperial centre for teaching applied science.[7] British pundits celebrated the growth of technical education in Hong Kong at the beginning of the twentieth century. Yet they wondered about the meaning of such global transfers without the communication of other parts of British culture.

Research in applied science also had an imperial reach. During the interwar period, the now-enlarged empire called for a wide variety of localised skills and knowledge to understand and improve practice. But, typically, a network of central institutions was devoted to building trade within the empire through applied science. Perhaps surprisingly, the new requirements seemed to reinforce rather than challenge the concept as it had developed in Britain. Applied science had been forged domestically in awareness of the sometimes substantial distance between the global and the local. Thus, the relationship between centres of colonial science, on the one hand, and the need for local expertise, on the other, mirrored metropolitan expectations. As for meanings adopted around the empire, we still need to understand better how connotations changed as the language circulated.[8]

---

[6] The American debates over the meanings of technology have been discussed by Wisnioski 2012 and Schatzberg 2018.
[7] On the idea of the circulating centres, see MacLeod 1980.
[8] See Seth 2009; Bennett and Hodge 2011; Lightman, McOuat and Stewart 2013; Clarke 2018. On the distinctive development of thinking about 'technology' in Canada, see Francis 2009.

As well as the centre of an empire, Britain was a European country with close ties to near neighbours. Yet the abundant studies of *technologie*, *technique*, *Technik*, and industrial science in France and Germany protect us from simply equating any British concept with such terms. Klein's work on the Prussian aggregation of 'useful sciences' reminds us how the linkage of science and practice in a different industrial policy regime could be otherwise construed. Certainly, this book has argued for the importance of the many two-way cultural interchanges underlying applied science across the continent. After all, the term itself grew from a translation by Coleridge from a pre-existing German usage. The French phrase *science appliquée aux arts industriels* contributed to the formation of the English term in the mid-nineteenth century.

Numerous transfers of concepts from Britain complemented British imports of terms. The 1898 establishment of the Institute of 'Applied Physics' in Göttingen, transgressing the border that traditionally encircled *Wissenschaft* in Germany, was attributed there to the university's close ties to the Anglo-American cultural world.[9] A French magazine dedicated to 'pure and applied science' was a well-known and valued innovation of the 1890s. However, Bensaude-Vincent has warned against finding a stabilised French counterpart to the British story presented here.[10] Instead, this book has suggested that historians continue addressing local constructions of the space between the laboratory and the market.

The implications should be more than historical. Bowker and Star ended their book on making categories by reflecting that the only good classification is a living classification.[11] For two centuries, applied science was acclaimed but also bitterly disputed amongst campaigners, politicians, and the community. Its limitations created spaces for other contested concepts with their own implications for national redemption. Therefore, the history of a single context should prompt us to reflect on change in the roles, organisation, and classification of knowledge. We cannot attribute this to the arbitration of just a few intellectuals. In Britain, the character of applied science developed through the jingle jangle of modern civilisation.

---

[9] Cramer 1908, 23. At the beginning of the twenty-first century, *Fachhochschulen* in German-speaking lands were relabelled *Hochschulen für angewandte Wissenschaften*, and the English translation 'University of Applied Sciences' is used often as a suffix.
[10] The magazine was the semi-popular *Revue Générale des Sciences Pures et Appliquées* (f. 1890). For the more general point, see Bensaude-Vincent 2020.
[11] Bowker and Star 1999, 326.

Archives Used

Bayerische Staatsbibliothek, Munich
  Papers of Justus Liebig
BBC Written Archives Centre, Caversham, Reading
  Contributors' files
British Library, London
  Babbage papers
  Balfour papers
City, University of London, Library, London
  Northampton Polytechnic Institute reports
Clothworkers Hall, Archives, London
  Company papers relating to Yorkshire College
Cornell University, Carl A. Kroch Library, Rare and Manuscript
  Collections, Ithaca, NY
  Papers of Henry Guerlac
Derbyshire Record Office, Matlock
  Papers of Doncaster Coalite Ltd
Glamorgan Archives, Cardiff
  Papers of the National Industrial Development Council of Wales
  and Monmouthshire
  Papers of Powell Duffryn Ltd
Guardian News and Media Archive, London
  Papers of Richard Fry
Guildhall Library, London
  City and Guilds examinations
House of Lords, Library, London
  Papers of Lloyd George
Institution of Engineering and Technology, IET Archives, London
  Papers of Arthur Fleming
  Papers of Walter Puckey
Imperial College London, College Archives and Corporate Records
  Unit, London
  City and Guilds Central Institution records
  Papers of Lyon Playfair
Imperial War Museum, Archives, London
  Papers of Henry Tizard
The Keep, Falmer
  Mass Observation Archive

263

Papers of J. G. Crowther
Papers of Richard Gregory
Papers of Richie-Calder
Kings College London, Archives and Special Collections, London
    King's College London Archives
Lambeth Palace Library, London
    William Rowe Lyall papers
London Metropolitan Archives, London
    British Empire Exhibition papers
    Papers of the Technical Education Board of the London County
    Council
London School of Economics Library, Archives and Special Collections,
    London
    Lord Passfield (Sidney Webb) papers
Manchester Central Library, Archives and Local History, Manchester
    Minutes of the Technical Education Committee;
    Papers of the Cotton Industry Research Association
McMaster University, The William Ready Collection of Archives and
    Research Collections, Hamilton
    Papers of George Catlin, including *The Realist*
National Aerospace Library/Royal Aeronautical Society, Farnborough
    Papers of George Cayley
The National Archives [TNA], Kew
    British public papers
National Library of Scotland, Edinburgh
    Blackwell archive
    Papers of R. B. Haldane
National Library of Wales, Aberystwyth
    Papers of Andrew Crombie Ramsay
Rice University, Woodson Research Center, Special Collections and
    Archives, Houston
    Papers of Julian Huxley
Royal Institution, Archive Collections, London
    Papers of William Henry Bragg
    Papers of William Lawrence Bragg
Royal Society, Library and Archives, London
    Papers of Patrick Blackett
    British Empire Exhibition papers
Royal Society of Arts, Archive, London
    The Society's papers
Science Museum, Science Museum Library and Research Centre, London
    Papers of Oswald John Silberrad
Teesside Archives, Middlesbrough
    Imperial Chemical Industries (ICI) archives
University of Birmingham, Cadbury Research Library, Birmingham
    Mason College and University archives
    Oliver Lodge papers

University of Cambridge, University Library, Cambridge
    Papers of Joseph Needham
    Papers of William Crookes
    University Archives
University of East Anglia, Archives Department, Norwich
    Papers of Solly Zuckerman
University of Edinburgh, Centre for Research Collections
    University of Edinburgh archives
University of Glasgow, Archive Services, Glasgow
    University of Glasgow archive
University of Illinois at Urbana Champaign, Rare Book and Manuscript
    Library, Urbana
    Papers of H. G. Wells
University of Liverpool Library, Special Collections and Archive,
    Liverpool
    Liverpool Royal Institution papers
University of London, Senate House Library, Special Collections,
    London
    University of London Senate Minutes
University of Maine, Raymond H. Fogler Library, Orono
    Floyd Gibbons Collection
University of Manchester Library, Special Collections Division,
    Manchester
    Archives of the University and UMIST
    Papers of Vivian Bowden
University of Oxford, Bodleian Library, Special Collections, Oxford
    Papers of Christopher Addison
University of Oxford, Museum of History of Science, Oxford
    Museum archives
University of Oxford, Oxford, Nuffield College Library, Oxford
    Papers of Private Conferences: 'Problems of Scientific and Industrial
    Research' (1944) and 'The Next Ten Years in British Economic
    Policy' (1955)
University of Reading, Museum of English Rural Life, Reading
    George Allen and Unwin Archive
University of Sheffield, The University Library, Special Collections,
    Sheffield
    Department of Applied Science archives
    Cecil Desch papers
University of Strathclyde, University Library, Special Collections, Glasgow
    University records
University of Warwick, Modern Records Centre, Coventry
    Papers of the Association of Scientific Workers
Victoria and Albert Museum, National Art Library, London
    Papers of Henry Cole
West Yorkshire Archive Service, Leeds
    Papers of Edward Baines

References

Archives, encyclopaedia entries, newspapers, and pre-1900 periodical articles are referenced in full in the notes, and are not included here.

Abbott, George C. 1970. 'British Colonial Aid Policy during the Nineteen Thirties'. *Canadian Journal of History* 5, no. 1: 73–89.

Adams, Mary (ed.). 1933. *Science in the Changing World*. London: George Allen & Unwin.

Adams, R. J. Q. 2007. *Balfour: The Last Grandee*. London: John Murray.

Addison, Christopher. 1939. *Four and a Half Years: A Personal Diary from June 1914 to January 1919*. Vol. 1. London: Hutchinson.

Adelman, Juliana. 2015. *Communities of Science in Nineteenth-Century Ireland*. London: Routledge.

Advisory Council on Scientific Policy. 1960. *Annual Report of the Advisory Council for Scientific Policy 1959–60*. Cmnd 1167. London: HMSO.

Agar, Jon. 2011. 'Thatcher, Scientist'. *Notes and Records of the Royal Society* 65, no. 3: 215–32.

2012. *Science in the Twentieth Century and Beyond*. London: Polity Press, 2012.

2016. 'The Defence Research Committee, 1963–1972'. In Don Leggett and Charlotte Sleigh (eds.), *Scientific Governance in Britain, 1914–79*. Manchester: Manchester University Press, 100–121.

2019. *Science Policy under Thatcher*. London: UCL Press.

Agar, Jon, and Brian Balmer. 1998. 'British Scientists and the Cold War: The Defence Research Policy Committee and Information Networks, 1947–1963'. *Historical Studies in the Physical and Biological Sciences* 28, no. 2: 209–52.

Allen, G. C. 1948. *Economic Thought and Industrial Policy*. An Inaugural Lecture Delivered at University College London, on March 4, 1948. London: H K Lewis.

Alter, Peter. 1987. *The Reluctant Patron: Science and the State in Britain 1850–1920*. Oxford: Berg.

Altick, Richard D. 1978. *The Shows of London*. Cambridge, MA: Harvard University Press.

AMRC. 2017. *Charities' Funding Contributes to UK Medical Research Excellence*. Available at www.amrc.org.uk/news/charities-funding-contributes-to-uk-medical-research-excellence (accessed March 2023).

Anderson, R. G. W. 1992. '"What Is Technology?"': Education through Museums in the Mid-Nineteenth Century'. *British Journal for the History of Science* 25 no. 2: 169–84.

Andrews, Kay. 1975. 'Politics, Professionalism and the Organisation of Scientists: the Association of Scientific Workers, 1917–1942'. DPhil thesis, Sussex University.

Anduaga, Aitor. 2009. *Wireless and Empire: Geopolitics, Radio Industry and Ionosphere in the British Empire, 1918–1939*. Oxford: Oxford University Press.

Anker, Peder. 2001. *Imperial Ecology: Environmental Order in the British Empire, 1895–1945*. Cambridge, MA: Harvard University Press.

Appleton, Edward V. 1945. 'The Work of the Department of Scientific and Industrial Research'. *Journal of the Royal Society of Arts* 93, no. 4694: 372–84.

1959. 'Notes on Science – Pure and Applied'. *British Affairs* 3, no. 4: 174–77.

Arapostathis, Stathis. 2012. 'Electrical Innovations, Authority and Consulting Expertise in Late Victorian Britain'. *Notes and Records of the Royal Society* 67, no. 1: 59–76.

Arapostathis, Stathis, and Graeme Gooday. 2013. 'Electrical Technoscience and Physics in Transition, 1880–1920'. *Studies in History and Philosophy of Science Part A* 44, no. 2: 202–11.

Archibald, Gail. 2006. 'How the "S" came to be in UNESCO'. In P. Petitjean, V. Zharov, G. Glaser, J. Richardson, B. de Padirac, and G. Archibald (eds.), *Sixty Years of Science at UNESCO, 1945–2005*. Paris: UNESCO, 36–40.

Argles, Michael. 1959. 'The Royal Commission on Technical Instruction, 1881–4: Its Inception and Composition'. *The Vocational Aspect of Education* 11, no. 23: 97–104.

Armytage, W. H. G. 1950. 'J. F. D. Donnelly: Pioneer in Vocational Education'. *The Vocational Aspect of Secondary and Further Education* 2, no. 4: 6–21.

1953. *Civic Universities: Aspects of a British Tradition*. London: Ernest Benn.

1965. *The Rise of the Technocrats: A Social History*. London: Routledge & Kegan Paul.

Arnold, Lorna. 1992. *Windscale 1957: Anatomy of a Nuclear Accident*. Basingstoke: Palgrave Macmillan.

Arnold, Matthew. 1868a. 'Report by Mr. Matthew Arnold'. Schools Inquiry Commission. [3966-V]. London: HMSO, Vol. 6, 441–726.

1868b. *Schools and Universities on the Continent*. London: Macmillan.

Aronova, Elena. 2012. 'The Congress for Cultural Freedom, Minerva, and the Quest for Instituting "Science Studies" in the Age of Cold War'. *Minerva* 50, no. 3: 307–37.

Ashby, Eric. 1958. *Technology and the Academics: An Essay on Universities and the Scientific Revolution*. London: Macmillan.

Ashby, Eric, and Mary Anderson. 1974. *Portrait of Haldane at Work on Education*. London: Macmillan.

Ashley, Percy. 1904. *Modern Tariff History. Germany – United States – France*. London: John Murray.

Ashworth, William J. 2017. *The Industrial Revolution: The State, Knowledge and Global Trade*. London: Bloomsbury.

Association of Scientific Workers. 1944. *Science in the Universities, Report of the Association of Scientific Workers to the University Grants Committee*. Oxford: Oxonian Press.

Atkins, Peter J. 2005. 'The Empire Marketing Board'. In D. J. Oddy and L. Petráňová (eds.), *The Diffusion of Food Culture in Europe from the Late Eighteenth Century to the Present*. Prague: Academic Press, 248–55.

Atkinson, Paul, and Sally Sheard (eds.). 2018. *Origins of the National Institute for Health Research: Transcript of a Witness Seminar Held at the University of Liverpool in London on 28 February 2018*. Liverpool: Department of Public Health and Policy, University of Liverpool.

Atkinson, Paul, Sally Sheard, and Tom Walley. 2019. 'All the Stars Were Aligned'? The Origins of England's National Institute for Health Research'. *Health Research Policy and Systems* 17, no. 95: doi.org/10.1186/s12961-019-0491-5 (accessed March 2023)

'Atomic Research in Britain: First Pictures Harwell Atomic Research Establishment'. 1948. *Illustrated London News*, 31 July: 132.

Austoker, Joan. 1989. 'Walter Morley Fletcher and the Origins of a Basic Biomedical Research Policy'. In Joan Austoker and Linda Bryder (eds.), *Historical Perspectives on the Role of the MRC*. Oxford: Oxford University Press, 23–33.

Aylen, Jonathan. 2015. 'First Waltz: Development and Deployment of Blue Danube, Britain's Post-War Atomic Bomb'. *The International Journal for the History of Engineering & Technology* 85, no. 1: 31–59.

Babbage, Charles. 1832. *On the Economy of Machinery and Manufactures*. London: Charles Knight.

Bacon, Reginald Hugh Spencer. 1929. *The Life of Lord Fisher of Kilverstone: Admiral of the Fleet*. 2 vols. London: Hodder and Stoughton.

Backhouse, Roger E., and Harro Maas. 2016. 'Marginalizing Maclaurin: The Attempt to Develop an Economics of Technological Progress at MIT, 1940–50'. *History of Political Economy* 48, no. 3: 423–47.

2017. 'A Road Not Taken: Economists, Historians of Science, and the Making of the Bowman Report'. *Isis* 108, no. 1: 82–106.

Bagrit, Leon. 1965. *The Age of Automation*. London: Weidenfeld.

Baker, Robert S. 2001. 'Science and Modernity in Aldous Huxley's Interwar Essays and Novels'. In C. C. Barfoot (ed.), *Aldous Huxley between East and West*. Amsterdam: Rodopi, 35–58.

Balconi, Margherita, Stefano Brusoni, and Luigi Orsenigo. 2010. 'In Defence of the Linear Model: An Essay'. *Research Policy* 39, no. 1: 1–13.

Balfour, Graham. 1903. *The Educational Systems of Great Britain and Ireland*. 2nd ed. Oxford: Clarendon Press.

Balls, W. Lawrence. 1926. 'The Classification of Research Work'. *Nature* 118, no. 2965: 298–99.

Balmer, Brian. 2015. 'An Open Day for Secrets: Biological Warfare, Steganography, and Hiding Things in Plain Sight'. In Brian Rappert and Brian Balmer (eds.), *Absence in Science, Security and Policy: From Research Agendas to Global Strategy*. Basingstoke: Palgrave Macmillan, 34–52.

Barker, Ernest. 1932. 'Universities in Great Britain'. In Walter M. Kotschnig and Prys (eds.), *The University in a Changing World: A Symposium*. Oxford: Oxford University Press, 85–120.

Barnes, Sarah V. 1996. 'England's Civic Universities and the Triumph of the Oxbridge Ideal'. *History of Education Quarterly* 36, no. 3: 271–305.

Barton, Ruth. 1998. 'Just before Nature: The Purposes of Science and the Purposes of Popularization in Some English Popular Science Journals of the 1860s'. *Annals of Science* 55, no. 1: 1–33.

2018. *The X Club: Power and Authority in Victorian Science*. Chicago: University of Chicago Press.

Bauer, M., J. Durant, A. Ragnarsdottir, and A. Rudolfsdottir. 1995. 'Science and Technology in the British Press, 1946–90'. 3 vols. Report on Wellcome Trust Grant no 037859/Z/93/Z.

Bauer, M., Kristina Petkova, Pepka Boyadjieva, and Galin Gornev. 2006. 'Long-Term Trends in the Public Representation of Science across the "Iron Curtain": 1946–1995'. *Social Studies of Science* 36, no. 1: 99–131.

Beard, Charles A. 1928. 'Introduction'. In Charles A. Beard (ed.), *Whither Mankind: A Panorama of Modern Civilization*. New York: Longmans Green, 1–24.

Beer, Gillian. 1996. '"Wireless": Popular Physics, Radio and Modernism'. In Francis Spufford and Jenny Uglow (eds.), *Cultural Babbage: Technology, Time and Invention*. London: Faber & Faber, 149–66.

Beinart, William, Karen Brown, and Daniel Gilfoyle. 2009. 'Experts and Expertise in Colonial Africa Reconsidered: Science and the Interpenetration of Knowledge'. *African Affairs* 108, no. 432: 413–33.

Belanger, Dian Olson. 1998. *Enabling American Innovation: Engineering and the National Science Foundation*. West Lafayette, IN: Purdue University Press.

Bell, John. 2017. *Life Sciences Industrial Strategy: A Report to the Government from the Life Sciences Sector*. Available at www.gov.uk/government/publications/life-sciences-industrial-strategy (accessed March 2023).

Belloc, Hilaire. 1926. *A Companion to Mr. Wells's 'Outline of History'*. London: Sheed and Ward.

Benn, Anthony Wedgwood. 1969. 'Government and Technology'. *Journal of the Royal Society for the Encouragement of Arts, Manufactures and Commerce* 117, no. 5151: 155–72.

Bennett, B. M., and J. M. Hodge (eds.). 2011. *Science and Empire: Knowledge and Networks of Science across the British Empire, 1800–1970*. Basingstoke: Palgrave Macmillan.

Bensaude-Vincent, Bernadette. 2009. 'A Historical Perspective on Science and Its "Others"'. *Isis* 100, no. 2: 359–68.

2020. 'At the Boundary between Science and Industrial Practices: Applied Science, Arts, and Technique in France'. *Science Museum Group Journal* 13 (Spring). https://journal.sciencemuseum.ac.uk/article/arts-and-technique-in-france/#abstract.

Berkowitz, Carin, and Bernard Lightman (eds.). 2017. *Science Museums in Transition: Cultures of Display in Nineteenth-Century Britain and America*. Pittsburgh: University of Pittsburgh Press.

Bernal, J. D. 1939. *The Social Function of Science*. London: Routledge.

Bingham, Adrian. 2004. *Gender, Modernity, and the Popular Press in Inter-War Britain*. Oxford: Clarendon Press.

Black, J. H. 1949. 'The Transmission of New Ideas: The Shirley Institute and the Cotton Industry'. *Journal of the Textile Institute Proceedings* 40, no. 7: 733–40.

Blackett, Patrick. 1968. 'Memorandum to the Select Committee on Science and Technology'. *Nature* 219, no. 5159: 1107–10.

Bland, Lucy, and Lesley Hall. 2010. 'Eugenics in Britain: The View from the Metropole'. In Alison Bashford and Philipa Levine (eds.), *Oxford Handbook of the History of Eugenics*. Oxford: Oxford University Press, 213–27.

Blatchford, Ian. 2021. 'Lyon Playfair: Chemist and Commissioner, 1818–1858'. *Science Museum Group Journal* 15 (Spring): doi.org/10.15180/211504.

Bloor, David. 2011. *The Enigma of the Aerofoil: Rival Theories in Aerodynamics, 1909–1920*. Chicago: University of Chicago Press.

Board of Education. 1905. *Preliminary Report of the Departmental Committee on the Royal College of Science and Royal School of Mines*. Cd 2610. London: HMSO.

　　1906. *Final Report of the Departmental Committee on the Royal College of Science, etc.* Cmd 2872. London: HMSO.

　　1915. *Scheme for the Organisation and Development of Scientific and Industrial Research*. Cd 8005. London: HMSO.

Board of Trade. 1929. *Final Report of the Committee on Industry and Trade*. Cmd 3282. London: HMSO.

　　1931. *Minutes of Evidence Taken before the Departmental Committee on the Patents and Designs Acts and Practice of the Patents Office*. London: HMSO.

Boden, Rebecca, Deborah Cox, and Maria Nedeva. 2006. 'The Appliance of Science? New Public Management and Strategic Change'. *Technology Analysis & Strategic Management* 18, no. 2: 125–41.

Boden, Rebecca, Deborah Cox, Maria Nedeva, and Kate Barker. 2004. *Scrutinising Science: The Changing UK Government of Science*. Basingstoke: Palgrave Macmillan.

Boden, Rebecca, Philip Gummett, Deborah Cox, and K. E. Barker. 1998. 'Men in White Coats … Men in Grey Suits: New Public Management and the Funding of Science and Technology Services to the UK Government'. *Accounting Auditing & Accountability Journal* 11, no. 3: 267–91.

Boon, Rachel. 2020. 'The Post Office Research Station at Dollis Hill, 1933–1965'. PhD thesis, University of Manchester.

Bowden, Vivian (Lord). 1965. 'Expectations for Science'. *New Scientist* 27, no. 463 (30 September): 849–53; and 28, no. 464 (7 October): 48–52.

Bowker, Geoffrey C., and Susan Leigh Star. 1999. *Sorting Things Out: Classification and Its Consequences*. Cambridge, MA: MIT Press.

Bowler, Peter J. 2009. *Science for All: The Popularization of Science in Early Twentieth-Century Britain*. Chicago: University of Chicago Press.

　　2010. *Reconciling Science and Religion: The Debate in Early-Twentieth-Century Britain*. Chicago: University of Chicago Press.

Bowler, Peter J., and Nicholas Whyte (eds.). 1997. *Science and Society in Ireland. The Social Context of Science and Technology in Ireland 1800–1950*. Belfast: Institute of Irish Studies, Queens University of Belfast.

Boyle, Alison. 2020. 'Stories and Silences in Modern Physics Collections: An Object Biography Approach'. PhD thesis, University College London.

Boyns, Trevor. 1987. 'Rationalisation in the Inter-War Period: The Case of the South Wales Steam Coal Industry'. *Business History* 29, no. 3: 282–303.

Bradley, Margaret, 2012. *Charles Dupin (1784–1873) and His Influence on France: The Contributions of a Mathematician, Educator, Engineer, and Statesman.* London: Cambria Press.

Brady, Robert A. 1932. 'The Meaning of Rationalization: An Analysis of the Literature'. *The Quarterly Journal of Economics* 46, no. 3: 526–40.

Bragg, W. H. 1928. 'Craftsmanship and Science'. *Supplement to Nature* 122, no. 3071: 353–63.

Brake, Laurel. 1997. 'Writing, Cultural Production, and the Periodical Press in the Nineteenth Century'. In J. B. Bullen (ed.), *Writing and Victorianism.* Harlow: Addison Wesley Longman, 57–72.

Brech, Edward, Andrew Thomson, and John F. Wilson. 2010. *Lyndall Urwick, Management Pioneer: A Biography.* Oxford: Oxford University Press.

Bret, Patrice. 2002. *L'état, l'armée, la science: l'invention de la recherche publique en France, 1763–1830.* Rennes: Presses universitaires de Rennes.

Brewster, David. 1837. 'Whewell's History of the Inductive Sciences'. *Edinburgh Review* 66: 110–51.

Briggs, Asa. 1957. *Where We Came In: The Industrial Revolution Reconsidered [A prospectus for Six Broadcast Talks. With Illustrations].* London: British Broadcasting Corporation.

1995. *The History of Broadcasting in the United Kingdom,* vol. 2: *The Golden Age of Wireless.* Rev. ed. Oxford: Oxford University Press.

'The British Air-Office: A Place of Wind-Towers, Whirling-Tables, and Gales Made to Order'. 1910. *The Illustrated London News* (2 April): 497.

British Association for the Advancement of Science. 1945. 'The Place of Science in Industry'. Transactions of a Conference Held by the Division for the Social and International Relations of Science, January 12–13, 1945. *The Advancement of Science* 3: 106–56.

British Empire Exhibition. 1924. *Pavilion of H. M. Government British Empire Exhibition, 1924.* London.

Brittain, Vera. 1928. *Women's Work in Modern England.* London: Noel Douglas.

Brock, William H. 2002. *Justus von Liebig: The Chemical Gatekeeper.* Cambridge University Press.

2008. *W. H. William Crookes 1832–1919 and the Commercialization of Science.* Aldershot: Ashgate.

Broks, Peter. 1996. *Media Science before the Great War.* Basingstoke: Macmillan Press.

Brooks, Harvey (chair). 1971. *Science, Growth and Society: A New Perspective. Report of the Secretary General's ad hoc Group on New Concepts of Science Policy.* Paris: OECD.

Brown, Andrew. 2006. *J. D. Bernal: The Sage of Science.* Oxford: Oxford University Press.

Brown, David. 2010. 'Morally Transforming the World or Spinning a Line? Politicians and the Newspaper Press in Mid-Nineteenth-Century Britain'. *Historical Research* 83, no. 220: 321–42.

Brownlie, David. 1927. 'Low-Temperature Carbonization: The Situation in Great Britain'. *Industrial & Engineering Chemistry* 19, no. 1: 39–45.

Buchanan, Alexandrina. 2013. *Robert Willis (1800–1875) and the Foundation of Architectural History*. Martlesham: Boydell & Brewer.

Buchanan, R. A. 1966. 'From Trade School to University: A Microcosm of Social Change'. In Gerald Walters (ed.), *A Technological University: An Experiment in Bath*. Bristol: Bath University Press, 12–26.

Buchwald, Jed Z., and Sungook Hong. 2003. 'Physics'. In David Cahan (ed.), *From Natural Philosophy to the Sciences: Writing the History of Nineteenth-Century Science*. Chicago: University of Chicago Press, 163–95.

Buckingham, John. 1952. 'The Scientific Research Department in the Time of the First Director, 1920–29'. *Journal of the Royal Naval Scientific Service* 7: 100–103.

Bud, Robert. 1980. 'The Discipline of Chemistry: The Origin and Early Years of the Chemical Society of London'. PhD thesis, University of Pennsylvania.

1993. *The Uses of Life: A History of Biotechnology*. Cambridge: Cambridge University Press.

1998. 'Penicillin and the New Elizabethans'. *British Journal for the History of Science* 31, no. 3: 305–33.

2007. *Penicillin: Triumph and Tragedy*. Oxford: Oxford University Press.

2008. 'Upheaval in the Moral Economy of Science? Patenting, Teamwork and the World War II Experience of Penicillin'. *History and Technology* 24, no. 2: 173–90.

2010. 'From Applied Microbiology to Biotechnology: Science, Medicine and Industrial Renewal'. *Notes and Records of the Royal Society* 64, Suppl. 1 (July): S17–S29.

2013a. 'Biological Warfare Warriors, Secrecy and Pure Science in the Cold War: How to Understand Dialogue and the Classifications of Science'. *Medicina Nei Secoli* 26, no. 2: 451–68.

2013b. 'Framed in the Public Sphere: Tools for the Conceptual History of "Applied Science": A Review Paper'. *History of Science* 51, no. 4: 413–33.

2013c. 'Infected by the Bacillus of Science'. In Peter Morris (ed.), *Science for the Nation: Perspectives on the History of the Science Museum*. London: Palgrave, 11–40.

2014a. '"Applied Science" in Nineteenth-Century Britain: Public Discourse and the Creation of Meaning, 1817–1876'. *History and Technology* 30, nos. 1–2: 30–36.

2014b. 'Responding to Stories: The 1876 Loan Collection of Scientific Apparatus and the Science Museum.' *Science Museum Group Journal* (Spring): doi.org/10.15180/140104.

2016. 'Science, Brands and the Museum'. *Journal of Science Communication* 15, no. 6. https://jcom.sissa.it/sites/default/files/documents/JCOM_1506_2016_C03.pdf (accessed March 2023).

2017. '"The Spark Gap Is Mightier Than the Pen": The Promotion of an Ideology of Science in the Early 1930s'. *Journal of Political Ideologies* 22, no. 2: 169–81.

2018a. 'Categorizing Science in Nineteenth and Early Twentieth-Century Britain'. In David Kaldewey and Désirée Schauz (eds.), *Basic and Applied Research: The Language of Science Policy in the Twentieth Century*. New York: Berghahn Books, 35–63.

2018b. 'Kantianism, Cameralism and Applied Science'. *Notes and Records of the Royal Society* 72, no. 3: 199–216.

2020. 'Science and the Press'. In Martin Conboy and Adrian Bingham (eds.), *The Edinburgh History of the British and Irish Press, vol. 3: Competition and Disruption, 1900–2017*. Edinburgh: Edinburgh University Press, 612–25.

Bud, Robert, Paul Greenhalgh, Frank James, and Morag Shiach (eds.). 2018. *Being Modern: The Cultural Impact of Science in the Early Twentieth Century*. London: UCL Press.

Bud, Robert, and Philip Gummett (eds.). 1999. *Cold War Hot Science: Applied Research in Britain's Defence Laboratories 1945–1990*. Reading: Harwood.

Bud, Robert, and Gerrylynn K. Roberts. 1984. *Science versus Practice: Chemistry in Victorian Britain*. Manchester: Manchester University Press.

Bunce, John Thackray. 1970. *Josiah Mason: A Biography with Sketches of the History of the Steel-Pen and Electro-Plating Trades*. 2nd ed. London: W. & R. Chambers [1890].

Burgess, Tyrrell. 1970. *Policy and Practice: The Colleges of Advanced Technology*. London: Allen Lane.

Burgon, John William. 1891. *Lives of Twelve Good Men*. New ed. London: Murray.

Burke, Edmund. 1800. *Thoughts and Details on Scarcity*. London: F. & C. Rivington.

Burke, Kenneth. 1960. *A Grammar of Motives*. First published 1945. Berkeley: University of California Press.

Bush, Vannevar. 2020. *Science – The Endless Frontier. A Report to the President on a Program for Postwar Scientific Research*. First issued 1945. Reprinted in celebration of the National Science Foundation's 70th anniversary. Washington, DC: National Science Foundation. www.nsf.gov/about/history/Endless Frontier_w.pdf.

Butler, Stuart, and Robert Bud. 2018. 'United Kingdom, Short Country Report'. History of Nuclear Energy and Society. www.honest2020.eu/sites/default/files/deliverables_24/UK.pdf.

Butterworth, H. 1968. 'The Science and Art Department 1853–1900'. PhD thesis, University of Sheffield.

Byatt, I. C. R., and A. V. Cohen. 1969. *An Attempt to Quantify the Economic Benefits of Scientific Research*. Department of Education and Science, Science Policy Studies no. 4. London: HMSO.

Cahan, David. 2004. *An Institute for an Empire: The Physikalisch-Technische Reichsanstalt, 1871–1918*. Cambridge: Cambridge University Press.

Calder, Peter Ritchie. 1934. *The Birth of the Future*. London: Arthur Barker.

Campbell, Colin, and Graham K. Wilson. 1995. *The End of Whitehall: Death of a Paradigm?* Oxford: Blackwell.

Campbell, John. 2020. *Haldane: The Forgotten Statesman Who Shaped Modern Britain*. Oxford: Oxford University Press.

Cannadine, David. 1984. 'The Present and the Past in the English Industrial Revolution 1880–1980'. *Past & Present* 103, no. 1: 131–72.

Cantor, G. N. 1996. 'The Scientist as Hero: Public Images of Michael Faraday'. In Michael Shortland and Richard Yeo (eds.), *Telling Lives in Science: Essays on Scientific Biography*. Cambridge: Cambridge University Press, 171–93.

2011. *Religion and the Great Exhibition of 1851*. Oxford: Oxford University Press.

Cantor, Leonard Martin, and Geoffrey Frank Matthews. 1977. *Loughborough from College to University: A History of Higher Education at Loughborough, 1906–1966*. Loughborough: Loughborough University of Technology.

Cardwell, D. S. L. 1972. *The Organisation of Science in England*. First published 1957. London: Heinemann Educational.

(ed.). 1974. *Artisan to Graduate: Essays to Commemorate the Foundation in 1824 of the Manchester Mechanics' Institution, Now in 1974 the University of Manchester Institute of Science and Technology*. Manchester: Manchester University Press.

Carlson, Bernard. 1997. 'Innovation and the Modern Corporation from Heroic Invention to Industrial Science'. In John Krige and Dominique Pestre (eds.), *Companion to the History of Twentieth-Century Science*. Reading: Harwood, 203–26.

Carson, Cathryn. 2010. *Heisenberg in the Atomic Age: Science and the Public Sphere*. Cambridge: Cambridge University Press.

Carter, C. F., and B. R. Williams. 1957. *Industry and Technical Progress: Factors Governing the Speed of Application of Science*. Oxford: Oxford University Press.

1958. *Investment in Innovation*. Oxford: Oxford University Press.

1959. *Science in Industry: Policy for Progress*. Oxford: Oxford University Press.

Carter, G. B. 2000. *Chemical and Biological Defence at Porton Down, 1916–2000*. 2nd ed. London: Stationery Office.

Carty, John J. 1916. 'The Relation of Pure Science to Industrial Research'. *Science* 44, no. 1137: 511–18.

Catlin, George Edward Gordon. 1972. *For God's Sake, Go!: An Autobiography*. Gerrards Cross: Smythe.

Catterall, Peter (ed.). 2011. *The Macmillan Diaries: The Cabinet Years, 1950–57*. London: Pan.

'The Centenary Journal'. 1966. *Journal of the Royal Aeronautical Society* 70, no. 661: 90–97.

Central Advisory Council for Science and Technology. 1968. *Technological Innovation in Britain*. Report of the Central Advisory Council for Science and Technology. London: HMSO.

Chairmen of the Advisory Council for Applied Research and Development (ACARD) and the Advisory Board for the Research Councils (ABRC). 1976. *The Science Base and Industry*. Second Joint Report. Cm 34. London: HMSO.

Chalaby, Jean K. 1996. 'Twenty Years of Contrast: The French and British Press during the Inter-War Period'. *European Journal of Sociology/Archives Européennes de Sociologie* 37, no. 1: 143–59.

Chambers, David Wade, and Richard Gillespie. 2000. 'Locality in the History of Science: Colonial Science, Technoscience, and Indigenous Knowledge'. *Osiris* 15, no. 1: 221–40.

Channell, David. 1982. 'The Harmony of Theory and Practice: The Engineering Science of W. J. M. Rankine'. *Technology and Culture* 23, no. 1: 39–52.

Chaptal, Jean-Antoine Claude. 1893. *Mes souvenirs sur Napoléon 1er*. Ed. A. Chaptal. Paris: Plon, Nourris.

Charles, G. W. 1951. 'The Merchant Venturers' Technical College, Bristol'. *The Vocational Aspect of Secondary and Further Education* 3, no. 6: 86–89.

Charnley, Berris. 2013. 'Experiments in Empire-Building: Mendelian Genetics as a National, Imperial, and Global Agricultural Enterprise'. *Studies in History and Philosophy of Science Part A* 44, no. 2: 292–300.

Chilvers, C. A. J. 2007. 'Something Wicked This Way Comes: The Russian Delegation at the 1931 Second International Congress of the History of Science and Technology'. DPhil thesis, Oxford University.

Cho, S. K. 2001. 'The Special Loan Collection of Scientific Apparatus, 1876: The Beginning of the Science Museum of London and the Popularization of Physical Science'. PhD thesis, Seoul National University (in Korean).

Christen-Lécuyer, Carole, and François Vatin (eds.). 2009. *Charles Dupin (1784– 1873): ingénieur, savant, économiste, pédagogue et parlementaire du Premier au Second Empire*. Rennes: Presses universitaires de Rennes.

Christensen, Clayton M. 1997. *The Innovator's Dilemma: When New Technologies Cause Great Firms to Fail*. Boston: Harvard Business School.

Church, A. G. 1920. 'Applied Science and Industrial Research'. *Nature* 105, no. 2640: 423.

Churchill, Randolph S. 1969. *Winston S. Churchill, vol. 2: Companion Part II, 1907–1911*. London: Heinemann.

Clarke, N. 1951. 'Impressions of a Discussion on Technical Universities'. *British Journal of Applied Physics* 2, no. 9: 245–48.

Clarke, Richard. 1973a. 'Mintech in Retrospect: I'. *Omega* 1, no. 1: 25–38. 1973b. 'Mintech in Retrospect: II'. *Omega* 1, no. 2: 137–63.

Clarke, Robin. 1968, 'How the 300 GeV Decision Was Made.' *Science Journal*, March: 4–7.

Clarke, Sabine. 2010. 'Pure Science with a Practical Aim: The Meanings of Fundamental Research in Britain, circa 1916-1950'. *Isis* 101, no. 2: 285–311.

2018. *Science at the End of Empire: Experts and the Development of the British Caribbean, 1940–62*. Manchester: Manchester University Press.

Class, Monica. 2012. *Coleridge and Kantian Ideas in England, 1796–1817: Coleridge's Responses to German Philosophy*. London: Bloomsbury.

Clayton, Sir Robert, and Joan Algar. 1988. *The GEC Research Laboratories 1919– 1984*. London: Peter Peregrinus in association with the Science Museum, London.

Cline, Peter. 2014. 'Winding Down the War Economy: British Plans for Peacetime Recovery, 1916–19'. In Kathleen Burk (ed.), *War and the State: The Transformation of British Government, 1914–1919*. London: Routledge, 157–81.

Cockcroft, John. 1954. 'Atomic Energy Research at Harwell'. *Atomic Scientists Journal* 4: 182–92.

1956. 'The United Kingdom Atomic Energy Authority and Its Functions'. *British Journal of Applied Physics* 7, no. 2: 43–51.

Cohen, I. Bernard. 1946. 'Authenticity of Scientific Anecdotes'. *Nature* 157, no. 3981: 196–97.

1987. 'Faraday and Franklin's "Newborn Baby"'. *Proceedings of the American Philosophical Society* 131, no. 2: 177–82.

Coleman, D. C. 1956. 'Industrial Growth and Industrial Revolutions'. *Economica* n.s. 23, no. 89: 1–22.

Coleridge, Samuel Taylor. 1852a. 'Blessed Are Ye That Sow beside all Waters: A Lay Sermon Addressed to the Higher and Middle Classes on the Existing Distresses and Discontents'. First published 1817. In Derwent Coleridge (ed.), *Lay Sermons*. 3rd ed. London: Edward Moxon, 127–267.

1852b. *On the Constitution of the Church and the State according to the Idea of Each*. 4th ed., ed. Henry Nelson Coleridge. First published 1830. London: Edward Moxon.

Collins, Peter. 2010. 'A Royal Society for Technology'. *Notes and Records of the Royal Society* 64, Suppl. 1: S43–S54.

2016. *The Royal Society and the Promotion of Science since 1960*. Cambridge: Cambridge University Press.

Collison, Robert. 1966. 'Samuel Taylor Coleridge and the "Encyclopaedia Metropolitana"'. *Journal of World History* 9, no. 1: 751–68.

Commission on the College of Science for Ireland. 1867. *Report of Commission on the College of Science for Ireland*. P.P. 1867 (219) LV. 777-80.

Committee of Enquiry into the Engineering Profession. 1980. *Engineering Our Future: Report of the Committee of Enquiry into the Engineering Profession*. Cmnd 7794. London: HMSO.

Committee of Enquiry into the Organisation of Civil Science. 1963. *Report of the Committee of Enquiry into the Organisation of Civil Science, under the Chairmanship of Sir Burke Trend*. Cmnd 2171. London: HMSO.

Committee on Science and the Public Welfare. 2020. 'Report of the Committee on Science and the Public Welfare'. Published as part of *Science – The Endless Frontier*. Reprinted in celebration of the National Science Foundation's 75th anniversary. Available at www.nsf.gov/about/history/EndlessFrontier_w.pdf, 77–149.

Conant, James Bryant. 1939. 'Lesson from the Past'. *Industrial and Engineering Chemistry* 31, no. 10: 1215–17.

1943. 'Science and Society in the Post-War World: Interchange of Ideas and Discoveries Necessary'. *Vital Speeches of the Day* 9, no. 13: 394–97.

Converse, Elliott. 2012. *History of Acquisition in the Department of Defense*. Washington, DC: Historical Office, Office of the Secretary of Defense.

Cooksey, David. 2006. *A Review of UK Health Research Funding*. Norwich: HMSO.

Cooper, Timothy. 2003. 'The Science-Industry Relationship: The Case of Electrical Engineering in Manchester, 1914–1960'. PhD thesis, Manchester University.

2011. 'Peter Lund Simmonds and the Political Ecology of Waste Utilization in Victorian Britain'. *Technology and Culture* 52, no. 1: 21–44.

Coopey, Richard. 1991a. 'The White Heat of Scientific Revolution'. *Contemporary Record* 5, no. 1: 115–27.

1991b. 'Ministry of Technology 1964–70'. *Contemporary Record* 5, no. 1: 128–48.

1993. 'Restructuring Civil and Military Science and Technology: The Ministry of Technology in the 1960s'. In Richard Coopey, Matthew Uttley, and Graham Spinardi (eds.), *Defence Science and Technology Adjusting to Change*. Reading: Harwood Academic, 65–84.

Cornish, Caroline. 2013. 'Curating Science in an Age of Empire: Kew's Museum of Economic Botany'. PhD thesis, Royal Holloway College.

Costa, Ettore. 2020. 'Whoever Launches the Biggest Sputnik Has Solved the Problems of Society? Technology and Futurism for Western European Social Democrats and Communists in the 1950s'. *History of European Ideas* 46, no. 1: 95–112.

Cottrell, Alan. 1980. 'Science Policy in Britain'. *Proceedings of the Royal Institution of Great Britain* 52: 179–91.

1992. 'Sunlight and Shadow in Applied Science'. *Interdisciplinary Science Reviews* 17: 120–26.

1993. 'Foreword'. In S. T. Keith, *The Fundamental Nucleus: A Study of the Impact of the British Atomic Energy Project on Basic Research*. Harwell: AEA Technology, i–iii.

1998. 'Harwell: The First 50 Years'. *Interdisciplinary Science Reviews* 23, no. 2: 139–45.

Council for Scientific Policy. 1966a. *Report on Science Policy*. Cmnd 3007. London: HMSO.

1966b. *Enquiry into the Flow of Candidates in Science and Technology into Higher Education, Interim Report*. Cmnd 2893. London: HMSO.

1967. *Second Report on Science Policy*. Cmnd 3420. London: HMSO.

1968. *Enquiry into the Flow of Candidates in Science and Technology into Higher Education*. Cmnd 3541. London: HMSO.

1972. *Third Report of the Council for Scientific Policy*. Cmnd 5117. London: HMSO.

Counsel, Robert E. 1987. 'The National Development Council of Wales and Monmouthshire 1932–55'. MA thesis, University of Wales, Cardiff.

Crabu, Stefano. 2018. 'Rethinking Biomedicine in the Age of Translational Research: Organisational, Professional, and Epistemic Encounters'. *Sociology Compass* 12, no. 10: doi.org/10.1111/soc4.12623.

Crafts, Nicholas, and Peter Fearon. 2013. *The Great Depression of the 1930s: Lessons for Today*. Oxford: Oxford University Press.

Cramer, August. 1908. 'Ansprache des Prorektors der Universität'. In *Zum Zehnjährigen Bestehen der Göttinger Vereinigung für Angewandte Physik und Mathematik*. Wiesbaden: Vieweg und Teubner, 21–24.

Crane, E. J. 1944. 'Growth of the Chemical Literature: Contributions of Certain Nations and the Effects of War'. *Chemical and Engineering News* 22, no. 17: 1478–81.

Cross, Charles. 1923–25. 'Fact and Phantasy in Industrial Science'. *Notices of the Proceedings at the Meetings of the Members of the Royal Institution* 24: 1–5.

Crossley, R. 2011. 'The First Wellsians: A Modern Utopia and Its Early Disciples'. *English Literature in Transition, 1880–1920* 54, no. 4: 444–69.

Crouch, Tom D. 2002. *A Dream of Wings: Americans and the Airplane, 1875–1905*. New York: W. W. Norton.

Crowther, J. G. 1959. *Six Great Engineers: De Lesseps, Brunel, Westinghouse, Parsons, Diesel, Hinton*. London: H. Hamilton.

1965. *Statesmen of Science*. London: Cresset Press.

Crowther, J. G., and Richard Whiddington. 1947. *Science at War*. London: HMSO.

Cullen, Clara. 2009. 'The Museum of Irish Industry, Robert Kane and Education for All in the Dublin of the 1850s and 1860s'. *History of Education* 38, no. 1: 99–113.

Daglish, Neil. 1998. '"Over by Christmas": The First World War, Education Reform and the Economy. The Case of Christopher Addison and the Origins of the DSIR'. *History of Education* 27, no. 3: 315–31.

Daniels, Mario, and John Krige. 2018. 'Beyond the Reach of Regulation? "Basic" and "Applied" Research in the Early Cold War United States'. *Technology and Culture* 59, no. 2: 226–50.

Davies, S. M. 2017. 'Organisation and Policy for Research and Development: The Health Department for England and Wales 1961 to 1986'. PhD thesis, London School of Hygiene and Tropical Medicine.

2021. 'Priorities in Medical Research: Elite Dynamics in a Pivotal Episode for British Health Research'. *British Journal for the History of Science* 54, no. 2: 195–211.

Davis, George Edward. 1904. *A Handbook of Chemical Engineering: Illustrated with Working Examples and Numerous Drawings from Actual Installations*. Manchester: Davis Bros.

Davis, Martin. 1990. 'Technology, Institutions and Status: Technological Education, Debate and Policy, 1944–1956'. In Penny Summerfield and Eric J. Evans (eds.), *Technical Education and the State since 1850: Historical and Contemporary Perspectives*. Manchester: Manchester University Press, 120–44.

De Clerq, Peter. 2002–3. 'The Special Loan Collection of Scientific Apparatus, South Kensington, 1876'. *Bulletin of the Scientific Instrument Society*, 72: 14–19; 73: 8–16; 73: 16–19; 76: 10–15.

De Vries, Marc. 2005. *80 Years of Research at the Philips Natuurkundig Laboratorium (1914–1994): The Role of the Nat. Lab. at Philips*. Amsterdam: Pallas Publications.

De Wit, Dirk. 1994. *The Shaping of Automation: A Historical Analysis of the Interaction between Technology and Organization, 1950–1985*. Hilversum: Uitgeverij Verloren.

Degrois, Denise. 1991. 'S. T. Coleridge et l'Encyclopaedia Metropolitana: impératifs pratiques et conflits idéologiques'. In Annie Becq (ed.), *L'encyclopédisme: actes du colloque de Caen, 12–16 janvier 1987*. Paris: Aux amateurs de livres, 431–35.

DeJager, Timothy. 1993. 'Pure Science and Practical Interests: The Origins of the Agricultural Research Council, 1930–1937'. *Minerva* 31, no. 2: 129–50.

Demant, V. A. 1932. *This Unemployment: Disaster or Opportunity?* London: Student Christian Movement Press.

Denton, Peter H. 2001. *The ABC of Armageddon: Bertrand Russell on Science, Religion, and the Next War, 1919–1938*. Albany: State University of New York Press.

Department for Business, Energy & Industrial Strategy. 2021. 'UK to Launch New Research Agency to Support High Risk, High Reward Science'. Press release. Available at www.gov.uk/government/news/uk-to-launch-new-research-agency-to-support-high-risk-high-reward-science (accessed August 2022).

Department of Health. Research and Development Directorate. 2006. *Best Research for Best Health: A New National Health Research Strategy*. London: Department of Health.

Department of Overseas Trade. 1938. *Guide to the Pavilion of His Majesty's Government in the United Kingdom. Empire Exhibition, Scotland, 1938*. London: HMSO.

Department of Science and Art. 1857. *Fourth Report of the Department of Science and Art*. [2240]. London: HMSO.

1857a. *Prospectus of the Metropolitan School of Science Applied to Mining and the Arts: 3d session, 1857–58*. London: HMSO.

1899. *Department of Science and Art: Forty-Sixth Report*. c. 9191. London: HMSO.

Dewey, Clive. 2007. *The Passing of Barchester*. London: Continuum.

Dickens, Elizabeth. 2011. '"Permanent Books": The Reviewing and Advertising of Books in the Nation and Athenaeum'. *The Journal of Modern Periodical Studies* 2, no. 2: 165–84.

Diebold, John. 1952 *Automation: The Advent of the Automatic Factory*. New York: Van Nostrand.

Dietz, Burkhard, Michael Fessner, and Helmut Maier (eds.). 1996. *Technische Intelligenz und 'Kulturfactor Technik'*. Münster: Waxmann.

Divall, Colin. 1990. 'A Measure of Agreement: Employers and Engineering Studies in the Universities of England and Wales, 1897–1939'. *Social Studies of Science* 20, no. 1: 65–112.

2006. 'Technological Networks and Industrial Research in Britain: The London, Midland & Scottish Railway, 1926–47'. *Business History* 48, no. 1: 43–68.

Divall, Colin, and Sean F. Johnston. 2000. *Scaling Up: The Institution of Chemical Engineers and the Rise of a New Profession*. Dordrecht: Kluwer Academic.

Dobbin, Frank R. 1993. 'The Social Construction of the Great Depression: Industrial Policy during the 1930s in the United States, Britain, and France'. *Theory and Society* 22, no. 1: 1–56.

1994. *Forging Industrial Policy: the United States, Britain, and France in the Railway Age*. 2nd ed. Cambridge: Cambridge University Press.

Dodgson, Mark. 1990. 'The Shock of the New: The Formation of Celltech and the British Technology Transfer System'. *Industry and Higher Education* 4, no. 2: 97–104.

Donaldson, Kirstin L. 2014. 'Experiment: A Manifesto of Young England, 1928–1931'. PhD thesis, University of York.

Donnelly, James. 1986. 'Representations of Applied Science: Academics and Chemical Industry in Late Nineteenth-Century England'. *Social Studies of Science* 16, no. 2: 195–234.

1987. 'Chemical Education and the Chemical Industry in England from the Mid-Nineteenth to the Early Twentieth Century'. PhD thesis, University of Leeds, 1987.

1989. 'The Origins of the Technical Curriculum in England during the Nineteenth and Early Twentieth Centuries'. *Studies in Science Education* 16, no. 1: 123–61.

1997. 'Getting Technical: The Vicissitudes of Academic Industrial Chemistry in Nineteenth-Century Britain'. *History of Education* 26, no. 2: 125–43.

'Dr. Henry Taylor Bovey, F.R.S.' 1912. *Nature* 88, no. 2207: 520–21.

Drayton, Richard Harry. 2000. *Nature's Government: Science, Imperial Britain, and the 'Improvement' of the World.* New Haven, CT: Yale University Press.

Drummond, Diane. 2020. 'Pure and Applied Science in the New University 1900–1914'. In James Mussell and Graeme Gooday (eds.), *A Pioneer of Connection: Recovering the Life and Work of Oliver Lodge.* Pittsburgh, PA: University of Pittsburgh Press, 431–35.

Drummond, Ian M. 1971. 'More on British Colonial Aid Policy in the Nineteen-Thirties'. *Canadian Journal of History* 6, no. 2: 189–95.

[DSIR]. 1916. *Report of the Committee of the Privy Council for Scientific and Industrial Research for the Year, 1915–16.* Cd 8336. London: HMSO.

1917a. *Report of the Committee of the Privy Council for Scientific and Industrial Research for the Year, 1916–17.* Cd 8718. London: HMSO.

1917b. *Report of the Fuel Research Board on Their Scheme of Research and on the Establishment of a Fuel Research Station.* London: HMSO.

1919. *Report of the Committee of the Privy Council for Scientific and Industrial Research for the Year, 1918–19.* Cmd 320. London: HMSO.

1920. *Report of the Committee of the Privy Council for Scientific and Industrial Research for the Year, 1919–20.* Cmd 905. London: HMSO.

1921. *Report of the Committee of the Privy Council for Scientific and Industrial Research for the Year, 1920–21.* Cmd 1491. London: HMSO.

1922. *First Report of the Adhesives Research Committee.* London: HMSO.

1923. *Report of the Committee of the Privy Council for Scientific and Industrial Research for the Year, 1922–23.* Cmd 1937. London: HMSO.

1924. *Report of the Committee of the Privy Council for Scientific and Industrial Research for the Year, 1923–24.* Cmd 2223. London: HMSO.

1925. *Report of the Committee of the Privy Council for Scientific and Industrial Research for the Year, 1924-25.* Cmd 2491. London: HMSO.

1927. *Report of the Committee of the Privy Council for Scientific and Industrial Research for the Year, 1925–26.* Cmd 2782. London: HMSO.

DSIR. 1927. *Co-operative Industrial Research: An Account of the Work of Research Associations under the Government Scheme.* London: HMSO.

1929. *Report for the Year, 1928–29.* Cmd 3471. London: HMSO

1931. *Report for the Year, 1929–30.* Cmd 3789. London: HMSO

1932. *Report for the Year, 1931–32.* Cmd 4254. London: HMSO.
1933. *Report for the Year, 1932–33.* Cmd 4483. London: HMSO.
1935. *Report for the Year, 1934–35.* Cmd 5013. London: HMSO.
1937. *Report for the Year, 1935–36.* Cmd 5350. London: HMSO.
1937. *Report for the Year, 1936–37.* Cmd 5647. London: HMSO.
1938. *Report for the Year, 1937–38.* Cmd 5927. London: HMSO.
DSIR Economics Committee. 1958. *Estimates of Resources Devoted to Scientific and Engineering Research and Development in British Manufacturing Industry, 1955.* London: HMSO.
Dudley, Fred A. 1942. 'Matthew Arnold and Science'. *PMLA* 57, no. 1: 275–94.
Dunbabin, John P. D. 1975. 'British Rearmament in the 1930s: A Chronology and Review'. *The Historical Journal* 18, no. 3: 587–609.
Duncan, Robert Kennedy. 1907. *The Chemistry of Commerce: A Simple Interpretation of Some New Chemistry in Its Relation to Modern Industry.* New York: Harper.
Eaglesham, E. J. 1963. 'The Centenary of Sir Robert Morant'. *British Journal of Educational Studies* 12, no. 1: 5–18.
Ede, Andrew. 2002. 'The Natural Defense of a Scientific People: The Public Debate over Chemical Warfare in Post-WWI America'. *Bulletin for the History of Chemistry* 27. no. 2: 128–35.
Edge, David. 1995. 'Reinventing the Wheel'. In Sheila Jasanoff, Gerald Markle, James Petersen, and Trevor Pinch (eds.), *Handbook of Science and Technology Studies.* Thousand Oaks, CA: Sage, 3–23.
Edgerton, David. 1996. 'The "White Heat" Revisited: The British Government and Technology in the 1960s'. *Twentieth Century British History* 7, no. 1: 53–82.
1997. 'The Decline of Declinism'. *Business History Review* 71, no. 2: 201–6.
2004. '"The Linear Model" Did Not Exist: Reflections on the History and Historiography of Science and Research in Industry in the Twentieth Century'. In Karl Grandin, Nina Wormbs, and Sven Widmalm (eds.), *The Science-Industry Nexus: History, Policy, Implications.* New York: Watson, 31–57.
2006. *Warfare State: Britain, 1920–1970.* Cambridge: Cambridge University Press.
2009. The "Haldane Principle" and Other Invented Traditions in Science Policy'. *History and Policy.* www.historyandpolicy.org/policy-papers/papers/the-haldane-principle-and-other-invented-traditions-in-science-policy (accessed March 2023).
2011. *Britain's War Machine: Weapons, Resources, and Experts in the Second World War.* Oxford: Oxford University Press.
2018. *The Rise and Fall of the British Nation: A Twentieth-Century History.* London: Penguin Books.
Edgerton, David, and Sally Horrocks. 1994. 'British Industrial Research and Development before 1945'. *Economic History Review* 47, no. 2: 213–38.
Edwards, Pamela. 2012. 'Coleridge on Politics and Religion: The Statesman's Manual, Aids to Reflection, On the Constitution of Church and State'. In Frederick Burwick (ed.), *The Oxford Handbook of Samuel Taylor Coleridge.* Oxford: Oxford University Press, 236–53.

Elliot, Walter. 1931. 'The World of the Empire Marketing Board'. *Journal of the Royal Society of Arts* 79, no. 4101: 736–48.

Empire Exhibition, Scotland. 1938. *Empire Exhibition: Official Guide*. Glasgow: Empire Exhibition.

English, Richard, and Michael Kenny. 1999. 'British Decline or the Politics of Declinism?' *The British Journal of Politics & International Relations* 1, no. 2: 252–66.

——— 2000. *Rethinking British Decline*. Basingstoke: Macmillan.

——— 2001. 'Public Intellectuals and the Question of British Decline'. *The British Journal of Politics & International Relations* 3, no. 3: 259–83.

'Equal to Six Trafalgar Squares: The Palace of Engineering'. 1924. *Illustrated London News*, 24 May: 18.

Eschenburg, Johann Joachim. 1792. *Lehrbuch der Wissenschaftskunde: Ein Grundriss Encyclopaedischer Vorlesung*. Berlin: Friedrich Nicolai.

Ewing, Alfred. 1932. 'An Engineer's Outlook'. *Supplement to Nature*, no. 3279: 341–50.

Fabian Society. 1923. *Is Civilisation Decaying? A Course of Six Lectures to Be Given at King's Hall, King St., Covent Garden, on Tuesdays at 8.30 p.m., Beginning 16th October, 1923*. London: Fabian Society.

Falby, Alison. 2008. *Between the Pigeonholes: Gerald Heard, 1889–1971*. Newcastle: Cambridge Scholars.

Farrall, Lyndsey Andrew. 2019. 'The Origins and Growth of the English Eugenics Movement, 1865–1925'. PhD thesis, University of Indiana (1969). Reissued by University College London, Department of Science and Technology Studies, as STS Occasional Paper, no. 9: www.ucl.ac.uk/sts/sts-research/sts-occasional-papers-series (accessed March 2023).

Farrar, W. V. 1972. 'The Society for the Promotion of Scientific Industry 1872–1876'. *Annals of Science* 29, no. 1: 81–86.

Favretto, Ilaria. 2000. '"Wilsonism" Reconsidered: Labour Party Revisionism 1952–64'. *Contemporary British History* 14, no. 4: 54–80.

Federation of British Industries (FBI). 1947. *Scientific and Technical Research in British Industry: A Statistical Survey*. London: FBI.

Fine, Gary Alan. 1995. 'Public Narration and Group Culture: Discerning Discourse in Social Movements'. In Hank Johnston and Bert Klandermans (eds.), *Social Movements and Culture*. 127–43. London: UCL Press.

Finkelstein, David. 2006. *Print Culture and the Blackwood Tradition, 1805–1930*. Toronto: University of Toronto Press.

Finlay, Mark R. 1991. 'The Rehabilitation of an Agricultural Chemist: Justus von Liebig and the Seventh Edition'. *Ambix* 38, no. 3: 155–66.

Finney, Patrick. 2000. 'The Romance of Decline: The Historiography of Appeasement and British National Identity'. *Electronic Journal of International History* 1: ISSN 1471-1443.

'The Five Best Brains'. 1930. *The Spectator*, no. 5320 (14 June): 19.

Fleming, A. P. M. 1930. 'The Profession of Engineering. Promising Fields in the Future'. *Journal of Careers* 9, no. 100: 9–14.

——— 1932. 'The Development of Invention: Bridging the Gap between the Birth of an Idea and its Industrial Application'. In R. J. Mackay (ed.), *Business and*

Science: Being Collected Papers Read to the Department of Industrial Co-operation at the Centenary Meeting of the British Association … 1931. London: British Association for the Advancement of Science, Department of Industrial Cooperation, 245–50.

1933. 'Research and Industrial Development. Is a New Organisation Required?' *State Service* 13: 31–32.

Fleming, A. P. M., and H. J. Brockelhurst. 1925. *A History of Engineering.* London: Black.

Flink, Tim, and David Kaldewey. 2018. 'The Language of Science Policy in the Twenty-First Century: What Comes after Basic and Applied Research?' In David Kaldewey and Désirée Schauz (eds.), *Basic and Applied Research: The Language of Science Policy in the Twentieth Century.* New York: Berghahn, 251–84.

Floud, Roderick, and Sean Glynn (eds.). 2000. *London Higher: The Establishment of Higher Education in London.* London: A&C Black.

Foden, Frank. 1951. 'The National Certificate'. *The Vocational Aspect of Secondary and Further Education* 3, no. 6: 38–46.

1961. 'A History of Technical Examinations in England to 1918 with Special Reference to the Examination Work of the City and Guilds of London Institute'. PhD thesis, University of Reading.

1970. *Philip Magnus: Victorian Educational Pioneer.* London: Vallentine & Mitchell.

Fontanon, Claudine, et al. (eds.). 1994. *Les professeurs du Conservatoire National des Arts et Métiers: dictionnaire biographique, 1794–1955.* Paris: Institut National de Recherche Pédagogique and Conservatoire National des Arts et Métiers.

Foreman-Peck, James, and Leslie Hannah. 1999. 'Britain: From Economic Liberalism to Socialism – And Back?' In James Foreman-Peck and Giovanni Federico (eds.), *European Industrial Policy: The Twentieth Century.* Oxford: Oxford University Press, 18–57.

'Foreword'. 1934. *Annals of Eugenics* 6, no. 1: 1.

Forgan, Sophie. 2003. 'Atoms in Wonderland: The Presentation of Atomic Science in Britain in Museums, Exhibitions and Print, 1945–1960'. *History and Technology* 19, no. 3: 177–96.

Forman, Paul. 2007. 'The Primacy of Science in Modernity, of Technology in Postmodernity, and of Ideology in the History of Technology'. *History and Technology* 23, nos. 1–2: 1–152.

Forman, Paul, John L. Heilbron, and Spencer Weart. 1975. 'Physics circa 1900: Personnel, Funding and Productivity of the Academic Establishments'. *Historical Studies in the Physical Sciences* 5.

Fort, Daniel G., et al. 2017. 'Mapping the Evolving Definitions of Translational Research'. *Journal of Clinical and Translational Science* 1, no. 1: 60–66.

Fox, Robert. 1974. 'Education for a New Age: The Conservatoire des Arts et Métiers, 1815–30'. In D. S. L. Cardwell (ed.), *Artisan to Graduate: Essays to Commemorate the Foundation in 1824 of the Manchester Mechanics' Institution, Now in 1974 the University of Manchester Institute of Science and Technology.* Manchester: Manchester University Press, 25–38.

Francis, R. Douglas. 2009. *The Technological Imperative in Canada: An Intellectual History.* Vancouver: UBC Press.

Freeden, Michael. 1979. 'Eugenics and Progressive Thought: A Study in Ideological Affinity'. *The Historical Journal* 22, no. 3: 645–71.

1996. *Ideologies and Political Theory: A Conceptual Approach.* Oxford: Clarendon Press.

Freeman, Christopher, and Luc Soete. 2009. 'Developing Science, Technology and Innovation Indicators: What We Can Learn from the Past'. *Research Policy* 38, no. 4: 583–89.

Friedmann, Georges. 1936. *La crise du progrès: esquisse d'histoire des idées, 1895–1935.* Paris: Gallimard.

Friday, James. 1974. 'The Braggs and Broadcasting'. *Proceedings of the Royal Institution of Great Britain* 47: 59–85.

Fry, R. H. 1956. 'The Breaking of the Crust'. *Manchester Guardian Survey of Industry, Trade and Finance* (March): 44–58.

'Fuel Research in Great Britain'. 1936. *Nature* 137, no. 3477: 1020–21.

Galison, Peter. 2004. 'Removing Knowledge'. *Critical Inquiry* 31, no. 1: 229–43.

2010. 'Secrecy in Three Acts'. *Social Research* 77, no. 3: 941–74.

Gallagher, Catherine. 2009. *The Body Economic: Life, Death, and Sensation in Political Economy and the Victorian Novel.* Princeton, NJ: Princeton University Press.

Gallie, W. B. 1955. 'Essentially Contested Concepts'. *Proceedings of the Aristotelian Society* n.s. 56: 167–98.

Galsworthy, John. 1923. *International Thought.* Cambridge: W. Heffer & Sons.

Galton, Francis. 1909. *Essays in Eugenics.* London: Eugenics Education Society.

Gambles, Anna. 1998. 'Rethinking the Politics of Protection: Conservatism and the Corn Laws, 1830–52'. *English Historical Review*, 113, no. 453: 928–52.

Garfield, Simon. 2022. *All the Knowledge in the World: The Extraordinary History of the Encyclopaedia.* London: Weidenfeld and Nicholson.

Gay, Hannah. 2000. 'Association and Practice: The City and Guilds of London Institute for the Advancement of Technical Education'. *Annals of Science* 57, no. 4: 369–98.

2007. *The History of Imperial College London 1907–2007: Higher Education and Research in Science, Technology and Medicine.* London: Imperial College Press.

'George Gore F.R.S.' 1909. *The Electrician* 63: 467.

Georges, R. A. 1969. 'Toward an Understanding of Storytelling Events'. *Journal of American Folklore* 82, no. 326: 313–28.

1994. 'The Concept of "Repertoire" in Folkloristics'. *Western Folklore* 53, no. 4: 313–23.

Georghiou, Luke, J. Stanley Metcalfe, Michael Gibbons, Tim Ray, and Janet Evans. 1986. 'Wealth from Knowledge Revisited'. In Luke Georghiou, Janet Evans, Tim Ray, J. Stanley Metcalfe, and Michael Gibbons (eds.), *Post–Innovation Performance Technological Development and Competition.* Basingstoke: Palgrave Macmillan, 9–30.

Gere, Cathy. 2009. *Knossos and the Prophets of Modernism.* Chicago: University of Chicago Press.

Gibbons, Michael. 1970. 'The CERN 300 GeV Accelerator: A Case Study in the Application of the Weinberg Criteria'. *Minerva* 8, no. 1: 180–91.

Gibbons, Michael, and Ron Johnston. 1974. 'The Roles of Science in Technological Innovation'. *Research Policy* 3, no. 3: 220–42.

Gibbons, Michael, Camille Limoges, Helga Nowotny, Simon Schwartzman, Peter Scott, and Martin Trow. 1994. *The New Production of Knowledge: The Dynamics of Science and Research in Contemporary Societies*. London: SAGE Publications.

Giffard, Hermione. 2020. 'Engine of Innovation: The Royal Aircraft Establishment, State Design and the Coming of the Gas Turbine Aero-Engine in Britain'. *Contemporary British History* 34, no. 2: 165–78.

Girard, Marion. 2008. *A Strange and Formidable Weapon: British Responses to World War I Poison Gas*. Lincoln: University of Nebraska Press.

Gispen, Kees. 1989. *New Profession, Old Order: Engineers and German Society, 1815–1914*. Cambridge: Cambridge University Press.

Glazebrook, Richard. 1905. 'Science and Industry'. *Times Engineering Supplement* 1, no. 31 (27 September): 245–46.

1917. *Science and Industry, the Place of Cambridge in Any Scheme for Their Combination*. Cambridge: Cambridge University Press.

1918. 'The National Industrial Research Laboratory'. *Engineering* 105 (8 March): 252–56.

1933. 'Early Days at the National Physical Laboratory'. Lecture delivered March 23, 1933. London: National Physical Laboratory.

Godin, Benoit. 2003. 'Measuring Science: Is There "Basic Research" without Statistics?' *Social Science Information* 42, no. 1: 57–90.

2006. 'The Linear Model of Innovation: The Historical Construction of an Analytical Framework'. *Science Technology and Human Values* 31, no. 6: 639–67.

2007. 'Science, Accounting and Statistics: The Input-Output Framework'. *Research Policy* 36, no. 9: 1388–1403.

2008. 'In the Shadow of Schumpeter: W. Rupert Maclaurin and the Study of Technological Innovation'. *Minerva* 46, no. 3: 343–60.

2015. *Innovation Contested: The Idea of Innovation over the Centuries*. London: Routledge.

2017. *Models of Innovation: The History of an Idea*. Cambridge, MA: MIT Press.

2019. *The Invention of Technological Innovation: Languages, Discourses and Ideology in Historical Perspective*. Cheltenham: Edward Elgar.

Godin, Benoit, and Joseph P. Lane. 2013. 'Pushes and Pulls: Hi(S)Tory of the Demand Pull Model of Innovation'. *Science, Technology, & Human Values* 38, no. 5: 621–54.

Godin, Benoit, and Désirée Schauz. 2016. 'The Changing Identity of Research: A Cultural and Conceptual History'. *History of Science* 54, no. 3: 276–306.

Golinski, Jan. 1983. 'Peter Shaw: Chemistry and Communication in Augustan England'. *Ambix* 30, no. 1: 19–29.

1992. *Science as Public Culture: Chemistry and Enlightenment in Britain, 1760–1820*. Cambridge: Cambridge University Press.

Gollin, Alfred M. 1981. 'England Is No Longer an Island: The Phantom Airship Scare of 1909'. *Albion* 13, no. 1: 43–57.

Gooday, Graeme. 2000. 'Lies, Damned Lies and Declinism: Lyon Playfair, the Paris 1867 Exhibition and the Contested Rhetorics of Scientific Education and Industrial Performance'. In I. Inkster (ed.), *The Golden Age: Essays in British Social and Economic History 1850–1870*. Aldershot: Ashgate, 105–20.

2012. '"Vague and Artificial": The Historically Elusive Distinction between Pure and Applied Science'. *Isis* 103, no. 3: 546–54.

Goodeve, Charles. 1972. 'Frank Edward Smith'. *Biographical Memoirs of Fellows of the Royal Society* 18: 532–33.

Gordon, W. R. 1934. 'The Utilisation of Coal'. *Journal of the Royal Society of Arts* 82, no. 4255: 756–80.

Gosden, P. H. J. H., and A. J. Taylor (eds.). 1975. *Studies in the History of a University, 1874–1974*. Leeds: E. J. Arnold.

Gowing, Margaret Mary. 1990. 'Lord Hinton of Bankside, O.M., F. Eng. 12 May 1901–22 June 1983'. *Biographical Memoirs of Fellows of the Royal Society* 36: 218–39.

Gowing, Margaret Mary, and Lorna Arnold. 1974. *Independence and Deterrence. Britain and Atomic Energy 1945–1952, vol. 2: Policy Execution*. London: Macmillan.

Grant, Alexander. 1884. *The Story of the University of Edinburgh during Its First Three Hundred Years*. London: Longmans Green.

Greg, Henry P. 1916. 'The Lancashire Sections. Research in the Cotton Trade'. *Journal of the Textile Institute* 7, no. 3: 244–49.

Gregory, Richard. 1916. *Discovery, or The Spirit and Service of Science*. London: Macmillan.

'The Grell Mystery'. 1913. *The Spectator*, no. 4444 (30 August): 22–23.

Grelon, André. 2000. 'Du bon usage du modèle étranger: la mise en place de l'École centrale des arts et manufactures'. *Bulletin de la Sabix* 26: 47–52.

Greysmith, D. 1990. 'The Empire as Infinite Resource: The Work of P. L. Simmonds (1814–1897)'. *Journal of Newspaper and Periodical History* 6, no. 1: 3–15.

Grove, Jack William. 1962. *Government and Industry in Britain*. London: Longmans.

Guagnini, Anna. 2010. 'The Fashioning of Higher Technical Education in Britain: The Case of Manchester, 1851–1914'. In Howard F. Gospel (ed.), *Industrial Training and Technological Innovation*. London: Routledge, 65–83.

Guise, George. 2014. 'Margaret Thatcher's Influence on British Science'. *Notes and Records of the Royal Society* 68, no. 3: 301–9.

Gummett, Philip. 1971. 'British Science Policy and the Advisory Council on Scientific Policy'. PhD thesis, University of Manchester.

1980. *Scientists in Whitehall*. Manchester: Manchester University Press.

Gummett, Philip, and Michael Gibbons. 1978. 'Redeployment and Diversification at Harwell'. *Omega* 6, no. 1: 65–69.

Gummett, Philip, and Geoffrey Price. 1977. 'An Approach to the Central Planning of British Science: The Formation of the Advisory Council on Scientific Policy'. *Minerva* 15, no. 2: 119–43.

Gummett, Philip, and Roger Williams. 1972. 'Assessing the Council for Scientific Policy'. *Nature* 240, no. 5380: 329–32.

Gunther, R. T. 1925. *Historic Instruments for the Advancement of Science: A Handbook to the Oxford Collections Prepared for the Opening of the Lewis Evans Collection of May 5, 1925.* London: Humphrey Milford.

Gusewelle, J. K. 1977. 'Science and the Admiralty during World War I: The Case of the BIR'. In Gerald Jordan (ed.), *Naval Warfare in the Twentieth Century, 1900–1945: Essays in Honour of Arthur Marder.* London: Taylor & Francis, 103–17.

Hackmann, Willem Dirk. 1984. *Seek and Strike: Sonar, Anti-Submarine Warfare and the Royal Navy 1914–54.* London: HMSO.

1988. 'Sonar, Wireless Telegraphy and the Royal Navy: Scientific Development in a Military Context, 1890–1939'. In Nicholaas A. Rupke (ed.), *Science, Politics and the Public Good: Essays in Honour of Margaret Gowing.* Basingstoke: Macmillan, 90–118.

Haigh, Alice. 2021. '"To Strive, to Seek, to Find": The Origins and Establishment of the British Post Office Engineering Research Station at Dollis Hill, 1908–1938'. PhD thesis, University of Leeds.

Haldane, J. B. S. 1929. 'The Place of Science in Western Civilisation'. *The Realist* 2, no. 2: 49–164.

Hamilton, C. I. 2014. 'Three Cultures at the Admiralty, c. 1800–1945: Naval Staff, the Secretariat and the Arrival of Scientists'. *Journal for Maritime Research* 16, no. 1: 89–102.

Hance, Nicholas John. 2006. *Harwell: The Enigma Revealed.* Buckland: Enhance Publishing.

Hansard HC Deb. 24 March 1868. Vol. 191, cols. 160–65. Available at https://api.parliament.uk/historic-hansard/commons/1868/mar/24/motion-for-a-select-committee (accessed March 2023).

29 April 1909. Vol. 4, cols. 473–619. Available at https://api.parliament.uk/historic-hansard/commons/1909/apr/29/revenue-and-expenditure-for-1909-10 (accessed March 2023).

2 August 1909. Vol. 8, cols. 1564–617. Available at https://api.parliament.uk/historic-hansard/commons/1909/aug/02/naval-and-military-aeronautics (accessed March 2023).

27 July 1925. Vol. 187, cols. 49–50W. Available at https://api.parliament.uk/historic-hansard/written-answers/1925/jul/27/scientific-research (accessed March 2023).

25 May 1938, Vol. 336, cols. 1233–348. Available at https://api.parliament.uk/historic-hansard/commons/1938/may/25/air-defences (accessed March 2023).

30 November 1945, Vol. 416, cols. 1846–47. Available at https://api.parliament.uk/historic-hansard/commons/1945/nov/30/scientific-manpower-and-resources (accessed March 2023).

21 July 1969. Vol. 787, cols. 1237–368. Available at https://api.parliament.uk/historic-hansard/commons/1969/jul/21/civil-science (accessed March 2023).

Hansard. HL Deb. 11 November 1852. Vol. 123, cols. 17–21. Available at https://api.parliament.uk/historic-hansard/lords/1852/nov/11/the-queens-speech (accessed March 2023).

15 November 1961. Vol. 235, cols. 648–728. Available at https://api.parliament.uk/historic-hansard/lords/1961/nov/15/technical-and-scientific-manpower-and (accessed March 2023).

Hård, Mikael. 1998. 'German Regulation: The Integration of Modern Technology into National Culture'. In Mikael Hård and Andrew Jamison (eds.), *The Intellectual Appropriation of Technology: Discourses on Modernity 1900–1939*. Cambridge, MA: MIT Press, 39–67.

Hartcup, Guy, and T. E. Allibone. 1984. *Cockcroft and the Atom*. Bristol: Adam Hilger.

Hartley, Harold. 1937. 'Agriculture as a Source of Raw Materials for Industry'. *Journal of the Textile Institute* 28: 151–72.

Harvey, Mark, and Andrew McMeekin. 2007. *Public or Private Economies of Knowledge? Turbulence in the Biological Sciences*. Cheltenham: Edward Elgar.

2010. 'Public or Private Economies of Knowledge: The Economics of Diffusion and Appropriation of Bioinformatics Tools'. *International Journal of the Commons* 4, no. 1: 481–506.

Harwood, Jonathan. 2005. *Technology's Dilemma Agricultural Colleges between Science and Practice in Germany, 1860–1934*. Bern: Peter Lang.

2006. 'Engineering Education between Science and Practice: Rethinking the Historiography'. *History and Technology* 22, no. 1: 53–79.

2010. 'Understanding Academic Drift: On the Institutional Dynamics of Higher Technical and Professional Education'. *Minerva* 48, no. 4: 413–27.

Hashimoto, Takehiko. 2000. 'The Wind Tunnel and the Emergence of Aeronautical Research in Britain'. In Peter Galison and Alex Roland (eds.), *Atmospheric Flight in the Twentieth Century*. Dordrecht: Springer, 223–39.

2007. 'Leonard Bairstow as a Scientific Middleman: Early Aerodynamic Research on Airplane Stability in Britain, 1909–1920'. *Historia Scientiarum*. 2nd Series 17, no. 2: 103–20.

Hastings, Adrian. 2001. *A History of English Christianity, 1920–2000*. 4th ed. London: SCM Press.

Hayes, Sarah. 2015. 'Industrial Automation and Stress, c. 1945–79'. In Mark Jackson (ed.), *Stress in Post-War Britain, 1945–1985*. London: Routledge, 75–93.

Haynes, Roslynn. 1980. *H. G. Wells, Discoverer of the Future: The Influence of Science on His Thought*. London: Macmillan.

Heard, Gerald. 1931. 'This Surprising World'. *The Listener* 6, no. 137 (26 August): 340.

1935. *Science in the Making*. London: Faber and Faber.

Hearnshaw, F. C. 1929. *Centenary History of King's College London 1828–1928*. London: G. G. Harrap.

Hempstead, Colin, and Gillian Cookson. 2000. *A Victorian Scientist and Engineer: Fleeming Jenkin and the Birth of Electrical Engineering*. Aldershot: Ashgate.

Hendry, John. 1991. 'Technological Decision Making in Its Organizational Context: Nuclear Power Reactor Development in Britain'. Cambridge University Engineering Department, Research Paper 4/91.

Hennessy, Peter. 2018. 'Searching for the Holy Grail of a Science and Innovation Strategy That Makes a Difference'. *FST Journal* 22, no. 2: 7–9.

Hennock, E. P. 1990. 'Technological Education in England, 1850–1926: The Uses of a German Model'. *History of Education* 19, no. 4: 299–331.

Hensley, Oliver D. (ed.). 1988. *The Classification of Research*. Lubbock: Texas Tech University Press.

Herklots, Hugh. 1928. *The New Universities: An External Examination*. London: Ernest Benn.

Hill-Andrews, Oliver. 2015. 'Interpreting Science: J. G. Crowther and the Making of Interwar British Culture'. PhD thesis, University of Sussex.

Hinton, Christopher. 1953. 'Atomic Energy Developments in Great Britain'. *Bulletin of the Atomic Scientists* 9, no. 10: 366–68, 390.

Hinton, James. 2013. *The Mass Observers: A History, 1937–1949*. Oxford: Oxford University Press.

HM Treasury. 1930. *Report of the Committee on the Staffs of Government Scientific Establishments*. London: HMSO.

Hodge, Joseph Morgan. 2007. *Triumph of the Expert: Agrarian Doctrines of Development and the Legacies of British Colonialism*. Athens: Ohio University Press.

2011. 'Science and Empire: An Overview of the Historical Scholarship'. In B. M. Bennett and J. M. Hodge (eds.), *Science and Empire: Knowledge and Networks of Science across the British Empire, 1800–1970*. Basingstoke: Palgrave Macmillan, 3–29.

Hogben, Lancelot. 1935. *Mathematics for the Million*. London: George Allen & Unwin.

1938. *Science for the Citizen*. London: George Allen & Unwin.

Hollinger, David. 1995. 'Science as a Weapon in Kulturkämpfe in the United States during and after World War II'. *Isis* 86, no. 3: 440–54.

Holmfeld, John D. 1970 'From Amateurs to Professionals in American Science: The Controversy over the Proceedings of an 1853 Scientific Meeting'. *Proceedings of the American Philosophical Society* 114, no. 1: 22–36.

Homburg, Ernst. 1986. 'De "Tweede Industriële Revolutie": Een Problematisch Historisch Concept'. *Theoretische Geschiedenis* 13, no. 3: 367–85.

Horn, C. A., and P. L. R. Horn. 1984. 'The City of London Guilds and Technical Education in Victorian England'. *Journal of Further and Higher Education* 8, no. 1: 75–89.

Horner, David. 1993. 'The Road to Scarborough: Wilson, Labour and the Scientific Revolution'. In Richard Coopey, Richard Fielding, and Nick Tiratsoo (eds.), *The Wilson Governments*. London: Pinter, 48–71.

Horrocks, Sally M. 1993. 'Consuming Science: Science, Technology and Food in Britain, 1870–1939'. PhD thesis, University of Manchester.

2000. 'A Promising Pioneer Profession? Women in Industrial Chemistry in Inter-War Britain'. *British Journal for the History of Science* 33, no. 3: 351–67.

Hottois, Gilbert. 2018. 'Technoscience: From the Origin of the Word to Its Current Uses'. In Sacha Loeve, Xavier Guchet, and Bernadette Bensaude Vincent (eds.), *French Philosophy of Technology: Classical Readings and Contemporary Approaches*. Cham: Springer, 121–38.

Houghton, Esther Rhoads. 1979. 'A "New" Editor of the "British Critic"'. *Victorian Periodicals Review* 12, no. 3: 102–5.

Hounshell, David A. 1980. 'Edison and the Pure Science Ideal in 19th-Century America'. *Science* n.s. 207, no. 4431: 612–17.

2000. 'Automation, Transfer Machinery, and Mass Production in the U.S. Automobile Industry in the Post–World War II Era'. *Enterprise & Society* 1, no. 1: 100–138.

House of Commons Innovation, Universities, Science and Skills Committee. 2008–9. Minutes of Evidence. Office for Strategic Coordination of Health Research, HC 655–I. Available at https://publications.parliament.uk/pa/cm200809/cmselect/cmdius/655i/9060801.htm (accessed March 2023).

House of Lords Select Committee on Science and Technology. 1988. *Priorities in Medical Research*. HL 54-1. Norwich: HMSO.

Hughes, Helen. 2021. 'Film and the British Atomic Project'. *Historical Journal of Film, Radio and Television* 42, no. 2: 219–43.

Hughes, Henry Stuart. 1967. *Consciousness and Society*. London: MacGibbon & Kee.

Hughes, Jeff. 1998. 'Plasticine and Valves: Industry, Instrumentation and the Emergence of Nuclear Physics'. In Jean-Paul Gaudillière and Ilana Löwy (eds.), *The Invisible Industrialist: Manufacturers and the Production of Scientific Knowledge*. Basingstoke: Macmillan, 58–101.

2016. 'Unity through Experiment? Reductionism, Rhetoric and the Politics of Nuclear Science, 1918–40'. In Harmke Kamminga and Geert Sommsen (eds.), *Pursuing the Unity of Science: Ideology and Scientific Practice from the Great War to the Cold War*. London: Routledge, 50–81.

2018. 'A Portrait of the Scientist as a Young Ham: Wireless, Modernity and Nuclear Physics'. In Robert Bud, Paul Greenhalgh, Frank James, and Morag Shiach (eds.), *Being Modern: The Cultural Impact of Science in the Early Twentieth Century*. London: UCL Press, 245–73.

Hull, Andrew J. 1994. 'Passwords to Power: A Public Rationale for Expert Influence on Central Government Policy-Making: British Scientists and Economists c. 1900–c. 1925'. PhD thesis, University of Glasgow.

1999. 'War of Words: The Public Science of the British Scientific Community and the Origins of the Department of Scientific and Industrial Research, 1914–16'. *British Journal for the History of Science* 32, no. 4: 461–81.

Hume, L. J. 1958. 'The Origins of the Haldane Report'. *Australian Journal of Public Administration* 17, no. 4: 344–52.

Hunt, Bruce J. 1983. '"Practice vs. Theory": The British Electrical Debate, 1888–1891'. *Isis* 74, no. 3: 341–55.

2005. *The Maxwellians*. Ithaca, NY: Cornell University Press.

Hutchinson, Eric. 1969. 'Scientists and Civil Servants: The Struggle over the National Physical Laboratory in 1918'. *Minerva* 7, no. 3: 373–98.

1972. 'A Fruitful Cooperation between Government and Academic Science: Food Research in the United Kingdom'. *Minerva* 10, no. 1: 19–50.

1975. 'The Origins of the University Grants Committee'. *Minerva* 13, no. 4: 583–620.

Huxley, Aldous. 1931. 'Sight-Seeing-in-Alien-Englands'. *Nash's Pall Mall Magazine*, June: 50–53, 118.

1932a. *Brave New World*. London: Chatto and Windus.

1932b 'Science: The Double-Edged Tool'. *The Listener* 7, no. 158 (20 January): 77–79 and 112.

Huxley, Julian (ed.). 1934. *Scientific Research and Social Needs*. London: Watts.
    1936. 'Science and Its Relation to Social Needs'. In *Scientific Progress*. London:
    George Allen & Unwin, 175–210.
Illinois Institute of Technology. 1968. *Technology in Retrospect and Critical Events
    in Science*. Chicago: Illinois Institute of Technology, Research Institute.
Ince, Martin. 1986. *The Politics of British Science*. Hemel Hempstead: Wheatsheaf.
Inge, William Ralph. 1930. *Christian Ethics and Modern Problems*. London:
    Hodder & Stoughton.
Inkster, Ian. 1991. *Science and Technology in History: An Approach to Industrial
    Development*. Basingstoke: Macmillan.
Innes, Hammond. 1939. *All Roads Lead to Friday*. London: Herbert Jenkins.
Irwin, Will. 1921. *'The Next War': An Appeal to Common Sense*. New York:
    Dutton.
Israel, Paul. 1992. *From Machine Shop to Industrial Laboratory: Telegraphy and the
    Changing Context of American Invention, 1830–1920*. Baltimore: Johns
    Hopkins University Press.
Ives, Eric William, L. D. Schwarz, and Diane K. Drummond. 2000. *The First
    Civic University: Birmingham 1880–1980: An Introductory History*.
    Birmingham: University of Birmingham Press.
Jacobson, Annie. 2015. *The Pentagon's Brain: An Uncensored History of DARPA,
    America's Top Secret Research Agency*. New York: Little, Brown, 2015.
James, C. L. R. 2003. 'A Visit to the Science and Art Museums'. First published
    1932. In Paul Buhle and Patrick Henry (eds.), *Letters from London: Seven
    Essays*. London: Signal Books, 1–14.
James, F. A. J. L. 2008. 'The Janus Face of Modernity: Michael Faraday in the
    Twentieth Century'. *The British Journal for the History of Science* 41, no. 4:
    477–516.
Jasanoff, Sheila. 2015. 'Future Imperfect: Science, Technology and the
    Imaginations of Modernity'. In Sheila Jasanoff and Sang-Hyun Kim (eds.),
    *Dreamscapes of Modernity: Sociotechnical Imaginaries and the Fabrication of
    Power*. Chicago: University of Chicago Press, 1–33.
Jenkin, John G. 2011. 'Atomic Energy Is "Moonshine": What Did Rutherford
    Really Mean?' *Physics in Perspective* 13, no. 2: 128–45.
Jenkins, E. W. 1973. 'The Board of Education and the Reconstruction
    Committee, 1916–1918'. *Journal of Educational Administration and History*
    5, no. 1: 42–51.
Jephcott, Sir Harry. 1965. 'The Glaxo Research Organization'. In Sir John
    Cockcroft (ed.), *The Organization of Scientific Establishments*. Cambridge:
    Cambridge University Press, 148–67.
Jevons, F. R. 1972. 'Preface'. In J. Langrish, M. Gibbons, W. G. Evans, and F. R.
    Jevons, *Wealth from Knowledge: Studies of Innovation in Industry*. London:
    Macmillan, ix–xiii.
    1976. 'The Interaction of Science and Technology Today, or, Is Science the
    Mother of Invention?' *Technology and Culture* 17, no. 4: 729–42.
Jewkes, John. 1960. 'How Much Science?' *The Economic Journal* 70, no. 277: 1–16.
Jewkes, John, David Sawers, and Richard Stillerman. 1958. *The Sources of
    Invention*. London: Macmillan.

Johnson, Paul Barton. 1969. *Land Fit for Heroes: The Planning of British Reconstruction, 1916–1919*. Chicago: University of Chicago Press.

Johnston, J. F. W. 1847. *Lectures on Agricultural Chemistry and Geology*. 2nd ed. Edinburgh: W. Blackwood.

Jones, Allan. 2012. 'Mary Adams and the Producer's Role in Early BBC Science Broadcasts'. *Public Understanding of Science* 21, no. 8: 968–83.

2020. 'Science in the Making: 1930s Citizen Science on the BBC'. *History of Education* 49, no. 3: 327–43.

Jones, Aubrey. 1985. *Britain's Economy: The Roots of Stagnation*. Cambridge: Cambridge University Press.

Jones, Edgar. 1996. *University College Durham: A Social History*. Aberystwyth: E. Jones.

Jones, Peter. 2004. 'Posting the Future: British Stamp Design and the "White Heat" of a Technological Revolution'. *Journal of Design History* 17, no. 2: 163–76.

Jones, Peter M. 2008. *Industrial Enlightenment: Science, Technology and Culture in Birmingham and the West Midlands, 1760–1820*. Manchester: Manchester University Press.

2016. *Agricultural Enlightenment: Knowledge, Technology, and Nature, 1750–1840*. Oxford: Oxford University Press.

Jones, Richard. 2019. 'A Resurgence of the Regions: Rebuilding Innovation Capacity across the Whole UK'. Available at www.softmachines.org/word press/wp-content/uploads/2019/05/ResurgenceRegionsRALJv22_5_19.pdf (accessed May 2020).

Jones, Richard, and James Wilsdon. 2018. *The Biomedical Bubble: Why UK Research and Innovation Needs a Greater Diversity of Priorities, Politics, Places and People*. London: NESTA.

Jordan, D. W. 1985. 'The Cry for Useless Knowledge: Education for a New Victorian Technology'. *IEE Proceedings A – Physical Science, Measurement and Instrumentation, Management and Education – Reviews* 132, no. 8: 587–601.

Joyce, James. 1932. *Ulysses*. Hamburg: Odyssey Press.

Joyce, Patrick. 1994. *Democratic Subjects: The Self and the Social in Nineteenth-Century England*. Cambridge: Cambridge University Press.

Kaiser, David. 2005. 'The Atomic Secret in Red Hands? American Suspicions of Theoretical Physicists during the Early Cold War'. *Representations* 90, no. 1: 28–60.

Kaldewey, David, and Desiree Schauz, 2018. *Basic and Applied Research: The Language of Science Policy in the Twentieth Century*. New York: Berghahn.

Kargon, Robert. 1977. *Science in Victorian Manchester: Enterprise and Expertise*. Manchester: Manchester University Press.

Kargon, Robert, and Scott G. Knowles. 2002. 'Knowledge for Use: Science, Higher Learning, and America's New Industrial Heartland, 1880–1915'. *Annals of Science* 59, no. 1: 1–20.

Katz, John. 1938. *The Will to Civilization: An Inquiry into the Principles of Historic Change* London: Secker & Warburg.

Katzir, Shaul. 2012. 'Who Knew Piezoelectricity? Rutherford and Langevin on Submarine Detection and the Invention of Sonar'. *Notes and Records of the Royal Society* 66, no. 2: 141–57.

2017. '"In War or In Peace": The Technological Promise of Science Following the First World War'. *Centaurus* 59, no. 3: 223–37.

Keanie, Andrew. 2012. 'Coleridge and Plagiarism'. In Frederick Burwick (ed.), *The Oxford Handbook of Samuel Taylor Coleridge*. Oxford: Oxford University Press, 435–54.

Keith, S. T. 1981. 'Inventions, Patents and Commercial Development from Governmentally Financed Research in Great Britain: The Origins of the National Research Development Corporation'. *Minerva* 19, no. 1: 92–122.

1982. 'The Role of Government Research Establishments: A Study of the Concept of Public Patronage for Applied Research and Development'. PhD thesis, Aston University.

Kelham, Brian B. 1967. 'The Royal College of Science for Ireland (1867–1926)'. *Studies: An Irish Quarterly Review* 56, no. 223: 297–309.

Keller, Vera. 2015. *Knowledge and the Public Interest, 1575–1725*. Cambridge: Cambridge University Press.

Kelly, A. 1976. 'Walter Rosenhain and Materials Research at Teddington'. *Philosophical Transactions of the Royal Society A: Mathematical, Physical and Engineering Sciences* 282: 5–36.

Kevles, Daniel J. 1977. 'The National Science Foundation and the Debate over Postwar Research Policy, 1942–1945: A Political Interpretation of Science – The Endless Frontier'. *Isis* 68, no. 1: 5–26.

1995. *In the Name of Eugenics: Genetics and the Uses of Human Heredity*. First published 1985. Cambridge, MA: Harvard University Press.

King, Alexander. 2001. 'Scientific Concerns in an Economic Environment: Science in OEEC–OECD'. *Technology in Society* 23, no. 3: 337–48.

Kirby, M. W. 1999. 'Blackett in the "White Heat" of the Scientific Revolution: Industrial Modernisation under the Labour Governments, 1964–1970'. *The Journal of the Operational Research Society* 50, no. 10: 985–93.

Kirk, F. A. 1998. 'Twentieth-Century Industry: Obsolescence and Change. A Case Study: The ICI Coal to Oil Plant and Its Varied Uses'. *Industrial Archaeology Review* 20, no. 1: 83–90.

Klancher, Jon P. 2013. *Transfiguring the Arts and Sciences: Knowledge and Cultural Institutions in the Romantic Age*. Cambridge: Cambridge University Press.

Klandermans, Bert, and Conny Roggeband. 2007. *Handbook of Social Movements across Disciplines*. New York: Springer Science & Business Media.

Klein, Ursula. 2016. *Nützliches Wissen: die Erfindung der Technikwissenschaften*. Göttingen: Wallstein Verlag.

2020. *Technoscience in History: Prussia, 1750–1850*. Cambridge, MA: MIT Press.

Kline, R. 1995. 'Construing "Technology" as "Applied Science": Public Rhetoric of Scientists and Engineers in the United States, 1880–1945'. *Isis* 86, no. 2: 194–221.

2015. *The Cybernetics Moment, or Why We Call Our Age the Information Age*. Baltimore: Johns Hopkins University Press.

Kline, R., and Thomas C. Lassman. 2005. 'Competing Research Traditions in American Industry: Uncertain Alliances between Engineering and Science at Westinghouse Electric, 1886–1935'. *Enterprise and Society* 6, no. 4: 601–45.

Kogan, Maurice, Mary Henkel, and Steve Hanney. 2006. *Government and Research: Thirty Years of Evolution*. 2nd ed. Dordrecht: Springer.

König, Wolfgang. 1996. 'Science-Based Industry or Industry-Based Science? Electrical Engineering in Germany before World War I'. *Technology and Culture* 37, no. 1: 70–101.

Koselleck, Reinhart. 2002. *The Practice of Conceptual History: Timing History, Spacing Concepts*. Trans. Samuel Presner Todd. Stanford, CA: Stanford University Press.

Kraft, Alison. 2004. 'Pragmatism, Patronage and Politics in English Biology: The Rise and Fall of Economic Biology 1904–1920'. *Journal of the History of Biology* 37, no. 2: 213–58.

Labour Party. 1938. *Labour's Plan for Oil from Coal*. London: Labour Party.

 1963. 'Labour and the Scientific Revolution, a Statement of Policy Approved by the Annual Conference of the Labour Party Scarborough 1963'. Appendix 2. In *The Labour Party. Report of the Sixty-Second Annual Conference*. London: Labour Party, 272–75.

Lander, Bryn, and Janet Atkinson-Grosjean. 2011. 'Translational Science and the Hidden Research System in Universities and Academic Hospitals: A Case Study'. *Social Science & Medicine* 72, no. 4: 537–44.

Langrish, J. 1988. Review of *Post-innovation Performance: Technological Development and Competition* by L. Georgiou et al. *Research Policy* 17, no. 2: 114–16.

 2023. *It's Not Physics. Part One: What's It All about Then?*. N.p.: Independently published.

Langrish, J., M. Gibbons, W. G. Evans, and F. R. Jevons. 1972. *Wealth from Knowledge: Studies of Innovation in Industry*. London: Macmillan.

Larédo, Philippe, and Philippe Mustar. 2004. 'Public Sector Research: A Growing Role in Innovation Systems'. *Minerva* 42, no. 1: 11–27.

Laucht, Christopher. 2012. 'Atoms for the People: The Atomic Scientists' Association, the British State and Nuclear Education in the Atom Train Exhibition, 1947–1948'. *British Journal for the History of Science* 45, no. 4: 591–608.

Layton, David. 1974. *Science for the People: The Origins of the School Science Curriculum in England*. London: George Allen & Unwin.

Leavis, F. R. 2013. *Two Cultures? The Significance of C. P. Snow*. Introduction by Stefan Collini. Cambridge: Cambridge University Press.

Lee, Kenneth. 1923. 'Industrial Research Associations'. *Nature* 112, no. 2825: 898–99.

Leggett, Donald. 2015. *Shaping the Royal Navy: Technology, Authority and Naval Architecture, c. 1830–1906*. Manchester: Manchester University Press.

 2016. 'Give Me a Laboratory and I Will Win You the War: Governing Science in the Royal Navy'. In Don Leggett and Charlotte Sleigh (eds.), *Scientific Governance in Britain, 1914–79*. Manchester: Manchester University Press, 27–44.

Leighton, Denys. 2000. 'Municipal Progress, Democracy and Radical Identity in Birmingham, 1838–1886'. *Midland History* 25: 115–42.

Lessing, L. 1963. 'The Three Ages of Science Writing'. *Chemical & Engineering News*, 41, no. 18: 88–92.

Lester, J. H. 1916. 'Textile Research'. *Journal of the Textile Institute* 7, no. 3: 202–16.

Lightman, Bernard. 2007. *Victorian Popularizers of Science: Designing Nature for New Audiences*. Chicago: University of Chicago Press.

Lightman, Bernard, Gordon McOuat, and Larry Stewart. 2013. *The Circulation of Knowledge between Britain, India, and China: The Early-Modern World to the Twentieth Century*. Leiden: Brill.

'"Lightning" in a Laboratory: Man-Made "Thunder-Storms"'. 1930. *Illustrated London News*. 15 March: 428.

Lindenfeld, David F. 1997. *The Practical Imagination: The German Sciences of State in the Nineteenth Century*. Chicago: University of Chicago Press.

Local Government Board. 1919. *Memorandum on the Provisions of the Ministry of Health Bill, 1919, as to the Work of the Medical Research Committee, (Clause 3 (1), proviso (i)*. Cmd 69. London: HMSO.

Lodge, Oliver. 1893. *Pioneers of Science*. London: Macmillan.

——— 1926. *Science and Human Progress: Halley Stewart Lectures 1926*. London: George Allen & Unwin.

Lord President of the Council. 1946. *Scientific Man-Power: Report of a Committee Appointed by the Lord President of the Council*. Cmd 6824. London: HMSO.

Lord Privy Seal. 1971. *A Framework for Government Research and Development*. Cmnd 4814. London: HMSO.

Lovell, Bernard. 1975. 'Patrick Maynard Stuart Blackett, Baron Blackett, of Chelsea. 18 November 1897–13 July 1974'. *Biographical Memoirs of Fellows of the Royal Society* 21: 1–115.

Lucier, Paul. 2012. 'The Origins of Pure and Applied Science in Gilded Age America'. *Isis* 103, no. 3: 527–36.

Lupton, M. C. 1964. 'The Mosely Education Commission to the United States, 1903'. *The Vocational Aspect of Secondary and Further Education* 16, no. 33: 36–49.

Macdonald, Lee T. 2018. *Kew Observatory and the Evolution of Victorian Science*. Pittsburgh, PA: University of Pittsburgh Press.

MacGregor, D. H. 1927. 'Rationalisation of Industry'. *Economic Journal* 37, no. 148: 521–50.

Maclaurin, W. Rupert. 1953. 'The Sequence from Invention to Innovation and Its Relation to Economic Growth'. *The Quarterly Journal of Economics* 67, no. 1: 97–111.

Maclean, George Edward. 1917. *Studies in Higher Education in England and Scotland: With Suggestions for Universities and Colleges in the United States*. Bulletin 1917, no. 16. Department of the Interior, Bureau of Education. Washington, DC: United States Government Printing Office.

MacLeod, Christine. 2007. *Heroes of Invention: Technology, Liberalism and British Identity, 1750–1914*. Cambridge: Cambridge University Press.

MacLeod, Christine, and Gregory Radick. 2013. 'Claiming Ownership in the Technosciences: Patents, Priority and Productivity'. *Studies in History and Philosophy of Science Part A* 44, no. 2: 188–201.

MacLeod, Roy M. 1969. 'Securing the Foundations'. *Nature* 224, no. 5218: 441–44.

1980. 'On Visiting the "Moving Metropolis": Reflections on the Architecture of Imperial Science'. *Historical Records of Australian Science* 5, no. 3: 1–16.

1993. 'The Chemists Go to War: The Mobilization of Civilian Chemists and the British War Effort, 1914–1918'. *Annals of Science* 50, no. 5: 455–81.

1994. 'Science for Imperial Efficiency and Social Change: Reflections on the British Science Guild, 1905–1936'. *Public Understanding of Science* 3, no. 2: 155–93.

1997. 'On Science and Colonialism'. In Peter Bowler and Nicholas Whyte (eds.), *Science and Society in Ireland: The Social Context of Science and Technology in Ireland 1800–1950*. Belfast: Institute of Irish Studies, the Queen's University of Belfast, 1–17.

2000. 'Sight and Sound on the Western Front: Surveyors, Scientists, and the "Battlefield Laboratory", 1915–1918'. *War & Society* 18, no. 1: 23–46.

2009a. 'The Scientists Go to War: Revisiting Precept and Practice, 1914–1919'. *Journal of War and Culture Studies* 2, no. 1: 37–51.

2009b. *Archibald Liversidge, FRS: Imperial Science under the Southern Cross*. Sydney: Sydney University Press.

MacLeod, Roy M., and E. Kay Andrews. 1969. 'The Committee of Civil Research: Scientific Advice for Economic Development 1925–30'. *Minerva* 7, no. 4: 680–705.

1970. 'The Origins of the DSIR: Reflections on Ideas and Men, 1915–1916'. *Public Administration* 48, no. 1: 23–48.

1971. 'Scientific Advice in the War at Sea, 1915–1917: The Board of Invention and Research'. *Journal of Contemporary History* 6, no. 2: 3–40.

Madge, Charles, and Tom Harrison. 1939. *Britain*. Harmondsworth: Penguin Books.

Magnello, Eileen. 1999a. 'The Non-Correlation of Biometrics and Eugenics: Rival Forms of Laboratory Work in Karl Pearson's Career at University College London, Part 1'. *History of Science* 37, no. 1: 79–106.

1999b. 'The Non-Correlation of Biometrics and Eugenics: Rival Forms of Laboratory Work in Karl Pearson's Career at University College London, Part 2'. *History of Science* 37, no. 2: 123–50.

2000. *A Century of Measurement: An Illustrated History of the National Physical Laboratory*. Bath: Canopus.

Magnus, Sir Philip. 1910. *Educational Aims and Efforts, 1880–1910*. London: Longmans Green.

Mah, Harold. 2000. 'Phantasies of the Public Sphere: Rethinking the Habermas of Historians'. *The Journal of Modern History* 72, no. 1: 153–82.

Malmsten, Neal R. 1977. 'British Government Policy toward Colonial Development, 1919–39'. *Journal of Modern History* 49, no. 2: D1249–D1287.

Mandler, Peter (ed.). 2006. *Liberty and Authority in Victorian Britain*. Oxford: Oxford University Press.

2015. 'Ben Pimlott Memorial Lecture 2014: The Two Cultures Revisited: The Humanities in British Universities since 1945'. *Twentieth Century British History* 26, no. 3: 400–423.

Manegold, Karl Heinz. 1970. *Universität, Technische Hochschule und Industrie: Ein Beitrag zur Emanzipation der Technik im 19. Jahrhundert unter besonderer Berücksichtigung der Bestrebungen Felix Kleins*. Berlin: Duncker & Humblot.

Manuel, Jeffrey T. 2021. 'Technical Literature and the Text-Searchable: The History of Technology and the Digitized Turn'. *Technology's Stories* 8, no. 3. doi.org/10.15763/jou.ts.2021.01.05.04.

Marsden, Ben. 1992. 'Engineering Science in Glasgow: W. J. M. Rankine and the Motive Power of Air'. PhD thesis, University of Kent.

2004. 'The Progeny of These Two "Fellows": Robert Willis, William Whewell and the Sciences of Mechanism, Mechanics and Machinery in Early Victorian Britain'. *British Journal for the History of Science* 37, no. 4: 401–34.

Marsh, R. W. 1953. 'The Past and the Future of the *Annals of Applied Biology*'. *Annals of Applied Biology* 40, no. 3: 435–48.

Martin, Ben. 2012. 'The Evolution of Science Policy and Innovation Studies'. *Research Policy* 41, no. 7: 1219–39.

Masefield, John. 1911. *The Street of To-Day*. London: J. M. Dent & Sons.

Mathers, Helen. 2005. *Steel City Scholars: The Centenary History of the University of Sheffield*. London: James & James.

Matsumoto, Miwao. 2006. *Technology Gatekeepers for War and Peace: The British Ship Revolution and Japanese Industrialisation*. Basingstoke: Palgrave Macmillan.

Matthew, H. C. G. 1973. *The Liberal Imperialists: The Ideas and Politics of a Post-Gladstonian Élite*. Oxford: Oxford University Press.

Mayer, Anna-Katherina. 1997. 'Moralizing Science: The Uses of Science's Past in National Education in the 1920s'. *British Journal for the History of Science* 30, no. 1: 51–70.

2000. '"A Combative Sense of Duty": Englishness and the Scientists'. In Christopher Lawrence and Anna-Katherina Mayer (eds.), *Regenerating England: Science, Medicine and Culture in Inter-War Britain*. Amsterdam: Rodopi, 67–106.

2005. 'When Things Don't Talk: Knowledge and Belief in the Interwar Humanism of Charles Singer (1876–1960)'. *British Journal for the History of Science* 38, no. 3: 325–47.

Mayr, Otto. 1976. 'The Science-Technology Relationship as a Historiographic Problem'. *Technology and Culture* 17, no. 4: 663–73.

Mazumdar, Pauline M. H. 1992. *Eugenics, Human Genetics and Human Failings: The Eugenics Society, Its Sources and Its Critics in Britain*. London: Routledge.

McAllister, John Francis Olivarius. 1987. 'Civil Science Policy in British Industrial Reconstruction, 1942–51'. DPhil thesis, University of Oxford.

McCance, Andrew. 1960. 'Cecil Henry Desch. 1874–1958'. *Biographical Memoirs of Fellows of the Royal Society* 5: 49–68.

McClellan, James E. (ed.). 2008. *The Applied-Science Problem: Papers from a Workshop at Stevens Institute of Technology Hoboken, NJ, May 6–8, 2005*. Hoboken, NJ: Jensen/Daniels.

McCraw, Thomas K. 2009. *Prophet of Innovation: Joseph Schumpeter and Creative Destruction*. Cambridge, MA: Harvard University Press.

McCulloch, Andrew. 2004. *The Feeneys of the Birmingham Post*. Birmingham: University of Birmingham Press.

McDonald, G. W., and Howard F. Gospel. 1973. 'The Mond-Turner Talks, 1927–1933: A Study in Industrial Co-operation'. *Historical Journal* 16, no. 4: 807–29.

McGee, Michael Calvin. 1980. 'The "Ideograph": A Link between Rhetoric and Ideology'. *Quarterly Journal of Speech* 66, no. 1: 1–16.

McGucken, William. 1978. 'On Freedom and Planning in Science: The Society for Freedom in Science, 1940–46'. *Minerva* 16, no. 1: 42–72.

McMahon, A. M. 1976. 'Corporate Technology: The Social Origins of the American Institute of Electrical Engineers'. *Proceedings of the IEEE* 64, no. 9: 1383–90.

McNinch, Syl. 1984. 'The Rise and Fall of RANN (Research Applied to National Needs)'. Ed. David E. Gould and Louise McIntire. Washington, DC: National Science Foundation.

Mees, Charles Edward Kenneth. 1920. *The Organization of Industrial Scientific Research*. New York: McGraw-Hill.

1952. 'Secrecy and Industrial Research'. *Nature* 170, no. 4336: 972.

Melville, Harry. 1962. 'D.S.I.R. Does It Pay Off?'. Manchester: Manchester Statistical Society.

Mercelis, J. 2022. '"Men Don't Like to Work under a Woman": Female Chemists in the Photographic Manufacturing Industry, ca. 1918–1950'. *Ambix* 69, no. 3: 291–319.

Mercier, Alain. 2018. *Le Conservatoire des Arts et Métiers: des origines à la fin de la Restauration, 1794–1830*. Ghent: Snoeck Publishers.

Mertens, Joost. 2000. 'From Tubal Cain to Faraday: William Whewell as a Philosopher of Technology'. *History of Science* 38, no. 3: 321–42.

Merton, Robert. 1936. 'Puritanism, Pietism and Science'. *Sociological Review* 28, no. 1: 1–30.

Merton, Robert, and Arnold Thackray. 1972. 'On Discipline Building: The Paradoxes of George Sarton'. *Isis* 63, no. 4: 473–95.

Meyer, Torsten, and Marcus Popplow. 2004. '"To Employ Each of Nature's Products in the Most Favorable Way Possible": Nature as a Commodity in Eighteenth-Century German Economic Discourse'. *Historical Social Research* 29, no. 4: 4–40.

Miller, David Philip. 2004. *Discovering Water: James Watt, Henry Cavendish and the Nineteenth-Century 'Water Controversy'*. Aldershot: Ashgate.

2019. *The Life and Legend of James Watt: Collaboration, Natural Philosophy, and the Improvement of the Steam Engine*. Pittsburgh, PA: University of Pittsburgh Press.

Miller, Ian. 2014. *Reforming Food Production: Agricultural Science and Education*. Manchester: Manchester University Press.

Miller, Steve. 2001. 'Public Understanding of Science at the Crossroads'. *Public Understanding of Science* 10, no. 1: 115–20.

Milligan, Ian. 2013. 'Illusionary Order: Online Databases, Optical Character Recognition, and Canadian History, 1997–2010'. *Canadian Historical Review* 94, no. 4: 540–69.

Ministry of Education. 1945. *Higher Technological Education: Report of a Special Committee Appointed in April 1944*. London: HMSO.

Minister of Fuel and Power. 1955. *A Programme of Nuclear Power*. Cmd 9389. London: HMSO.

Ministry of Reconstruction. 1918. *Report of the Machinery of Government Committee*. Cd 9230. London: HMSO.

Ministry of Technology. 1970. *Industrial Research and Development in Government Laboratories: A New Organisation for the Seventies*. London: HMSO.

Mokyr, Joel. 2004. *Gifts of Athena: Historical Origins of the Knowledge Economy*. Princeton, NJ: Princeton University Press.

Mond, Sir Alfred. 1927. *Industry and Politics*. London: Macmillan.

Moore, D. C. 1965. 'The Corn Laws and High Farming'. *The Economic History Review* 18, no. 3: 544–61.

Moore, Olive. 1943. '"Vigilantibus non dormientibus": Whiston A. Bristow'. *Scope. Magazine for Industry*, July: 23–28.

Morant, Robert. 1910. 'Prefatory Memorandum'. In *Regulations for Technical Schools, Schools of Art and Other Forms of Provision of Further Education in England and Wales*. Cd 5329. London: HMSO, 1910, iv–v.

——— 1911. 'Prefatory Memorandum'. In *Statement of Grants Available from the Board of Education in Aid of Technological and Professional Work in Universities in England and Wales*. Cd. 5762. London: HMSO, iii–iv.

Morrell, J. B. 1973. 'The Patronage of Mid-Victorian Science in the University of Edinburgh'. *Science Studies* 3, no. 4: 353–88.

Morrell, J. B., and Arnold Thackray. 1981. *Gentlemen of Science: Early Years of the British Association for the Advancement of Science*. Oxford: Oxford University Press.

Morris, Peter J. T. (ed.). 2010. *Science for the Nation: Perspectives on the History of the Science Museum*. Basingstoke: Palgrave Macmillan.

Morse, Elizabeth J. 1992. 'English Civic Universities and the Myth of Decline'. *History of Universities* 11: 177–204.

Mortimer, James E., and Valerie E. Ellis. 1980. *A Professional Union: The Evolution of the Institution of Professional Civil Servants*. London: George Allen & Unwin.

Moseley, Henry. 1843. *The Mechanical Principles of Engineering and Architecture*. London: Longman.

——— 1857. *Illustrations of Mechanics*. First ed. 1839. New York: Harper.

Moseley, Russell. 1976. 'Science, Government and Industrial Research: The Origins and Development of the National Physical Laboratory, 1900–1975.' DPhil thesis, University of Sussex.

——— 1978. 'The Origins and Early Years of the National Physical Laboratory: A Chapter in the Pre-History of British Science Policy'. *Minerva* 16, no. 2: 222–50.

——— 1980. 'Government Science and the Royal Society: The Control of the National Physical Laboratory in the Inter-War Years'. *Notes and Records of the Royal Society* 35, no. 2: 167–93.

Mosley, S. 2016. 'Selling the Smokeless City: Advertising Images and Smoke Abatement in Urban-Industrial Britain, circa 1840–1960'. *History and Technology* 32, no. 2: 201–11.

Moura, Mário Graça. 2017. 'Schumpeter and the Meanings of Rationality'. *Journal of Evolutionary Economics* 27, no. 1: 115–38.

Mowery, David. 1986. 'Industrial Research in Britain, 1900–1950'. In Bernard Elbaum and William Lazonick (eds.), *Decline of the British Economy*. Oxford: Clarendon Press, 189–222.

Müller, Ernst, and Falko Schmieder. 2018. 'Begriffsgeschichte und Wissenschaftsgeschichte: Bestandsaufnahme und Forschungsperspektiven'. *Geschichte und Gesellschaft* 44, no. 1: 79–106.

Multhauf, Robert P. 1971. 'The French Crash Program for Saltpeter Production, 1776–94'. *Technology and Culture* 12, no. 2: 163–81.

Murphy, Kate. 2016. *Behind the Wireless: A History of Early Women at the BBC*. Basingstoke: Palgrave Macmillan.

Musgrave, P. W. 1964. 'The Definition of Technical Education: 1860–1910'. *The Vocational Aspect of Secondary and Further Education* 16, no. 34: 105–11.

Mussell, James, and Graeme Gooday. 2020. *A Pioneer of Connection: Recovering the Life and Work of Oliver Lodge*. Pittsburgh, PA: University of Pittsburgh Press.

Nahum, Andrew. 1999. 'The Royal Aircraft Establishment from 1945 to Concorde'. In Robert Bud and Philip Gummett (eds.), *Cold War Hot Science: Applied Research in Britain's Defence Laboratories 1945–1990*. Reading: Harwood, 29–58.

National Academy of Sciences. 1967. *Applied Science and Technological Progress: A Report to the Committee on Science and Astronautics, U.S. House of Representatives*. Washington, DC: Government Printing Office.

National Physical Laboratory. 1936. *Report for the Year 1935*. London: HMSO.

Nelson, Richard R. 1989. 'What Is Private and What Is Public about Technology?' *Science, Technology, & Human Values* 14, no. 3: 229–41.

'The New Municipal Technical School'. 1900. *Journal of the Manchester Geographical Society* 16: 293–95.

'News and Views'. 1936. *Nature* 118, no. 2982: 922–25.

Niblett, C. A. 1980. 'Images of Progress: Three Episodes in the Development of Research Policy in the UK Electrical Engineering Industry'. PhD thesis, University of Manchester.

Nicholson, Rafaelle M., and John W. Nicholson. 2012. 'Martha Whiteley of Imperial College, London: A Pioneering Woman Chemist'. *Journal of Chemical Education* 89, no. 5: 598–601.

Nieto-Galan, Agustí. 2016. *Science in the Public Sphere: A History of Lay Knowledge and Expertise*. London: Routledge.

Nobel Industries Limited, 1924. *Nobel Industries Limited: A Record of Their Exhibits at the British Empire Exhibition, 1924*. London: Nobel Industries Limited.

Norris, John Pilkington. 1869. *The Education of the People, Our Weak Points and Our Strength, Occasional Essays*. London: Simpkin Marshall.

'Notes'. 1912. *Nature* 88, no. 2201: 322–26.

Nuffield College. 1944. *Problems of Scientific and Industrial Research: A Statement*. Oxford: Oxford University Press.

Nurse, Paul. 2023. *Independent Review of the UK's Research, Development and Innovation Organisational Landscape: Final Report and Recommendations*. Available at www.gov.uk/government/publications/research-development-and-innovation-organisational-landscape-an-independent-review (accessed March 2023).

Nutting, P. G. 1919. 'Institutes of Applied Science'. *Journal of the Franklin Institute* 187, no. 4: 487–93.

Nye, Mary Jo. 2004. *Blackett: Physics, War, and Politics in the Twentieth Century*. Cambridge, MA: Harvard University Press.

2011. *Michael Polanyi and His Generation: Origins of the Social Construction of Science*. Chicago: University of Chicago Press.

Office of the Minister for Science. 1961. *Report of the Committee on the Management and Control of Research and Development*. London: HMSO.

Office for National Statistics. 2017. 'Business Enterprise Research and Development, UK: 2016'. Available at www.ons.gov.uk/economy/govern mentpublicsectorandtaxes/researchanddevelopmentexpenditure/bulletins/ businessenterpriseresearchanddevelopment/2016 (accessed August 2022).

2019. 'Gross Domestic Expenditure on Research and Development, UK: 2017'. Available at www.ons.gov.uk/releases/ukgrossdomesticexpenditureon researchanddevelopment2017 (accessed March 2023).

2022. 'Comparison of ONS Business Enterprise Research and Development Statistics with HMRC Research and Development Tax Credit Statistics'. Available at www.ons.gov.uk/economy/governmentpublicsectorandtaxes/ researchanddevelopmentexpenditure/articles/comparisonofonsbusinessen terpriseresearchanddevelopmentstatisticswithhmrcresearchanddevelopment taxcreditstatistics/2022-09-29. (accessed March 2023).

2023. 'Gross Domestic Product at Market Prices: Current Price: Seasonally Adjusted £m'. Available at www.ons.gov.uk/economy/grossdomesticpro ductgdp/timeseries/ybha/pn2 (accessed March 2023).

Olby, Robert. 1991. 'Social Imperialism and State Support for Agricultural Research in Edwardian Britain'. *Annals of Science* 48, no. 6: 509–26.

Olewski, Todd. 2018. 'Between Bench and Bedside: Building Clinical Consensus at the NIH, 1977–2013'. *Journal of the History of Medicine and Allied Sciences* 73, no. 4: 464–500.

Olivier, Théodore. 1850. 'Monge et l'École polytechnique'. *Revue Scientifique et Industrielle* 7: 64–68.

1851. *Mémoires de géométrie descriptive, théorique et appliqué*. Paris.

Orlow, Dietrich. 2000. *Common Destiny: A Comparative History of the Dutch, French, and German Social Democratic Parties, 1945–1969*. New York: Berghahn Books.

Ormerod, Henry A. 1953. *The Liverpool Royal Institution: A Record and a Retrospect*. Liverpool: University Press.

Ormerod, Richard. 2003. 'The Father of Operational Research'. In Peter Hore (ed.), *Patrick Blackett: Sailor, Scientist and Socialist*. London: F. Cass, 187–200.

Ortolano, Guy. 2009. *The Two Cultures Controversy: Science, Literature and Cultural Politics in Postwar Britain*. Cambridge: Cambridge University Press.

Osler, W. 1915. 'Science and War'. *Lancet*, no. 4806: 795–801.

1919. 'The Old Humanities and the New Science: The Presidential Address Delivered before the Classical Association at Oxford, May, 1919'. *BMJ* 2, no. 3053: 1–7.

O'Sullivan, O. P., R. M. Duffy, and B. D. Kelly. 2019. 'Culturomics and the History of Psychiatry: Testing the Google Ngram Method'. *Irish Journal of Psychological Medicine* 36, no. 1: 23–27.

Overy, Richard. 2009. *The Morbid Age*. London: Penguin Books.

Palladino, Paolo. 1990. 'The Political Economy of Applied Research: Plant Breeding in Great Britain, 1910–1940'. *Minerva* 28, no. 4: 446–68.

1993. 'Between Craft and Science: Plant Breeding, Mendelian Genetics, and British Universities, 1900–1920'. *Technology and Culture* 34, no. 2: 300–323.

Parker, Miles. 2016. 'The Rothschild Report (1971) and the Purpose of Government-Funded R&D: A Personal Account'. *Palgrave Communications* 2 (2 August). doi.org/10.1057/palcomms.2016.53.

Parliamentary and Scientific Committee. 1947. *Colleges of Technology and Technological Manpower*. London: Parliamentary and Scientific Committee.

Pasteur, Louis. 1871. 'Pourquoi la France n'a pas trouvé d'hommes supérieurs au moment du Péril'. Reprinted in *Quelques Réflexions sur la Science en France*. Paris: Gauthier-Villars, 25–40.

Paterson, C. C. 1945. *A Confidential History of the Research Laboratories*. London: General Electric.

Paterson, C. C., J. C. Smuts, R. E. B. Crompton, and F. B. Jewett. 1931. 'Speeches at the Royal Albert Hall, 23rd September, 1931, at the Opening of the Faraday Exhibition'. *Journal of the Institution of Electrical Engineers* 69, no. 419: 1385–87.

Pattison, Michael. 1983. 'Scientists, Inventors and the Military in Britain, 1915–19: The Munitions Inventions Department'. *Social Studies of Science* 13, no. 4: 521–68.

Pavitt, K., and W. Walker. 1976. 'Government Policy towards Industrial Innovation: A Review'. *Research Policy* 5, no. 1: 11–97.

Pearson, George C. 1885. 'Editor's Notice'. In George C. Pearson (ed.), *Propaedeia Prophetica or, The Use and Design of the Old Testament Examined by William Rowe Lyall*. New ed. London: Kegan Paul Trench, 5–6.

Pechenick, Eitan Adam, Christopher M. Danforth, and Peter Sheridan Dodds. 2015. 'Characterizing the Google Books Corpus: Strong Limits to Inferences of Socio-Cultural and Linguistic Evolution'. *PLoS One* 10, no. 10. doi.org/10.1371/journal.pone.0137041.

Percy, (Lord) Eustace. 1930. *Education at the Crossroads*. London: Evans Bros.

Perkin, Harold. 1972. 'University Planning in Britain in the 1960s'. *Higher Education* 1: 111–20.

Petch, N. J. 1992. 'Sir Harold Montague Finniston, 15 August 1912–2 February 1991'. *Biographical Memoirs of Fellows of the Royal Society* 38: 131–44.

Petitjean, P., V. Zharov, G. Glaser, J. Richardson, B. de Padirac, and G. Archibald (eds.). 2006. *Sixty Years of Science at UNESCO, 1945–2005*. Paris: UNESCO.

Phillips, Denise. 2012. *Acolytes of Nature: Defining Natural Science in Germany, 1770–1850*. Chicago: University of Chicago Press.

Playfair, Lyon. 1852. *Industrial Instruction on the Continent*. London: HMSO.

1857. *On Scientific Institutions in Connexion with the Department of Science and Art Department*. London: Chapman and Hall.

Pocock, Rowland F. 1977. *Nuclear Power: Its Development in the United Kingdom*. Old Woking: Unwin Brothers.

Polanyi, Michael. 1956. 'Pure and Applied Science and Their Appropriate Forms of Organization'. *Dialectica* 10, no. 3: 231–42.

1958. *Personal Knowledge*. London: Routledge & Kegan Paul.

Political and Economic Planning, Industries Group. 1936. *Report on the British Coal Industry*. London: PEP.

Pollock, Frederick. 1957. *Automation: A Study of Its Economic and Social Consequences*. Trans. W. O. Henderson and W. H. Challoner. New York: Praeger.

Pratt, John. 1997. *The Polytechnic Experiment: 1965–1992*. Buckingham: Open University Press and SRHE.

Preece, C. 1982. 'The Durham Engineering Students of 1838'. *Transactions of the Architectural and Archaeological Society of Durham and Northumberland* n.s. 6: 71–74.

President of the United States. 1964. *Economic Report of the President*. Transmitted to the Congress January 1964. Washington, DC: United States Government Printing Office.

Prickett, Stephen. 1979. 'Coleridge and the Idea of the Clerisy'. In Walter B. Crawford (ed.), *Reading Coleridge: Approaches and Applications*. Ithaca, NY: Cornell University Press, 252–73.

Pursell, Carroll. 1974. '"A Savage Struck by Lightning": The Idea of a Research Moratorium, 1927–37'. *Lex et Scientia* 10: 146–58.

Pyenson, Lewis. 2007. *The Passion of George Sarton: A Modern Marriage and Its Discipline*. Philadelphia: American Philosophical Society.

Quayle, J. Rodney, and Geoffrey W. Greenwood. 2003. 'Leonard Rotherham CBE. 31 August 1913 – 23 March 2001'. *Biographical Memoirs of Fellows of the Royal Society* 49: 431–46.

Rayleigh, Lord (chairman). 1916. *The Neglect of Science: Report of Proceedings at a Conference Held … 3rd May, 1916*. London: Harrison & Sons.

1930. *Lord Balfour in His Relation to Science*. Cambridge: Cambridge University Press.

Reader, W. J. 1970, 1975. *Imperial Chemical Industries: A History*. 2 vols. Oxford: Oxford University Press.

1977. 'Imperial Chemical Industries and the State, 1926–1945'. In Barry Supple (ed.), *Essays in British Business History*. Oxford: Oxford University Press, 1977, 227–43.

Recueil des lois, décrets, ordonnances … du Conservatoire National des Arts et Métiers, et à la création des cours publics de cet établissement etc. 1889. Paris: Imprimerie Nationale.

Reeks, Margaret. 1920. *Register of the Associates and Old Students of the Royal School of Mines, and History of the Royal School of Mines*. London: Royal School of Mines Old Students' Association.

Reich, Leonard S. 1987. 'Edison, Coolidge, and Langmuir: Evolving Approaches to American Industrial Research'. *The Journal of Economic History* 47, no. 2: 341–51.

Reid, Wemyss. 1899. *Memoirs and Correspondence of Lyon Playfair, First Lord Playfair of St. Andrews*. London: Cassel.

Reingold, Nathan. 1987. 'Vannevar Bush's New Deal for Research: Or the Triumph of the Old Order'. *Historical Studies in the Physical and Biological Sciences* 17, no. 2: 299–344.

Resnikoff, Jason 2022. *Labor's End: How the Promise of Automation Degraded Work*. Urbana: University of Illinois Press.

Review of Frank Froest, *The Grell Mystery*. 1913. *The Spectator*, no. 4444 (30 August): 22–23.

Reynolds, L. A., and E. M. Tansey (eds.). 2008. *Superbugs and Superdrugs: A History of MRSA*. London: Wellcome Trust Centre for the History of Medicine at UCL.

Richter, M. 1986. 'Conceptual History (Begriffsgeschichte) and Political Theory'. *Political Theory* 14, no. 4: 604–37.

Roberts, Gerrylynn Kuszen. 1973. 'The Royal College of Chemistry (1845–1853): A Social History of Chemistry in Early-Victorian England.' PhD thesis, Johns Hopkins University.

Roberts, Gerrylynn Kuszen, and A. E. Simmons. 2009. 'British Chemists Abroad, 1887–1971: The Dynamics of Chemists' Careers'. *Annals of Science* 66, no. 1: 103–28.

Roberts, K. O. 1969. 'The Separation of Secondary Education from Technical Education 1899–1903'. *The Vocational Aspect of Education* 21: 101–5.

Roderick, Gordon Wynne, and Michael Dawson Stephens. 1972. *Scientific and Technical Education in 19th Century England: A Symposium*. Newton Abbot: David and Charles.

Rodrick, Anne B. 2004. *Self Help, and Civic Culture: Citizenship in Victorian Birmingham*. Aldershot: Ashgate.

Roe, Nicholas. 2002. *Samuel Taylor Coleridge and the Sciences of Life*. Oxford: Oxford University Press.

Rolls-Hansen, Niels. 2006. 'The Lysenko Effect: The Politics of Science'. *Journal of the History of Biology* 39, no. 1: 202–4.

2015. 'On the Philosophical Roots of Today's Science Policy: Any Lessons from the "Lysenko Affair"?' *Studies in East European Thought* 67, nos. 1–2: 91–109.

Ronayne, J. 1984. *Science in Government*. London: Edward Arnold.

Root, John David. 1980. 'The Philosophical and Religious Thought of Arthur James Balfour (1848–1930)'. *Journal of British Studies* 19, no. 2: 120–41.

Rose, Hilary, and Steven Rose. 1969. *Science and Society*. Harmondsworth: Allen Lane.

Rosenhain, Walter. 1914. *Metallurgy: An Introduction to the Study of Physical Metallurgy*. New York: D. Van Nostrand.

1915–16. 'The National Physical Laboratory: Its Work and Aims'. *Journal of the West Scotland Iron & Steel Institute* 23: 213–63.

1930. 'The Development of Materials for Aircraft Purposes'. *Proceedings of the Royal Aeronautical Society* 34, no. 236: 631–48.

Ross, S. 1962. 'Scientist: The Story of a Word'. *Annals of Science* 18, no. 2: 65–85.

Ross, William. 2002. *H. G. Wells's World Reborn: The Outline of History and Its Companions*. Selinsgrove, PA: Susquehanna University Press.

Rotherham L., and A. B. Mcintosh. 1956. 'Applied Research and Development within the Industrial Group of the U.K. Atomic Energy Authority'. *Nature* 178, no. 4532: 524–27.

Rothschild, Victor (Lord). 1972. 'Forty-Five Varieties of Research (and Development)'. *Nature* 239, no. 5372: 373–78.

Rothwell, Roy. 1992. 'Successful Industrial Innovation: Critical Factors for the 1990s'. *R&D Management* 22, no. 3: 221–39.

Rowland, H. A. 1886. 'Address of Professor Rowland'. In *Report of the Electrical Conference at Philadelphia in September 1884*. Washington, DC: United States Government Printing Office, 12–28.

Royal Commission on the Coal Industry. 1926. *Report of the Royal Commission on the Coal Industry (1925) with Minutes of Evidence and Appendices*. Cmd 2600. London: HMSO.

Royal Commission on Secondary Education. 1895. *Minutes of Evidence Taken before the Royal Commission on Secondary Education*. Vol II. c-7862-1. London: HMSO.

Royal Commission on Technical Instruction. 1884. *Second Report of the Royal Commission on Technical Instruction*, vols. 1–4. c. 3981. London: HMSO.

Royal Society, British Empire Exhibition Committee. 1924. *British Empire Exhibition 1924: Handbook to the Exhibition of Pure Science: Galleries 3 and 4 British Government Pavilion*. London: Royal Society.

1925. *Phases of Modern Science*. London: A. and F. Denny.

Rubin, Charles T. 2005. 'Daedalus and Icarus Revisited'. *The New Atlantis*, no. 8, Spring: 73–91.

Rushforth, Alexander. 2012. 'Translational Science: The Very Idea. Transformations in Contemporary Academic Health Research?' PhD thesis, University of Surrey.

Russell, A. S. 1933. 'Science Notes: An Engineer's Outlook'. *The Listener* 10, no. 240 (16 August): 252.

Russell, Bertrand. 1924. *Icarus; or, The Future of Science*. London: K. Paul, Trench, Trubner.

1928. 'Science'. In C. A. Beard (ed.), *Whither Mankind? A Panorama of Modern Civilization*. New York: Longmans Green, 63–82.

1956. 'Science and Human Life'. In J. R. Newman (ed.), *What Is Science? Twelve Eminent Scientists and Philosophers Explain Their Various Fields to the Layman*. London: Victor Gollancz, 6–23.

Russell, E. J. 1919. 'The Development of Agricultural Research and Education in Great Britain'. *Nature* 103, no. 2586: 227–29.

Russell, Peter (pseud. for John Maynard Smith). 1962. 'The Social Control of Science'. *New Left Review* 1, no. 16: 10–20.

Sainsbury, David (Lord). 2005. *The Race to the Top: A Review of Government's Science and Innovation Policies* Norwich: HMSO.

Salin, Edgar. 1956. 'Industrielle Revolution'. *Kyklos* 9, no. 3: 299–317.

Sanderson, Michael. 1972. *The Universities and British Industry, 1850–1970*. London: Routledge and Kegan Paul.

1978. 'The Professor as Industrial Consultant: Oliver Arnold and the British Steel Industry, 1900–14'. *The Economic History Review* 31, no. 4: 585–60.

Sarton, George. 1931. *History of Science and the New Humanism*. New York: Henry Holt.

Sayers, R. S. 1950. 'The Springs of Technical Progress in Britain, 1919–39'. *Economic Journal* 60: 275–91.

Schaffer, Simon. 1992. 'Late Victorian Metrology and Its Instrumentation: A Manufactory of Ohms'. In Robert Bud and Susan E. Cozzens (eds.), *Invisible Connections: Instruments, Institutions, and Science.* Bellingham, WA: SPIE, 23–56.

2019. 'Ideas Embodied in Metal: Babbage's Engines Dismembered and Remembered'. In Joshua Nall, Liba Taub, and Francis Willmoth (eds.), *The Whipple Museum of the History of Science: Objects and Investigations, to Celebrate the 75th Anniversary of R. S. Whipple's Gift to the University of Cambridge.* Cambridge: Cambridge University Press, 119–58.

Schatzberg, Eric. 2006. 'Technik Comes to America: Changing Meanings of Technology before 1930'. *Technology and Culture* 47, no. 3: 486–512.

2018. *Technology: Critical History of a Concept.* Chicago: University of Chicago Press.

Schauz, Désirée. 2020. *Nützlichkeit und Erkenntnisfortschritt: Eine Geschichte des modernen Wissenschaftsverständnisses.* Göttingen: Wallstein.

Schefold, Bertram. 2004. 'Edgar Salin and His Concept of "Anschauliche Theorie" ("Intuitive Theory") during the Interwar Period'. *Annals of the Society for the History of Economic Thought* 46: 1–16.

Scheinfeldt, Tom. 2003. 'Sites of Salvage: Science History between the Wars'. DPhil thesis, University of Oxford.

Schleifer, Ronald. 2000. *Modernism and Time: The Logic of Abundance in Literature, Science, and Culture, 1880–1930.* Cambridge: Cambridge University Press.

Schmid, C. C. E. 1810. *Allgemeine Encyklopädie und Methodologie der Wissenschaften.* Jena: Akad. Buchhandlung.

Schmoch, Ulrich. 2007. 'Double-Boom Cycles and the Comeback of Science-Push and Market-Pull'. *Research Policy* 36, no. 7: 1000–1015.

Schools Inquiry Commission. 1867. *Report Relative to Technical Education.* [3898]. London: HMSO.

Schumpeter, Joseph. 1928. 'The Instability of Capitalism'. *Economic Journal* 38: 361–86.

'Science in Civilisation'. 1923. *Nature* 112, no. 2825: 889–91.

Searle, Geoffrey Russell. 1971. *The Quest for National Efficiency: A Study in British Politics and Political Thought, 1899–1914.* Oxford: Blackwell.

Searle, John, 2010. *Making the Social World: The Structure of Human Civilization.* Oxford: Oxford University Press.

Sebag-Montefiore, Ruth. 1994–96. 'A Quest for a Grandfather: Sir Philip Magnus, 1st Bart., Victorian Educationalist'. *Jewish Historical Studies* 34: 141–59.

Sebestik, J. 1983. 'The Rise of the Technological Science'. *History and Technology* 1, no. 1: 25–43.

1984. 'De la technologie à la technonomie: Gerard-Joseph Christian'. *Cahiers STS* 2: 56–69.

1986. 'The Introduction of Technological Education at the Conservatoire des Arts et Métiers'. In Melvin Kranzberg (ed.), *Technological Education – Technological Style.* San Francisco: San Francisco Press, 26–40.

Secord, James A. 2004. 'Knowledge in Transit'. *Isis* 95, no. 4: 654–72.

Secretary of State for the Colonies. 1925. *Report of the East Africa Commission.* Cmd 2387. London: HMSO.

Select Committee. 1867–68. 'On the Provisions for Giving Instruction in Theoretical and Applied Science to the Industrial Classes'. *P.P.* 1867–68 (432) XV.1.

Self, Robert. 1994. 'Treasury Control and the Empire Marketing Board: The Rise and Fall of Non-Tariff Preference in Britain, 1924–1933'. *Twentieth Century British History* 5, no. 2: 153–82.

Seppel, Marten, and Keith Tribe (eds.). 2017. *Cameralism in Practice.* Martlesham: Boydell Press.

Seth, Suman. 2009. 'Putting Knowledge in Its Place: Science, Colonialism, and the Postcolonial'. *Postcolonial Studies* 12, no. 4: 373–88.

Seward, A. C. 1917. 'Preface'. In A. C. Seward (ed.), *Science and the Nation.* Cambridge: Cambridge University Press, v–vii.

Sewell, William H., Jr. 2005. *Logics of History: Social Theory and Social Transformation.* Chicago: University of Chicago Press.

Sexton, James. 1996. '*Brave New World* and the Rationalization of Industry'. In Jerome Meckier (ed.), *Critical Essays on Aldous Huxley.* New York: Hall, 88–102.

Seymour-Ure, Colin. 1975. 'The Press and the Party System between the Wars'. In Gillian Peale and Chris Cooke (eds.), *The Politics of Reappraisal 1918–1939.* Basingstoke: Palgrave Macmillan, 232–57.

Shapin, Steven. 2010. *The Scientific Life: A Moral History of a Late Modern Vocation.* Chicago: University of Chicago Press.

Shapiro, Fred R. 1985. 'Neologisms in Coleridge's Notebooks'. *Notes and Queries* 32, no. 3: 346–47.

Sharot, Stephen. 1979. 'Reform and Liberal Judaism in London: 1840–1940'. *Jewish Social Studies* 41, nos. 3–4: 211–28.

Sharp, Paul Richard. 1971. '"Whiskey Money" and the Development of Technical and Secondary Education in the 1890s'. *Journal of Educational Administration and History* 4, no. 1: 31–36.

Sharp, Robert. 2003. 'Oswald Silberrad: The Work of a Forgotten Chemist'. *Ambix* 50, no. 3: 302–9.

Shaw, Peter (ed.). 1733. *The Philosophical Works of Francis Bacon.* London.

Shelley, Mary. 2017. *Frankenstein.* Ed. David H. Guston, Ed Finn, and Jason Scott Robert. Cambridge, MA: MIT Press.

Sheppard, Deri. 2017. 'Robert Le Rossignol, 1884–1976: Engineer of the "Haber" Process'. *Notes and Records of the Royal Society* 71, no. 3: 263–96.

Sheppard, F. H. W. (ed.). 1975. *Survey of London, vol. 38: South Kensington Museums Area.* London: London County Council.

Sherington, G. E. 1974. 'R. B. Haldane, The Reconstruction Committee and the Board of Education, 1916–18'. *Journal of Educational Administration and History* 6, no. 2: 18–25.

Sherwin, C. W., and R. S. Isenson. 1967. 'Project Hindsight'. *Science* n.s. 156, no. 3782: 1571–77.

Shiach, Morag. 2018. 'Woolf's Atom, Eliot's Catalyst and Richardson's Waves of Light: Science and Modernism in 1919'. In Robert Bud, Paul Greenhalgh, Frank James, and Morag Shiach (eds.), *Being Modern: The Cultural Impact of Science in the Early Twentieth Century*. London: UCL Press, 58–76.

Shils, Edward. 1956. *The Torment of Secrecy: The Background and Consequences of American Security Policies*. Glencoe, IL: Free Press.

Shinn, C. H. 1980. 'The Beginnings of the University Grants Committee'. *History of Education* 9, no. 3: 233–43.

1986. *Paying the Piper: The Development of the University Grants Committee 1919–1946*. London: Falmer Press.

Short, P. J. 1974. 'The Municipal School of Technology and the University, 1890–1914'. In D. S. L. Cardwell (ed.), *Artisan to Graduate: Essays to Commemorate the Foundation in 1824 of the Manchester Mechanics' Institution, Now in 1974 the University of Manchester Institute of Science and Technology*. Manchester: Manchester University Press, 157–64.

Siemens, Sir Charles William. 1889. 'Remarks on the House of Applied Science'. Presented to the Iron and Steel Institute, 1877. In E. F. Bamber (ed.), *The Scientific Works of C. William Siemens*. Vol. 3. London: John Murray, 154–55.

Simcock, A. V. 1985. *Robert T. Gunther and the Old Ashmolean*. Oxford: Museum of the History of Science.

Simon, Francis Eugene. 1960. *The Neglect of Science: Essays Addressed to Laymen*. Oxford: Basil Blackwell. First published 1951.

Singer, Charles. 1960. 'How "A History of Technology" Came into Being'. *Technology and Culture* 1, no. 4: 302–11.

Skelton, Matthew. 2001. 'The Paratext of Everything: Constructing and Marketing H. G. Wells's *The Outline of History*'. *Book History* 4, no. 1: 237–75.

Slosson, Edwin E. 1928. 'The Coming of the New Coal Age'. In *Annual Report of the Board of Regents of the Smithsonian Institution 1927*. Washington, DC: United States Government Printing Office, 243–53.

Smalley, George W. 1909. *The Life of Sir Sydney H. Waterlow, Bart., London Apprentice, Lord Mayor, Captain of Industry, and Philanthropist*. London: E. Arnold.

Smeaton, W. A. 1997. 'History of Science at University College London'. *British Journal for the History of Science* 30, no. 1: 25–28.

Smith, Crosbie. 1998. *The Science of Energy: A Cultural History of Energy Physics in Victorian Britain*. London: Athlone.

Smith, Olivia. 1986. *The Politics of Language, 1791–1819*. Oxford: Clarendon Press.

Snelders, H. A. M. 1981. 'James F. W. Johnston's Influence on Agricultural Chemistry in the Netherlands'. *Annals of Science* 38, no. 5: 571–84.

Snow, C. P. 1936. 'What We Need from Applied Science'. *The Spectator*, no. 5656 (20 November): 904.

1954. *The New Men*. London: Macmillan.

1993. *The Two Cultures*. With Introduction by Stefan Collini. Cambridge: Cambridge University Press.

Snyder, Alice D. (ed.). 1934. *S. T. Coleridge's Treatise on Method: As Published in the Encyclopaedia Metropolitana*. London: Constable & Co.

1940. 'Coleridge and the Encyclopedists'. *Modern Philology* 38: 173–91.

Soddy, Frederick. 1920. 'Applied Science and Industrial Research'. *Nature* 105, no. 2640: 422.

Soloway, Richard A. 1995. *Demography and Degeneration: Eugenics and the Declining Birthrate in Twentieth-Century Britain*. Chapel Hill: University of North Carolina Press.

Soubiran, Sébastien. 2002. 'De l'utilisation contingente des scientifiques dans les systèmes d'innovation des marine française et britannique entre les deux guerres mondiales. Deux exemples: la conduite du tir des navires et la télémécanique'. PhD thesis, Université de Paris 7, Denis Diderot.

Staley, Richard. 2018. 'The Interwar Period as a Machine Age: Mechanics, the Machine, Mechanisms, and the Market in Discourse'. *Science in Context* 31, no. 3: 263–92.

Staudenmaier, John M. 1985. *Technology's Storytellers: Reweaving the Human Fabric*. Cambridge, MA: MIT Press.

Stead, H. 1976. 'The Costs of Technological Innovation'. *Research Policy* 5, no. 1: 2–9.

Steiner, Helmuth (ed.). 1989. *J. D. Bernal's The Social Function of Science, 1939–1989*. Berlin: Akademie-Verlag.

Stewart, W. 1989. *Higher Education in Post-War Great Britain*. Basingstoke: Macmillan.

Stokes, D. E. 1997. *Pasteur's Quadrant: Basic Science and Technological Innovation*. Washington, DC: Brookings Institution Press.

Stranges, Anthony N. 1985. 'From Birmingham to Billingham: High-Pressure Coal Hydrogenation in Great Britain'. *Technology and Culture* 26, no. 4: 726–57.

Stuart, Joseph T. 2008. 'The Question of Human Progress in Britain after the Great War'. *British Scholar* 1, no. 1: 53–78.

Sturchio, Jeffrey. 2020. 'Experimenting with Research: Kenneth Mees, Eastman Kodak and the Challenges of Diversification'. *Science Museum Group Journal*, 13 (Spring). http://dx.doi.org/10.15180/201311 (accessed March 2023).

Sumner, James. 2014. 'Defiance to Compliance: Visions of the Computer in Postwar Britain'. *History and Technology* 30, no. 4: 309–33.

2015. *Brewing Science, Technology and Print, 1700–1880*. London: Routledge.

Supple, Barry. 1989. 'The British Coal Industry between the Wars'. *Refresh* 9: 5–8.

Swade, Doron. 2000. *The Cogwheel Brain: Charles Babbage and the Quest to Build the First Computer*. London: Little, Brown.

Swiatecka, M. Jagdiga. 1980. *Idea of the Symbol: Some Nineteenth Century Comparisons with Coleridge*. Cambridge: Cambridge University Press.

Swinney, Geoffrey N. 2016. 'George Wilson's Map of Technology: Giving Shape to the "Industrial Arts" in Mid-Nineteenth-Century Edinburgh'. *Journal of Scottish Historical Studies* 36, no. 2: 165–90.

Sykes, F. H. 1913. 'Military Aviation'. *The Journal of the Aeronautical Society* 17, no. 67: 127–39.

Tansey, E. M. 1989. 'The Wellcome Physiological Research Laboratories 1894–1904: The Home Office, Pharmaceutical Firms, and Animal Experiments'. *Medical History* 33, no. 1: 1–41.

1990. 'The Early Scientific Career of Sir Henry Dale FRS (1875–1968)'. PhD thesis, University of London.

Tawney. R. H. 1943. 'The Abolition of Economic Controls, 1918–1921'. *Economic History Review* 13, nos. 1–2: 1–30.

Taylor, A. J. P. 1965. *English History 1914–1945*. Oxford: Clarendon Press.

Teague, S. John. 1980. *The City University: A History*. London: City University.

Technical Instruction Act. 1889. c.76. London: HMSO.

Thackray, Arnold. 1974. 'Natural Knowledge in Cultural Context: The Manchester Model'. *American Historical Review* 79, no. 3: 672–709.

Thackray, Arnold, Jeffrey Sturchio, P. Thomas Carroll, and Robert Bud. 1985. *Chemistry in America, 1876–1976: Historical Indicators*. Dordrecht: Reidel.

'Things to Come'. 1937. *The Economist* (11 September): 511.

Thomas, Edward. 1936. 'Computing Progress in Chemistry'. *Science* n.s. 83, no. 2146: 159–61.

Thomas, Mark. 1983. 'Rearmament and Economic Recovery in the Late 1930s'. *Economic History Review* 36, no. 4: 552–79.

Thompson, F. M. L. (ed.). 1993. *The Cambridge Social History of Britain 1750–1950*. Cambridge: Cambridge University Press.

Thompson, Joseph. 1886. *The Owens College: Its Foundation and Growth; and Its Connection with the Victoria University*. Manchester: Cornish.

Thompson, Peter. 2022. 'From Gas Hysteria to Nuclear Fear: A Historical Synthesis of Chemical and Atomic Weapons'. *Historical Studies in the Natural Sciences* 52, no. 2: 223–64.

Thomson, A. Landsborough. 1973. 'Origin of the British Legislative Provision for Medical Research'. *Journal of Social Policy* 2, no. 1: 41–54.

Thomson, J. J. 1917. 'Address of the President, Sir J. J. Thomson, O.M., at the Anniversary Meeting, November 30, 1916'. *Proceedings of the Royal Society of London. Series A* 93, no. 647: 90–98.

Thorsheim, P. 2017. *Inventing Pollution: Coal, Smoke, and Culture in Britain since 1800*. 2nd ed. Athens: Ohio University Press.

Thurston, R. H. 1902. 'Scientific Research: The Art of Revelation and of Prophecy'. *Science* n.s.16, no. 402: 401–24.

Tilley, Helen. 2011. *Africa as a Living Laboratory: Empire, Development, and the Problem of Scientific Knowledge, 1870–1950*. Chicago: University of Chicago Press.

Titley, Arthur. 1920. 'Presidential Address'. *Transactions of the Newcomen Society* 1, no. 1: 65–76.

1941. 'Beginnings of the Society: A Note on the Attainment of Its Majority'. *Transactions of the Newcomen Society* 22, no. 1: 37–39.

Tizard, H. T. 1929. 'Science and the New Industrial Revolution'. *Journal of the Textile Institute Proceedings* 20 (April): 79–93.

Tomlinson, Jim. 1994. *Government and the Enterprise since 1900: The Changing Problem of Efficiency*. Oxford: Clarendon.

2000. *The Politics of Decline: Understanding Post-War Britain*. Harlow: Longmans.

Tout, T. F. 1915. 'The University'. In H. M. McKechnie (ed.), *Manchester in 1915*. Manchester: Manchester University Press/Longman, 38–50.

Travis, Anthony S. 2006. 'Decadence, Decline and Celebration: Raphael Meldola and the Mauve Jubilee of 1906'. *History and Technology* 22, no. 2: 131–52.

——— 2013. 'Modernising Industrial Organic Chemistry'. In Anthony S. Travis, Harm G. Schröter, Ernst Homburg, and Peter J. T. Morris (eds.), *Determinants in the Evolution of the European Chemical Industry, 1900–1939: New Technologies, Political Frameworks, Markets and Companies*. Dordrecht: Springer Science & Business Media, 171–98.

Trentmann, Frank. 2008. *Free Trade Nation: Commerce, Consumption, and Civil Society in Modern Britain*. Oxford: Oxford University Press.

Trischler, Helmuth, and Robert Bud. 2018. 'Public Technology: Nuclear Energy in Europe'. *History and Technology* 34, nos. 3–4: 187–212.

Trischler, Helmuth, and Rüdiger Vom Bruch. 1999. *Forschung für den Markt: Geschichte der Fraunhofer-Gesellschaft*. Munich: C. H. Beck.

Tully, James (ed.). 1989. *Meaning and Context: Quentin Skinner and His Critics*. Cambridge: Polity Press.

Turchetti, Simone. 2012. *The Pontecorvo Affair: A Cold War Defection and Nuclear Physics*. Chicago: University of Chicago Press.

Turner, Dorothy Mabel. 1933. *The Book of Scientific Discovery: How Science Has Enabled Human Welfare*. London: G. G. Harrap & Co.

Turney, Jon. 1998. *Frankenstein's Footsteps: Science, Genetics and Popular Culture*. New Haven, CT: Yale University Press.

Tyndall, John. 1873. *Six Lectures on Light, Delivered in America in 1872–1873*. New York: Appleton.

Uekötter, F. 2021. 'The Revolt of the Chemists: Biofuels, Agricultural Overproduction, and the Chemurgy Movement in New Deal America'. *History and Technology* 37, no. 4: 429–45.

United States Congress, Joint Economic Committee. 1956. *Automation and Technological Change, United States*. Congress, Senate, Eighty-Fourth Congress, Second Session. Washington, DC: United States Government Printing Office.

United States Congress, Subcommittee on Economic Stabilization of the Joint Committee on the Economic Report. 1955. *Automation and Technological Change*. Hearings. October 14, 15, 17, 18, 24, 25, 26, 27, and 28, Eighty-Fourth Congress, First Session. Washington, DC: Government Printing Office.

United States Department of Commerce. 1967. *Technological Innovation: Its Environment and Management*. Report of the Panel on Invention and Innovation. Washington, DC: United States Department of Commerce.

United States House of Representatives, Subcommittee of the Committee on Government Operations. 1958. *Research and Development*. Hearing. 15 January. Eighty-Fifth Congress, Second Session. Washington, DC: United States Government Printing Office.

United States National Science Foundation. 1980. *Categories of Scientific Research*. Papers Presented at a National Science Foundation Seminar,

Washington, DC, December 8, 1979. Washington DC: United States Government Printing Office.

University Grants Committee. 1921. *Report of the University Grants Committee.* 3 February. Cmd 1163. London: HMSO.

1953. *University Development. Report on the Years 1947 to 1952.* London: HMSO.

Urwick, Lyndall F. 1930. *The Meaning of Rationalisation.* London: Nisbet and Co.

Valone, David A. 2001. 'Hugh James Rose's Anglican Critique of Cambridge: Science, Antirationalism, and Coleridgean Idealism in Late Georgian England'. *Albion* 33, no. 2: 218–42.

Van der Kloot, William. 2005. 'Lawrence Bragg's Role in the Development of Sound-Ranging in World War I'. *Notes and Records of the Royal Society* 59, no. 3: 273–84.

Van Dijk, Teun A. 2013. 'Ideology and Discourse'. In Michael Freeden, Lyman Tower Sargent, and Marc Stears (eds.), *The Oxford Handbook of Political Ideologies.* Oxford: Oxford University Press, 191–212.

Van Miert, Dirk. 2017. 'Structuring the History of Knowledge in an Age of Transition: The Göttingen Geschichte between Historia Literaria and the Rise of the Disciplines'. *History of Humanities* 2, no. 2: 389–416.

Varcoe, Ian. 1970. 'Scientists, Government and Organised Research in Great Britain 1914–16: The Early History of the DSIR'. *Minerva* 8, no. 2: 192–216.

1981. 'Co-operative Research Associations in British Industry, 1918–34'. *Minerva* 19, no. 3: 433–63.

Varley, E. A. 2002. *The Last of the Prince Bishops: William Van Mildert and the High Church Movement of the Early Nineteenth Century.* Cambridge: Cambridge University Press.

Vatin, François. 2004. 'Machinisme, Marxisme, Humanisme: Georges Friedmann avant et après-guerre'. *Sociologie du Travail* 46, no. 2: 205–23.

Veblen, Thorstein. 1906. 'The Place of Science in Modern Civilization'. *American Journal of Sociology*, 11, no. 5: 585–609.

Vernon, Keith. 1994. 'Microbes at Work: Micro-Organisms, the D.S.I.R. and Industry in Britain, 1900–1936'. *Annals of Science* 51, no. 6: 593–613.

1997. 'Science for the Farmer? Agricultural Research in England 1909–36'. *Twentieth Century British History* 8, no. 3: 310–33.

2004. *Universities and the State in England, 1850–1939.* London: Routledge.

2009. 'Civic Universities and Community Engagement in Inter-War England'. In Peter Cunningham with Susan Oosthuizen and Richard Taylor (eds.), *Beyond the Lecture Hall: Universities and Community Engagement from the Middle Ages to the Present Day.* Cambridge: University of Cambridge Faculty of Education, 31–48.

Vickers, Neil. 2004. *Coleridge and the Doctors.* Oxford: Oxford University Press.

Vig, Norman. 1968. *Science and Technology in British Politics.* Oxford: Pergamon.

Viklund, Daniel. 1955. 'The Automatic Factory: What Does It Mean?' *Institution of Production Engineers Journal* 34, no. 12: 817–19.

Volkov, Shulamit. 2012. *Walther Rathenau: The Life of Weimar's Fallen Statesman.* New Haven, CT: Yale University Press.

Vonnegut, Kurt. 1953. *Player Piano*. London: Macmillan.

Wakefield, Andre. 2009. *The Disordered Police State: German Cameralism as Science and Practice*. Chicago: University of Chicago Press.

Walsh, James Jackson. 1998. 'Postgraduate Technological Education in Britain: Events Leading to the Establishment of Churchill College, Cambridge, 1950–1958'. *Minerva* 36, no. 2: 147–77.

Walsh, Vivien. 1984. 'Invention and Innovation in the Chemical Industry: Demand-Pull or Discovery-Push?' *Research Policy* 13, no. 4: 211–34.

Walters, Gerald (ed.). 1966. *A Technological University: An Experiment in Bath*. Bristol: Bath University Press.

Ward, Jacob. 2018. 'Information and Control: Inventing the Communications Revolution in Post-War Britain'. PhD thesis, University College London.

Waters, D. W. 1980. 'Seamen, Scientists, Historians, and Strategy: Presidential Address, 1978'. *British Journal for the History of Science* 13, no. 3: 189–210.

Weart, Spencer R. 1988. *Nuclear Fear: A History of Images*. Cambridge, MA: Harvard University Press.

Webb, Sidney. 1901. *Twentieth Century Politics: A Policy of National Efficiency*. London: Fabian Society.

1904. *London Education*. London: Longmans Green.

Weedon, Brenda. 2008. *The Education of the Eye: History of the Royal Polytechnic Institution 1838–1881*. London: University of Westminster.

Weidlein, Edward R., and William A. Hamor. 1936. *Glances at Industrial Research, during Walks and Talks in Mellon Institute*. New York: Reinhold.

Weiss, John Hubbel. 1982. *The Making of Technological Man: The Social Origins of French Engineering Education*. Cambridge, MA: MIT Press.

Wellerstein, Alex. 2021. *Restricted Data: The History of Nuclear Secrecy in the United States*. Chicago: University of Chicago Press.

Wellmon, Chad. 2015. *Organizing Enlightenment: Information Overload and the Invention of the Modern Research University*. Baltimore: Johns Hopkins University Press.

Wells, H. G. 1911. *Tono Bungay*. London: Macmillan.

1914. *The World Set Free*. London: Macmillan.

1920. *The Outline of History; Being a Plain History of Life and Mankind*. London: Cassell.

1921. *The Salvaging of Civilization*. London: Cassell.

1926. *The World of William Clissold*. London: Benn.

1932. *The Work, Wealth and Happiness of Mankind*. 2nd ed. London: William Heinemann.

Werner, P., and F. L Holmes. 2002. 'Justus Liebig and the Plant Physiologists'. *Journal of the History of Biology* 35, no. 3: 421–41.

Werskey, Gary. 1978. *The Visible College: A Collective Biography of British Scientists and Socialists of the 1930s*. London: Allen Lane.

Westman, Robert S. 2011. *The Copernican Question: Prognostication, Skepticism, and Celestial Order*. Berkeley: University of California Press.

Whitehead, Alfred North. 1926. *Science and the Modern World*. Cambridge: Cambridge University Press.

Whitehead, Thomas P. 1978. 'Before and after the Rothschild Report'. In Gordon McLachlan (ed.), *Five Years After: A Review of Health Care Research Management after Rothschild*. Oxford: Oxford University Press for the Nuffield Provincial Hospitals Trust, 9–50.

Whitfield, Jakob. 2013. 'Metropolitan Vickers, the GasTurbine, and the State: A Socio-Technical History, 1935-1960'. PhD thesis, University of Manchester.

Wiener, Norbert. 1950. *The Human Use of Human Beings: Cybernetics and Society*. Boston: Houghton Mifflin.

1965. *Cybernetics, or Control and Communication in the Animal and Machine*. First published 1948. Cambridge, MA: MIT Press.

Willetts, David. 2019. *The Road to 2.4%: Transforming Britain's R&D Performance*. London: Policy Institute, King's College London.

Williams, Raymond. 1976. *Keywords: A Vocabulary of Culture and Society*. Abingdon: CroomHelm.

Williams, Roger. 1973. 'Some Political Aspects of the Rothschild Affair'. *Science Studies* 3, no. 1: 31–46.

1983. 'British Nuclear Power Policies'. In P. Tempest (ed.), *Energy Economics in Britain*. Dordrecht: Springer, 35–58.

Williamson, A. W. 1870. *Plea for Pure Science: Being the Inaugural Lecture at the Opening of the Faculty of Science in University College, London, October 4, 1870*. London: Taylor and Francis.

Williamson, J. W. 1920a. 'Applied Science and Industrial Research'. *Nature* 105, no. 2639: 387–88.

1920b. 'Applied Science and Industrial Research'. *Nature* 105, no. 2643: 518.

Willis, Kirk. 1995. 'The Origins of British Nuclear Culture, 1895–1939'. *Journal of British Studies* 34, no. 1: 59–89.

Wilson, Daniel. 2010. 'Machine Past, Machine Future: Technology in British Thought, c. 1870-1914'. PhD thesis, Birkbeck, University of London.

Wilson, George. 1855. *What Is Technology? An Inaugural Lecture*. Edinburgh: Sutherland and Knox.

Wilson, Harold. 1963. 'Labour and the Scientific Revolution'. In *The Labour Party. Report of the Sixty-Second Annual Conference*. London: Labour Party, 133–40.

Wilson, Jessie A. 1860. *Memoir of George Wilson*. Edinburgh: Edmonston and Douglas.

Wise, George. 1985. *Willis R. Whitney, General Electric, and the Origins of U.S. Industrial Research*. New York: Columbia University Press.

Wisnioski, Matthew H. 2012. *Engineers for Change: Competing Visions of Technology in 1960s America*. Cambridge, MA: MIT Press.

Wolfe, Audra J. 2018. *Freedom's Laboratory: The Cold War Struggle for the Soul of Science*. Baltimore: Johns Hopkins University Press.

Wolffe, John. 1994. *God and Greater Britain: Religion and National Life in Britain and Ireland, 1843-1945*. London: Routledge.

Wood, A. B. 1965. 'From Board of Invention to Royal Naval Scientific Service'. *Journal of the Royal Naval Scientific Service* 20, no. 4: 201–81.

Wood, Henry Trueman. 1913. *A History of the Royal Society of Arts*. London: Murray.

Woolf, Steven H. 2008. 'The Meaning of Translational Research and Why It Matters'. *JAMA* 299. no. 2: 211–13.

Woolf, Virginia. 1928. *Mr Bennett and Mrs Brown*. London: Hogarth Press.

Worboys, Michael. 1979. 'Science and British Colonial Imperialism'. PhD thesis, Sussex University.

1990. 'The Imperial Institute, the State and the Development of the Natural Resources of the Colonial Empire, 1887–1923'. In John McKenzie (ed.), *Imperialism and the Natural World*. Manchester: Manchester University Press, 164–86.

1996. 'British Colonial Science Policy 1918–1939'. In Patrick Petitjean (ed.), *Les sciences coloniales: figures et institutions/Colonial Sciences: Researchers and Institutions*. Paris: Orstom, 99–111.

Wynne, Bryan. 1992. 'Misunderstood Misunderstanding: Social Identities and the Public Uptake of Science'. *Public Understanding of Science* 1, no. 3: 281–304.

Yavetz, Ido. 1993. 'Oliver Heaviside and the Significance of the British Electrical Debate'. *Annals of Science* 50, no. 2: 135–73.

1994. 'A Victorian Thunderstorm: Lightning Protection and Technological Pessimism in the Nineteenth Century'. In Yaron Ezrahi, Everett Mendelsohn, and Howard P. Segal (eds.), *Technology, Pessimism, and Postmodernism*. Dordrecht: Kluwer, 53–75.

Yeo, Richard. 1985. 'An Idol of the Marketplace: Baconianism in Nineteenth-Century Britain'. *History of Science* 23, no. 3: 251–98.

2001. *Encyclopaedic Visions: Scientific Dictionaries and Enlightenment Culture*. Cambridge: Cambridge University Press.

Zeitlin, Jonathan. 2008. 'Re-Forming Skills in British Engineering, 1900–40: A Contingent Failure'. *Historical Studies in Industrial Relations* 25–26, no. 1: 19–77.

Zheludev, I. S., and L. V. Konstantinov. 1980. 'Nuclear Power in the USSR'. *IAEA Bulletin* 22, no. 2: 34–45.

Ziman, John. 1968. *Public Knowledge: An Essay Concerning the Social Dimension of Science*. Cambridge: Cambridge University Press.

1994. *Prometheus Bound: Science in a Dynamic Steady State*. Cambridge: Cambridge University Press.

Zuckerman, Solly. 1966. *Scientists and War: The Impact of Science on Military and Civil Affairs*. London: H. Hamilton.

1978. *From Apes to Warlords: The Autobiography (1904–1946) of Solly Zuckerman*. London: Hamilton.

1988. *Monkeys, Men, and Missiles: An Autobiography, 1946–1988*. London: Collins.

Index

Page numbers in *italics* refer to illustrations and those followed by n indicate footnotes. Universities whose names begin with 'University of' are indexed under the name of the town or city, e.g. University of Sheffield is indexed as 'Sheffield University'.

Printed in the United States
by Baker & Taylor Publisher Services